# New Media in New Europe-Asia

This volume offers an in-depth investigation of the role of new media in the political, social and cultural life in the region of Europe-Asia. By focusing on new media, which is understood primarily as internet-enabled networked social practice, the book puts forward a political and cultural redefinition of the region which is determined by the recognition of the diversity of new media uses in the countries included in the study. This book focuses on the period prior to the advent of 'world internet revolutions', and it registers the region at its pivotal moment—at the time of its entry into the post-broadcast era. Does the Internet aid democratization or is it conditioned by socio-political norms? Has the Internet changed politics or has it had to fit existing political structures? Has the use of digital technologies revolutionized election campaigns? How is hyperlinked society different from society prior to the advent of the web? How do ordinary people actually use the Internet. These and other pressing questions—crucial to understanding the post-socialist world—are investigated in the current volume.

This book is based on a special issue of *Europe-Asia Studies*.

**Jeremy Morris** is Senior Lecturer in Russian at the University of Birmingham. His current research is focused on ethnographic approaches to understanding 'actually lived experience' in the former Soviet Union, particularly in relation to the diverse economy and new media.

**Natalia Rulyova** is Lecturer in Russian at the University of Birmingham. She has research interests in Russian media studies, post-Soviet television and Russian-language new media.

**Vlad Strukov** is Associate Professor in Digital Culture at the University of Leeds. He is the founding and principal editor of *Digital Icons: Studies in Russian, Eurasian and Central European New Media*. His research interests include contemporary film, animation, digital media, especially the internet, and popular culture; digital and web-induced arts.

**Routledge Europe-Asia Studies Series**

A series edited by Terry Cox

*University of Glasgow*

The **Routledge Europe-Asia Studies Series** focuses on the history and current political, social and economic affairs of the countries of the former 'communist bloc' of the Soviet Union, Eastern Europe and Asia. As well as providing contemporary analyses it explores the economic, political and social transformation of these countries and the changing character of their relationships with the rest of Europe and Asia.

**Challenging Communism in Eastern Europe**
1956 and its Legacy
*Edited by Terry Cox*

**Globalisation, Freedom and the Media after Communism**
The Past as Future
*Edited by Birgit Beumers,*
*Stephen Hutchings and*
*Natalia Rulyova*

**Power and Policy in Putin's Russia**
*Edited by Richard Sakwa*

**1948 and 1968 – Dramatic Milestones in Czech and Slovak History**
*Edited by Laura Cashman*

**Perceptions of the European Union in New Member States**
A Comparative Perspective
*Edited by Gabriella Ilonszki*

**Symbolism and Power in Central Asia**
Politics of the Spectacular
*Edited by Sally N. Cummings*

**The European Union, Russia and the Shared Neighbourhood**
*Edited by Jackie Gower and*
*Graham Timmins*

**Russian Regional Politics under Putin and Medvedev**
*Edited by Cameron Ross*

**Russia's Authoritarian Elections**
*Edited by Stephen White*

**Elites and Identities in Post-Soviet Space**
*Edited by David Lane*

**EU Conditionality in the Western Balkans**
*Edited by Florian Bieber*

**Reflections on 1989 in Eastern Europe**
*Edited by Terry Cox*

**Russia and the World**
The Internal-External Nexus
*Edited by Natasha Kuhrt*

# New Media in New Europe-Asia

*Edited by*
**Jeremy Morris, Natalia Rulyova and
Vlad Strukov**

LONDON AND NEW YORK

First published 2015
by Routledge
2 Park Square, Milton Park, Abingdon, Oxon, OX14 4RN, UK

and by Routledge
605 Third Avenue, New York, NY 10017

First issued in paperback 2020

*Routledge is an imprint of the Taylor & Francis Group, an informa business*

*British Library Cataloguing in Publication Data*
A catalogue record for this book is available from the British Library

ISBN 13: 978-0-367-73957-7 (pbk)
ISBN 13: 978-0-415-73709-8 (hbk)

Typeset in Times New Roman
by RefineCatch Limited, Bungay, Suffolk

**Publisher's Note**
The publisher accepts responsibility for any inconsistencies that may have
arisen during the conversion of this book from journal articles to book chapters,
namely the possible inclusion of journal terminology.

**Disclaimer**
Every effort has been made to contact copyright holders for their permission to
reprint material in this book. The publishers would be grateful to hear from any
copyright holder who is not here acknowledged and will undertake to rectify
any errors or omissions in future editions of this book.

# Contents

CONTENTS

# Citation Information

The following chapters were originally published in *Europe-Asia Studies*, volume 64, issue 8 (October 2012). When citing this material, please use the original page numbering for each article, as follows:

**Chapter 2**
*Mundane Citizenship: New Media and Civil Society in Bulgaria*
Maria Bakardjieva
*Europe-Asia Studies*, volume 64, issue 8 (October 2012) pp. 1356–1374

**Chapter 3**
*The Role of Social Networking Sites in Civic Activism in Russia and Finland*
Boris Gladarev & Markku Lonkila
*Europe-Asia Studies*, volume 64, issue 8 (October 2012) pp. 1375–1394

**Chapter 4**
*From Blogging Central Asia to Citizen Media: A Practitioners' Perspective on the Evolution of the* neweurasia *Blog Project*
Cai Wilkinson & Yelena Jetpyspayeva
*Europe-Asia Studies*, volume 64, issue 8 (October 2012) pp. 1395–1414

**Chapter 5**
*Blog Medvedev: Aiming for Public Consent*
Dmitry Yagodin
*Europe-Asia Studies*, volume 64, issue 8 (October 2012) pp. 1415–1434

**Chapter 6**
*Blogging for the Sake of the President: The Online Diaries of Russian Governors*
Florian Toepfl
*Europe-Asia Studies*, volume 64, issue 8 (October 2012) pp. 1435–1459

**Chapter 7**
*Political Challengers or Political Outcasts?: Comparing Online Communication for the Communist Party of the Russian Federation and the British Liberal Democrats*
Sarah Oates
*Europe-Asia Studies*, volume 64, issue 8 (October 2012) pp. 1460–1485

CITATION INFORMATION

**Chapter 9**

*Contesting Bulgaria's Past Through New Media: Latin, Cyrillic and Politics*
Orlin Spassov
*Europe-Asia Studies*, volume 64, issue 8 (October 2012) pp. 1486–1504

**Chapter 10**

*Ukrainian Nation Branding Off-line and Online: Verka Serduchka at the Eurovision Song Contest*
Galina Miazhevich
*Europe-Asia Studies*, volume 64, issue 8 (October 2012) pp. 1505–1523

**Chapter 11**

*Blogging the Other: Construction of National Identities in the Blogosphere*
Natalia Rulyova & Taras Zagibalov
*Europe-Asia Studies*, volume 64, issue 8 (October 2012) pp. 1524–1545

**Chapter 12**

*Learning How to Shoot Fish on the Internet: New Media in the Russian Margins as Facilitating Immediate and Parochial Social Needs*
Jeremy Morris
*Europe-Asia Studies*, volume 64, issue 8 (October 2012) pp. 1546–1564

**Chapter 13**

*Co-opting Transmedia Consumers: User Content as Entertainment or 'Free Labour'? The Cases of* S.T.A.L.K.E.R. *and* Metro 2033
Natalia Sokolova
*Europe-Asia Studies*, volume 64, issue 8 (October 2012) pp. 1565–1583

**Chapter 14**

*Spatial Imagining and Ideology of Digital Commemoration (Russian Online Gaming)*
Vlad Strukov
*Europe-Asia Studies*, volume 64, issue 8 (October 2012) pp. 1584–1604

Please direct any queries you may have about the citations to
clsuk.permissions@cengage.com

# Notes on Contributors

**Maria Bakardjieva** is based at the Faculty of Communication and Culture, University of Calgary, Canada. She holds a doctorate in Sociology from the Bulgarian Academy of Sciences and a doctorate in Communication from Simon Fraser University, Canada. She is the author of *Internet Society: The Internet in Everyday Life* (Sage, 2005) and co-editor of *How Canadians Communicate* (University of Calgary Press, 2004 and 2007). Her research has examined internet use practices across different social and cultural contexts with a focus on the ways in which users understand and actively appropriate new media. She has also published on the topics of online community, e-learning and research ethics. Her current projects look at the interactions between traditional and new media with a view to identifying opportunities for citizen participation in the public sphere.

**Boris Gladarev** works as a Research Fellow at the Center for Independent Social Research and as a research fellow at the European University (St Petersburg). His research interests include actor–network analysis, sociology of social movements and social activism.

**Yelena Jetpyspayeva** is a social media marketing specialist, marketing manager and managing editor for the NewEurasia.net/ru project, and a Freedom House country consultant on the internet.

**Markku Lonkila** is Professor of Sociology at the University of Tampere. His research interests include Russian studies, comparative research, analysis of personal networks and the use of social media in civic activism.

**Galina Miazhevich** is the Gorbachev Media Research Fellow at the University of Oxford. Her research interests include post-Soviet identity, media, communication and transnationalism, the relationship between language and power, and gender issues. She is working on several projects dealing with mass-media representations of terrorism and the discourse of 'security threat', the interaction between the 'new' and 'old' media in post-communist societies and issues of press freedom. In particular her postdoctoral research focuses on the dramatic rise in post-communist xenophobia by exploring the state media's treatment of extremisms in Ukraine and Belarus.

**Jeremy Morris** is Senior Lecturer in Russian at the Centre for Russian and East European Studies (CREES), University of Birmingham. He graduated with degrees in literature from the Universities of Lancaster, and Queen Mary & Westfield College. His DPhil (Sussex, 2003) on Russian nonconformist literature from the 1970s to the present focused on the writer Evgeny Popov. His research interests include contemporary Russian literature and popular culture as well as ethnographic approaches to understanding 'actually lived experience' in the former Soviet Union, particularly in relation to the diverse economy and new media.

**Sarah Oates** is a Professor at the Philip Merrill College of Journalism at the University of Maryland. She is the author of *Revolution Stalled: The Political Limits of the Internet in the Post-Soviet Sphere* (Oxford University Press, forthcoming December 2012) and *Terrorism, Elections, and Democracy: Political Campaigns in the United States, Great Britain, and Russia* (with Lynda Lee Kaid and Mike Berry, Palgrave-Macmillan, 2010). A major theme in her work is the way in which the media and the internet can support or subvert democracy in places as diverse as Russia, the United States and the United Kingdom. Founder of the Google Forum to bring together academics with Google UK, she currently serves as an expert for the European Commission's Digital Futures project.

**Graham Roberts** is Vice-President of the British-French Association for the Study of Russian Culture. Graham Roberts teaches Russian and International Business at Université Paris Ouest Nanterre La Défense, just outside Paris. He has published on a wide range of topics, including images of masculinity in Soviet cinema, the internationalisation strategy of French hypermarket chain Auchan, and the relationship between packaging and branding in post-socialist Russia. His current projects include an edited volume on material culture in Russia from Peter the Great to Putin, and a monograph, entitled *Consumer Culture, Branding and Identity in the New Russia: From Five-Year Plan to 4x4.*

**Natalia Rulyova** is Lecturer in Russian at the Centre for Russian and East European Studies (CREES), University of Birmingham. Her research focuses on genre and literary practices in contemporary Russian-language media. She is currently leading an AHRC-funded project entitled 'Interdisciplinary Genre Studies Network' (2012–2013). With Birgit Beumers and Stephen Hutchings, she co-edited *Television and Culture in Putin's Russia: Remote Control* (Routledge, 2009) and *The Post-Soviet Russian Media: Conflicting Signals* (Routledge, 2009). Her other research interests lie in the area of translation studies and the work of Joseph Brodsky. Her PhD dissertation, Joseph Brodsky: Translating Oneself, which she completed at the University of Cambridge in 2002, examines the poet's self-translations.

**Robert Saunders** is a Professor in the Department of History and Political Science and Chair of the Science, Technology and Society (STS) program at Farmingdale State College-SUNY, where he teaches courses on comparative religions, international politics, and Russian history. The author of three books and numerous articles, his research explores the impact of information and communications technologies (ICTs), mass media, and popular culture on national identity, geopolitics, and international relations.

**Natalia Sokolova** is an Associate Professor at the University of Samara, Russia. She is the Director of the MA programme 'Digital Media and Internet'. Her academic interests include critical theory of media, new media, visual studies and gender studies. Her publications include *Popular Culture Web 2.0: Mapping Modern Media-Landscape* (University of Samara, 2009) and articles (in Russian) on aesthetics and politics of popular culture. She is currently researching Russian transmedia with a particular focus on the role of audiences in its production.

**Orlin Spassov** is Doctor of Sociology and Associate Professor at the University of Sofia 'St. Kliment Ohridski', Faculty of Journalism and Mass Communication. Until 2011 he was Head of the Department of Radio and Television. Dr Spassov teaches Media and Communication Studies, and Internet Culture. He is executive director of the Media Democracy Foundation (fmd.bg). Spassov is the author of *Transition and the Media: Politics of Representation* (2000, in Bulgarian), and editor of 13 books, including *Media and Politics* (2011), *New Media—New Mobilizations* (2011, in Bulgarian), *New Youth and New Media* (2009, in Bulgarian), *Quality Press in South East Europe* (2004) and *New Media in South East Europe* (2003). His current research concerns transformations of the public sphere caused by the traditional and new media, and subcultural activities on the internet.

**Vlad Strukov** is Associate Professor in Digital Culture at the University of Leeds. His research is on film, animation, art, media, television, digital culture, theories of globalisation and post-broadcast media. His publications include such books as *Shocking Chic: Glamour and Celebrity in Contemporary Russian Culture* and *From Central to Digital: Television in Russia*. He is the founding and principal editor of the journal entitled *Studies in Russian, Eurasian and Central European New Media* (*Digital Icons*, www.digitalicons.org). He is a frequent contributor to the media, including Al Jazeera, BBC, Calvert 22, Channel 4, Radio 4, and many others. He has worked on collaborative projects with Tate Modern (London), Leeds International Film festival and other arts organisations.

**Florian Toepfl** is a Marie Curie Postdoctoral Fellow at the Department of Media and Communications, London School of Economics and Political Science, where he is currently working on a project titled 'Mediating Semi-Authoritarianism: The Power of the Internet in Russia'. He received his PhD in Political Science from the University Passau, Germany, in 2009. Since then, he has been a postdoctoral fellow at the Harriman Institute at Columbia University, New York, and a postdoctoral researcher at the LMU University, Munich. He also had visiting fellowships at the Aleksanteri Instititute in Helsinki and at the Higher School of Economics in St Petersburg.

**Cai Wilkinson** is a Lecturer in International Relations whose research focuses on societal security issues in Central Asia and Russia and the use of interpretive methods in Critical Security Studies. Her PhD (Birmingham, 2009) was entitled Interpreting Security: Grounding the Copenhagen School in Kyrgyzstan and explored understandings of security in post-Akaev Kyrgyzstan with reference to the concepts of securitisation and societal security within a broadly interpretivist methodology. As part of this, she explored the role of the media in creating and sustaining securitisation narratives at different spatial scales and analytical levels. Her current research examines the contestation of international human rights norms in Kyrgyzstan and Russia. She has previously published in *Security Dialogue and Central Asian Survey* and has contributed chapters to volumes on Central Asia and research methods.

**Dmitry Yagodin** is a graduate student at the University of Tampere. He is currently working on his PhD dissertation which examines political blogging on the Runet.

**Taras Zagibalov** is a Natural Language Researcher in a social media monitoring company Brandwatch, UK. His first degree was in Chinese linguistics and translation. His DPhil (Sussex, 2010) on Multilingual Automatic Sentiment Analysis focused on techniques for automatic processing of textual data from online sources. He is currently doing research in different areas of natural language processing used in commercial applications.

# Introduction: The Role of New Media in the Political, Social and Cultural Life in the Region of Europe-Asia

JEREMY MORRIS, NATALIA RULYOVA & VLAD STRUKOV

SINCE THE DISSOLUTION OF THE SOVIET UNION IN 1991, Eurasia—as a geographical, political and cultural entity—has been relentlessly re-drawn, re-defined and re-conceptualised by actors in the region as well as in the communities on other continents. In fact, we write this introduction to the volume[1] at the time when the political crisis in Ukraine continues to unfold: the Russian Federation has already annexed Crimea, there is growing unrest in the eastern part of the country which, with financial and political support from the US and EU, is preparing for presidential elections. The crisis started in 2013 when tens of thousands of protesters took to the streets of central Kiev and other cities to express their anger at the government's decision to abandon plans to sign an association agreement with the EU and instead join the Eurasian Economic Community Customs Union, whose members are Belarus, Kazakhstan and Russia. The Customs Union was originally launched as a step towards forming a broader EU-style economic alliance of the former Soviet states, thereby extending the EU model of economic cooperation across the Eurasian space, with Russia being the central node in this new conglomerate of countries. Thus, while the crisis in Ukraine had its own social, economic and political causes (endemic corruption, shrinking economy, ineffective government, etc.),[2] it was fuelled by the confrontation between two geopolitical entities—the EU and the Eurasian Economic Community—each of which continues to evolve and shape the configurations of world trade, economy, communication and everyday life. In the Western media, the conflict over Ukraine has been described as a new Cold War between Russia and the West; however, it is clear that in the twenty-first century the confrontation is not between two ideological systems—capitalism and

---

[1] This edited volume *New Media in New Europe-Asia* is an expanded version of the special issue *New Media in New Europe-Asia* published in October 2012.

[2] This is according to the statement made by US Vice-President Joe Biden during his visit to Kiev in April 2014 (for more information, see BBC 2014).

communism—but rather between competing visions for globalising factors in the region that are often acted out in networked-media, which, by definition, transcends the boundaries of nation-states.

The crisis in Ukraine has re-ignited century-long conflicts, imperial ambitions and post-colonial phobias in the bordering states.[3] Unlike in the previous decades, the Russian incursion in Ukraine has been almost instantaneously documented thanks to social media. In this regard, the protest movement and social unrest in Ukraine have been similar to those in Egypt, Turkey and elsewhere in the world where internet penetration has reached significant levels.[4] Such 'mediation' of events in social media creates an unprecedented sense of participation even in those communities that are not directly linked to the conflict. Such instantaneous grass-roots documentation also poses challenges to professional journalism which finds it increasingly difficult to compete with independent bloggers, video-sharing sites and other types of digital citizenscape. On the one hand, social media provided a platform for critical reflection over the situation in Ukraine, particularly over the annexation of Crimea, whereby discussions in Facebook and VKontakte revealed people's actual attitude to the events which often were in conflict with the outcome of the so-called 'referendum' in Crimea. On the other hand, social media have been a space where users staged 'digital wars' against each other and hate speech has dominated the discourse. The crisis in Ukraine has become a major political and diplomatic event since 9/11 and the Arab Spring and will perhaps inform the political and social process on the continent and elsewhere in the world in the decades to come. It has been a pivotal moment in the development of networked media in the region as digital revolution sweeps across the continent.

The events in Ukraine have also demonstrated that government media have extended their outreach to the blogosphere, social media, *YouTube*, thus creating a more complex and transcending media system than ever before. In previous years there had been much speculation about how the Russian government would act at the time of crisis and whether, for example, it would block access to social media. The experience of 2013–14—as well as the Russian protest movement of 2011–12 (Strukov *et al.* 2013)—has demonstrated that this is not their preferred option although the situation remains uncertain. As a matter of fact, Russian social media such as odnoklassniki.ru have been used by protesters in Kiev and elsewhere in Ukraine for social mobilisation. Instead of blocking access to social media as, for example, the Turkish government did in 2014, the

---

[3]For example, *The Economist* reported that there is a growing sentiment in Poland to view western parts of Ukraine as under the Polish influence; this sentiment has historical origins as this part of Ukraine had been part of Poland until 1939 (*The Economist* 2014). Just as Eastern Ukraine depends on remittances from temporary workers in Russia, Western Ukraine relies on Ukrainians working in Poland. And Bloomberg reported Hungary's Prime Minister Viktor Orban repeated calls for Ukraine to grant autonomy to ethnic Hungarians living in the country; he made these calls even after the annexation of Crimea and Western sanctions against Russia (Simon 2014).

[4]At the end of 2012, internet penetration reached 64.9% in Poland, 58.4% in Italy, 47.7% in Russia, 45.7% in Turkey, 45.0% in Kazakhstan, 44.1% in Romania, 40.1% in China, 34.7% in Ukraine and 5.0% in Turkmenistan (Internet World Stats 2012).

Russian government decided to utilise its powerful media empire, which now includes *RT* that airs its programmes in English, Spanish and Arabic, ORT and other multi-region television channels, blogs and social media, to advance its own version of the events in Ukraine.[5] (In this respect, the Russian media follow closely in the footsteps of their US and EU counterparts, whilst Ukraine lacks its own international media presence—in most cases in Western and Russian mainstream media, the voices of Ukrainian people are available in an indirect, mediated form.) The Ukrainian government retaliated by closing down Russian-speaking channels, and thus, the conflict in the region evolved into a fully-fledged confrontation on the ground, in diplomatic circles, on air and in the arena of digital networked media.

Whilst the outcome of the political crisis in Ukraine is impossible to predict, and perhaps Putin's celebration of the annexation of Crimea is premature, it is undeniable that we have recently lived through a change of paradigm in terms of the media structure. In the region we shifted from the mass media paradigm, which dominated the twentieth century, to the new media paradigm, which has become influential in the twenty-first century. There are several theories which attempt to account for this transformation. One way of theorising this new paradigm is by looking at it as a shift from 'broadcast' to 'post-broadcast' era. The strength of this approach is in acknowledging the reducing role of mass media. Broadcasting corporations now have to adapt to a new environment dominated by the internet-enabled media and develop new ways of producing news stories: (1) each story has to be relatively short for an online user to have a snippet of events; (2) stories have to contain a visual element, often in the form of a video/photograph, etc.; (3) ordinary citizens enabled by new technological devices, such as smart phones and cameras, produce their own news posted online on *Twitter, Facebook*, etc.[6] Media practitioners and media scholars have provided their reflection on the shift (for example, Gillmor 2004; Castells 2009). However, the concept of the 'post-broadcast era' does not fully explain the current paradigm because it assumes that broadcasting is no longer there. In reality, as it has been evidenced in the crisis in Ukraine, broadcasting companies still dominate the world news market. Also, the element of '-casting' has

---

[5]There has been an even more aggressive campaign in Russia itself against those who take an opposing view on the events. For example, the opposition leader Alexei Navalny's VKontakte account was frozen for 'extremism' on 20 March 2014, and the editor of the independent news website Lenta.ru was forced out to be replaced by someone more compliant. At the same time there is evidence to suggest some Russian media outlets have remained independent from the government; for example, VKontakte refused to disclose personal information of its Ukrainian users to Russian security forces, the FSB. In April 2014, Russian media company rbc.ru published photocopies of the documents sent by the FSB to VKontakte administration. Rbc. ru also published Pavel Durov's, the founder of the social network, statement, in which he defended the right of the company to withhold information from security forces (*rbc.ru* 2014). This is an unprecedented event in the era of wide-spread surveillance over personal accounts of citizens and prominent politicians (uncovered by the British newspaper *The Guardian*) as well as corporate involvement in online censorship (*Google* in China). Since then there are signs that the government will move to take indirect control over VKontakte, and Durov has fled Russia. As of May 2014 there are moves by the government to force bloggers with more than 3,000 visitors to register as media outlets.

[6]For full discussion see Strukov (2012).

become more important in the new era dominated by visual narratives. Recent research into *YouTube* suggests that 'the Internet is transitioning to a predominantly video-based medium', and broad– or narrowcasting oneself and one's stories has already become a widespread phenomenon in the twenty-first century (Strangelove 2010, p. 11). The same has been observed by Gunther Kress who argues that as images including moving images are becoming increasingly important in addition to writing and colour, we need to develop new ways of analysing online content; Kress develops a multimodal social semiotic approach to communication (Kress 2010).

Another aspect of the shift to new media is the increasing segmentation of audiences due to more narrowcasting online. The existing research into audience segmentation has been mainly focused on 'virtual communities within organizations or communities of practice' (Foster *et al.* 2011). Audiences have been segmented by researchers based on how active they are in their use of social media (Williams and Merten 2008), on consumption patterns (Hammond *et al.* 1996; Baltas and Doyle 1998), on the kind of social media activities that they are engaged in (Li and Bernoff 2008). Research into audience segmentation is often sponsored and utilised by businesses to help them market their brands successfully online. So, this approach does not reflect the full scope of change in the production and use of new media. Manuel Castells suggests another more encompassing definition of the change occurring in the media landscape. He identifies the new form of online communication as mass self-communication: 'a new form of interactive communication [. . .], characterized by the capacity from many to many, in real time or chosen time, and with the possibility of using point-to-point communication, narrowcasting or broadcasting' (Castells 2009, p. 38). His observations are based on an analysis of new media developments across the world:

> the common culture of the global network society is a culture of protocols of communication enabling communication between different cultures on the basis not of shared values but of the sharing of the value of communication. This is to say: the new culture is not made of content but of process, as the constitutional democratic culture is based on procedure, not on substantive programmes. (p. 38)

This last citation about the changing culture of communication—from the culture based on content to the culture based on the value of communication—helps to shed light on the current conflict between different modes of communication and values attached to them. In a contemporary media environment, as Castells observes,

> the sender/addressee has to interpret the message she receives from multiple modes of communication and multiple channels of communication by engaging their own code in interaction with the code of the message originated by the sender and processed by subcodes of modes and channels. (p. 132)

Castells describes this as a 'self-selected meaning' and proceeds to associate it with the 'creative audience' (p. 132). The concept of the 'creative audience' has received a new connotation since the advent of internet-enabled media: contemporary internet users are no longer just audiences, they are prosumers, i.e. they are consumers and producers at the same time. In relation to the region in question, it is important to enquire whether such 'creative audiences', 'active audiences', constitute an active social group capable of political participation. Similarly, can governments utilise such creative audiences for their purpose? And ultimately, how do these new practices and concepts enrich our understanding of the region?

This edited volume addresses many of these questions in detail and offers an in-depth investigation of the role of new media in the political, social and cultural life in the region of Europe-Asia which is defined loosely as Eastern and Central Europe, Russia and Central Asia. This region shares a common cultural and ideological past which has been shaped by histories of imperial expansion, the rise of national identities and rapid entry into the global community in the 1980s and 1990s. As this volume shows, the political and geopolitical, to a significant degree, structure new media use. Thus the common thread of incomplete democratisation, the attempts at more traditional agenda management, can be observed across the post-socialist space and come to define political media use. Similarly, geopolitically, the recent events in Ukraine serve as a reminder of the linguistic ties that bind new media use, with *Twitter* and Russophone social networking becoming a geopolitical space of conflict and dialogue, as much as the traditional media. At the same time, linguistic areas of exclusion continue to grow, thus creating new boundaries in the era of transnational flows of information, capital and human resources.

By focusing on new media, which is understood primarily as internet-enabled networked social practice, that encompasses social networking sites, blogs, email, social gaming and mobile telecommunications, this volume puts forward a political re-definition of the region which is determined by the recognition of the diversity of new media uses in the countries included in the study. It is becoming more and more apparent since protests in Russia in 2012 and the Ukraine revolution in 2014, that media-use in most of the region is characterised by 'dissent', whereby protests are directed against political regimes, local governments, cultural traditions or forces of globalisation. In this regard, new media in Europe-Asia display the same potential for social intervention and political reform that has been manifested in other regions, particularly in the Arab world. While cognizant of the 2011–12 protest movements in Arab countries, China, Russia and other places, the contributions to this volume focus on the period prior to the advent of 'internet revolutions'. This is because research presented here aims to capture new media use in the region at its pivotal moment—at the time of the paradigm shift from mass media to networked media, and thus the volume documents this transitional moment and relates it to the history of change of post-communist states.

First of all, the volume is concerned with the involvement of new media in political processes and considers political activity in the region in a medium-focused fashion, i.e. by looking at blogs, but also broader multi-media use whereby media convergence exemplifies new mixes of political powers in given countries. Second, the volume engages with media practitioners in the region by giving voice to Cai Wilkinson and Jelena Jetpyspayeva, neweurasia bloggers, who share their experiences of working on a Central Asian blogging platform. The engagement with media practitioners speaks directly to 'policy practitioners': governmental organisations, NGOs and communities that are involved in negotiating policy online.[7] Third, the volume broadens the spectrum of enquiry by identifying new areas of the political as such. For example, political processes are revised in the realm of popular culture, particularly in the context of identity construction, as observed in the Eurovision contest (Miazhevich). *New Media in New Europe-Asia* places significant emphasis on new developments in the media system in the region: the calls of digital modernity are examined so as to reveal the development of social networks and their impact (or a lack of it) on both civil society and everyday life as well as a whole new dimension in cultural politics brought by computer games (Sokolova and Strukov).

This collection is part of a larger debate concerning new media applications in the region. For a number of reasons, the discussion has been dominated by focus on the Russian Federation in spite of the general interest in new media in various countries of the region. First, traditionally Western academia has singled out Russia in its approach to Eastern Europe, Central Asia and the Caucasus, i.e. the so-called 'communist space'. Second, the number of unique users of internet and mobile phones in Russia is higher than in any other country in the region. In fact, as of December 2011, Russia was the second largest country in Europe in terms of the number of internet users—70 million (Internet World Stats 2012). The statistics exclude the practice of so-called 'shared use', that is when people access the internet from such public spaces as internet cafes or use the internet collectively at home (see Morris' chapter); and therefore the number of people using the internet in Russia is certainly even higher. Indeed, in Morris' chapter there is concern to check the myth of a methodologically individualistic portrait of use—as if the lone gamer, or networker sits physically or socially atomised, while maintaining at distance a network of status equals and by definition one's own social capital. Similarly, Sokolova points to promising new avenues of research on the problematic meaning of a consumer of media and a producer. Perhaps there is something of a collective hang-over from the socialist period that might make 'prosuming'—the joint production, or labouring at a task, from fanzines to gaming—more likely in the new media of Eurasia.

At the same time, in terms of internet penetration Russia is still behind other post-communist states, for example, Estonia and Slovenia, with penetration below 50%. Finally, because of the legacy of Russian Empire and the Soviet Union, the Russian language has a status of the language of intercultural communication, and is in fact spoken by diasporas in many countries in the world. Therefore, Runet—the Russophone segment of the internet—is an entity significantly larger than the internet in the Russian Federation per se. This came to particular prominence in the 2013–14 Ukraine events where significant transnational 'dialogue' online was observed. Thus, the focus of new media studies on Russia is accounted for by the simple numerical factor as well as by the structure of Western academia and knowledge, in general.

---

[7]Luc Levy, the Ministry of Foreign and European Affairs (France) who works on the relationship between diplomacy, soft power and the internet, found the discussion informative and helpful for the understanding of the blogosphere, in particular. In his interview with the editors, which can be found at the project's website (http://eurasia.vladstrukov.com/), he highlighted the importance of the discussion to the governmental policy in France.

Among various approaches to the study of new media in the region, three centres are particularly distinct: (1) *The Future of Russian: Language Culture in the Era of New Technology* research project, based at the University of Bergen and led by Dr Ingunn Lunde (www.uib.no/rg/future_r), is an example of a linguistic and communicative approach to the study of new media in the region. It investigates the development of and the relationship between the standard language and its vernacular uses in new media; (2) Berkman Centre for Internet and Society at Harvard University (Drs Karina Alexanyan, Bruce Etling, John Kelly *et al.*; http://cyber.law.harvard.edu/) has utilised quantitative methods in its examination of the Russian blogosphere, including Twitter, and has produced analytical maps that represent the usage dynamics on the internet; and (3) *Digital Icons: Studies in Russian, Eurasian and Central European New Media* (edited by Vlad Strukov, Henrike Schmidt, Ellen Rutten *et al.*; www.digitalicons.org; ISSN 2043-7633) publishes research that examines the impact of digital and electronic technologies on politics, economics, society, culture and the arts in the region. The journal's research agenda is chiefly driven by the Cultural Studies approach with its preoccupation with issues of power, alternative cultural practice and protest activity.

This collection of Europe-Asia Studies has two chief objectives. First of all, research presented in *New Media in New Europe-Asia* is informed by different disciplines—linguistics, political science, sociology, ludology, and so forth—placing itself firmly in the tradition of area studies. At the same time, few of our contributors would see their work as originating in that tradition and the works here show a trajectory crossing disciplinary boundaries from Anthropology (Morris), to International Relations (Wilkinson). Second, it pushes the thematic and geographical boundaries of the field. It explores new areas of social activity, for example, ecological activism on social media platforms, and cultural production, such as online gaming. The issue goes beyond the Russo-centric approach to the study of new media in the region by presenting research on countries that tend to be overlooked, for example, Bulgaria and Kazakhstan. At the same time the aim of this volume is not to provide an overview of new media use in all countries of the region, nor is it to survey all types and forms of new media, for example, mobile phones, telephony, tablets, and so forth. Rather its aim is to generate a discussion of new media use in the region that goes beyond national boundaries, recognised social groups and communities, and established cultural practices. As a result, the choice of case studies is determined by the urgent necessity to present general trends of new media use in the region that are informed by local cultural traditions and social practice and provide a bigger picture of digital communication in the region. To achieve this objective, the editors consciously selected works that utilise a comparative approach to the study of new media, whether by comparing two countries, for example, Russia and China (Rulyova), or by contrasting political entities, for example, political parties in Russia and the United Kingdom (Oates), or by exploring diverging social practices, for example, use of Cyrillic and Latin encoding on Bulgarian internet and ensuing political tendencies (Spassov), or by investigating the merger of offline and online activities and emergence of transmedial practices (Sokolova and Strukov). The authors of this volume explore both the democratising (horizontal) effect of social media networks (Bakardjieva, Lonkila and Gladarev) and the ways in which governments use the internet in order to increase control, surveillance and consolidation of their power (vertical or top-down), for example the use of new media by regional governors and local authorities (Toepfl), and the Russian President Medvedev's blogging (Yagodin), and the increasing importance of branding for both countries (Saunders) and cities (Roberts). The editors hope that this volume will generate a discussion that is interdisciplinary, multi-platform and transnational.

Digital modernity, as any other modernity, privileges social and political progress and celebrates technological and intellectual advancements. The consumer society in the West, as well as in emerging economies of Eastern Europe and Asia, promotes the consumption of new gadgets, ideas and social practices. In this volume the editors hope to avoid the pitfalls of some new media studies, namely obsessive concentration on the most recent events and phenomena, known as 'presentism'. Furthermore, each chapter situates new media use in the region in a wider historical context whether it concerns the rise of civic society, transformation of class systems, journalistic practice, cultural heritage or popular media. While the volume frequently deals with ephemeral phenomena, the lasting effect of the presented research is in its intrinsic links with the theories of democratisation, cultural memory, identity and practice. In addition, the volume advances methodological approaches to new media, particularly regarding the analysis of transmedial phenomena, visual apparatus of online environments and social milieu of users of digital technologies.

One of the most striking insights that the volume brings to a framing of new media debates with an Area-Studies purview is how 'pervasive communication' (Fischer *et al.* 2008, p. 529) affects social relationships in societies with a history of centralised control and ideological straightjacketing. This enquiry is often visible in the study of the ambivalent and unpredictable nature and meaning of 'democratisation', the conundrum of virtual-real civic engagement, and the potential for decentralised networks to 'out-compete' vertical hierarchies (Juris 2004, p. 341). In the case of societies emergent or still beholden to relatively centralised, authoritarian patterns of social control, it can sometimes appear that observers bring a digital naivety to their analysis of new media's potential to enact or foster lasting change. With that in mind, it is particularly salutary that contributors to this volume recognise the minefield of imputing seismic, or even incremental social change to new media activities.

Notwithstanding such caution, even in those political economies that appear most ossified the contributors are adept in seeking out and explaining the ways in which new media do effect change of one sort or another. As an extension of transitional communication methods for agitation and propaganda, one would expect political traditions such as that in Russia to make effective use of such a malleable and attractive platform as the blog. As such, both Yagodin and Toepfl analyse the interpellation of Russian netizens by successful and well-spun elite blogs; these chapters should be seen in the tradition of effective and *affective* personalisation of political communication, where an attempt is made to find a socio-cultural technological fix to narrow the gulf between ruler and ruled, to placate, to elicit loyalty, and of course to 'manage' a regime's image. In this sense, any analysis of political leaders' blogs must be seen in a historical context alongside such hegemonic techno-tactics within traditional broadcast communication as the Roosevelt's fireside speeches, Churchill's radio broadcasts[8] and so on. But this comparison then begs the question: can any attempt through the noisy blogosphere to engage with the voting public compete with conflicting messages or the sheer noise of other sources (Jenkins and Thorburn 2004)?

Oates' analysis of the Communist Party's use of the internet re-examines the view that may have been drawn from the blogging articles that Russia's political elite are new media savvy manipulators of image and opinion. The communist party's online presence is a timely reminder that new media communication can be as circular, isolated and, ironically, private as traditional political discourse. Despite the huge potential for message dissemination and social-networking opportunities for this opposition party, Oates shows that new media use was far more closely tied to 'national political culture than to cyber-culture in general'.

Turning from elite and political economy use of new media, the importance of a focus on the practical interpretation of civic activism as supported and enhanced by the internet is shown by Gladarev and Lonkila. This grounded approach is taken up by Bakardjieva too in the Bulgarian context. She is careful to separate new technological facilitation from utopian teleology: there is nothing inherently 'democratising' about new media. Nonetheless, the facilitating communicative potential of the internet for expanding and enriching civic engagement at the level of 'mundane citizenship' and 'subactivism' are illustrated well by both the Bulgarian and Russian cases. Here, grassroots actors are shown to be empowered in their counter-hegemonic, often localised struggles. The logic of the particular informational and communitarian potential of the new media is shown in both cases, but Gladarev and Lonkila carefully note that the building and reinforcing of an emergent collective identity and virtual consensus within movements is not the same as a discursive public space a lá Habermas. In the Russian case, the reach and impact of civic activism through new media is limited and particular—these groupings are inclusive only to the extent that immediate aims are shared. While this activity is self-directed and managed and involves the open circulation of information, it is still a far cry from Castells' vision of the building of horizontal ties among diverse interests and groups and collaboration and coordination in democratic praxis (2001, pp. 54–55). It is particularly apt then for Bakardjieva to question the potential for subactivism to break out of the confines of the private sphere and to percolate into the more visible and institutionalised spheres of activity characterising subpolitics and formal politics.

In a particularly interesting examination of how a single issue—the Bulgarian usage of Latin script online—is illustrative of an emerging tension between elites and ordinary people, Spassov shows how the domestication of technology—the wider socio-cultural fact of transliteration (Latin is necessary for mobile phone text-messaging)—dramatises a series of ideological struggles around national identity. As in the Russian cases, we have a backdrop of strong ideological control not only over the written language but even who has the right to expertise and even *speech* on the issue. The post-socialist era crisis of linguistic standards is symptomatic of wider social contesting of identity and legitimacy. The binary Latin-Cyrillic shows up the lack of middle ground for value judgements on the past-present Socialist-EU debates. This is a classic case of new media as heralding the return of the 'oppressed'. When the hoi polloi are let out into

---

[8]In both Churchill's and Medvedev's cases, the personal communication may ironically be carried out by a team of spin doctors, political managers or others.

cyberspace all kinds of opinions, imaginings and, with them, demons are released too. And elite management is often neither here nor there.

Thus a further re-evaluation of the misused term 'democratisation' emerges. Perhaps mundane use, popular culture and popular opinion (which may or may not overlap with the sphere of politics proper) are more fruitful sites of the democratising potential of new media. In quite diverse ways, various chapters indicate a turning on its end of the inter-pretation, diffusion and remaking of meaning: from the ethnography on 'ordinary' use by 'ordinary' people by Morris, prosumption of online content by gamers and fantasy fans in Sokolova's and Strukov's cases, and the grass-roots creation of meaning in the interpreta-tion of celebrity and otherness in Miazhevich's and Rulyova and Zagibalov's chapters. Mundane use of new media can bring to the foreground overlooked affordances often pigeon-holed as 'mere' popular culture: online chat forums on Eurovision or ethnic stere-otypes, mass online gaming; or an issue entirely absent from an elite focus: the internet as facilitating existing parochial sociality and leisure activities that are not necessarily peculiar to virtuality.

Morris questions how 'transformative' of everyday lived experience is the perva-sive background culture of virtualised information for those who in Russia inhabit marginal urban spaces that are neither an urban centre nor a 'black hole' of informa-tional capitalism. His chapter offers an ethnographic reflection on perhaps the key question all the authors of this volume engage with: the place-specific 'culture of real virtuality' in Eurasia. In each case the authors offer insight into the question of new media's replacement of stable markers of space and identity with a shifting set of alle-giances and masks, a 'patchwork of interests and experiences' (Castells 1996, p. 199), replacing more 'traditional' identities based on physical space, ethnicity, social class and gender. Rulyova and Zagibalov advance our knowledge of how national identities are constructed through the representation of the other in the blogosphere by presenting their results of the quantitative and qualitative analysis of Russian and Chinese blog-ging posts. They identify specific 'masks' and genres in which bloggers tend to discuss the other online.

While Rulyova and Zagibalov focus on issues of national identity and ethnic stere-otypes, Sokolova and Strukov explore more fundamental questions of cultural tradition, transformations in cultural production and ensued economic and political tensions, and, more generally, the epistemological turn caused by digital technologies. Sokolova traces changes in creative process by analysing practices of transmedial storytelling and the role of fans as a new social group that possesses significant amount of cultural capital and potential to impact social and political processes. Strukov's chapter is concerned with the theory of digital economy and the transformational impact of digital technolo-gies on cultural experiences in networked ludic spaces. He calls for a renegotiation of the Russian mediascape in its post-broadcast phase for it to include new ludic spaces of production of cultural symbols and value. His argument builds on many interpretations of space presented in the volume which he considers primarily as a locus of cultural imagining and a place of commemoration.

Continuing and complementing discussions of cultural specificity, Saunders and Roberts, in their respective chapters, link cultural and national identity to political and media incorporation of the 'former Second world' into the global space. For Saunders, the main question is the particularities of the region's 'international mediatization'

(Hitchcock 2003). As demonstrated elsewhere in the volume, the state is seen to embrace the use of new media technologies to inform but also manipulate external audiences in soft-power externally orientated production, particularly in social media. At the same time, others in the region use cyberspace to contest state-affiliated representations. Finally, Saunders assesses cultural production from outside new Europe-Asia in terms of an external (re-)defining of the region through parody, satire and pastiche, and also examines how such popular culture influences and informs national cultures (and the Jamesonian 'commodification of culture') in the post-Second World. Roberts, on the other hand, analyses how the branding of a city has been re-shaped by the use of social media by focusing on one case study, the project 'I love Moscow'. Roberts finds that 'city branding is moving into a new phase, one in which the use of social media turns those living in the city from passive viewer to active participant; from spectator to spectacle'.

In order to focus discussion, the editors have divided the volume into four parts. Each part develops a major theme, while all are concerned with the transformative potential of new media in the region. The first part focuses on the development of civic activism in new media; the second examines how new media have influenced on political processes; the third part explores the formation of online and offline identities, and the fourth is focused on consumption and production in relation to new media technologies. In each part there are contributions by scholars based both in the West and in post-socialist countries. This exchange is possible because the field of Eurasian new media studies has grown in the past few years thanks to the contributions of many scholars residing in the region and beyond.

## References

Baltas, G. and Doyle, P. (1998) 'An Empirical Analysis of Private Brand Demand Recognising Heterogeneous Preferences and Choice Dynamics', *The Journal of the Operational Research*, Vol. 49, No. 8, August, 790–798.

BBC (2014) 'Ukraine Crisis: Biden Says Russia Must "Start Acting" ', *BBC*. Online. Available at: http://www.bbc.co.uk/news/world-europe-27106630 (accessed 22 April 2014).

Castells, M. (1996) *The Rise of the Network Society, The Information Age: Economy, Society and Culture, Vol. I*. Cambridge, MA; Oxford, UK: Blackwell.

—— (2001) *The Internet Galaxy: Reflections on the Internet, Business and Society*, Oxford: Oxford University Press.

—— (2009) *Communication Power*, Oxford: OUP.

*The Economist* (2014) 'A Village in Western Ukraine', *The Economist*. Online. Available at: http://www.economist.com/blogs/easternapproaches/2014/03/poland-and-ukraine-0 (accessed 17 April 2014).

Fischer, M., Lyon, S. and Zeitlyn, D. (2008) 'The internet and the future of social research', in, Fielding, N., Lee, R.M. and Blank, G. (eds) *The Sage Handbook of Online Research Methods*, London: Sage, pp. 519–536.

Foster, M., West, B. and Francescucci, A. (2011) 'Exploring Social Media User Segmentation and Online Brand Profiles', *Journal of Brand Management*, 19, 4–17.

Gillmor, D. (2004) *We the Media: Grassroots Journalism by the People, for the People*, Sebastopol, Canada: O'Reilly Media.

Hammond, K., Ehrenberg, A. S. C. and Goodhardt, G. J. (1996) 'Market Segmentation for Competitive Brands', *European Journal of Marketing*, Vol. 30, No. 12, 39–49.

Hitchcock, P. (2003) *Imaginary States: Studies in Cultural Transnationalism*, Champaign, IL: University of Illinois Press.

Internet World Stats: Usage and Population Statistics (2012) 'Internet Users in the World Distribution by World Regions—2011' [report]. Online. Available at: http://www.internetworldstats.com/stats.htm (accessed 17 April 2014).

Jenkins H. and Thorburn, D. (2004) *Democracy and the Media*, Cambridge, MA: MIT Press.

Juris J. S. (2004) 'Networked social movements: global movements for global justice' in, Castells, M. (ed.) *The Network Society: A Cross-cultural Perspective*, Cheltenham, UK: Edward Elgar, pp. 341–362.

Kress, G. (2010) *Multimodality: A Social Semiotic Approach to Communication*, London: Routledge Falmer.

Li, C. and Bernoff, J. (2008) *Groundswell: Winning in a World Transformed by Social Technologies,* Boston, MA: Harvard Business Press.

*rbc.ru* (2014) 'Durov rasskazal o trebovanii FSB peredat' dannye po gruppam Evromaidana' (2014) *rbc. ru*. Online. Available at: <http://top.rbc.ru/politics/17/04/2014/918521.shtml> (accessed 17 April 2014).

Simon, Z. (2014) 'Ukraine is Urged to Extend Autonomy for Ethnic Hungarians', *Bloomberg*. Online. Available at: http://www.bloomberg.com/news/2014-05-12/ukraine-is-urged-to-extend-autonomy-for-ethnic-hungarians.html (accessed 4 June 2014).

*The Economist* (2014) 'A Village in Western Ukraine', *The Economist.* Online. Available at: http://www.economist.com/blogs/easternapproaches/2014/03/poland-and-ukraine-0 (accessed 17 April 2014).

Strangelove, M. (2010) *Watching YouTube: Extraordinary Videos by Ordinary People*, Toronto, Buffalo, London: University of Toronto Press.

Strukov, V. (2012) 'BBC's Video Hub: Working in the Post-Broadcast Era; Interview with Zoya Trunova', *Studies in Russian, Eurasian and Central European New Media* (Digital Icons), 7. Online. Available at: http://www.digitalicons.org/issue07/vlad-strukov/ (accessed 24 April 2014).

Strukov, V., *et al.* (2013) *Russian Protest Movement (R)e-Visited* [Special Issue], *Studies in Russian, Eurasian and Central European New Media* (Digital Icons), 9. Online. Available at: http://www.digitalicons.org/issue09/ (accessed 22 April 2014).

Williams, A. L. and Merten, M. J. (2008) 'A Review of Online Social Networking Profiles by Adolescents: Implications for Future Research and Intervention', *Adolescence*, Vol. 43, No. 170, 253–274.

# Mundane Citizenship: New Media and Civil Society in Bulgaria

## MARIA BAKARDJIEVA

*Abstract*

This essay examines the new forms of civic and political engagement that the increasing accessibility of internet-based media has precipitated in the Bulgarian context. It discusses the results of three case studies which focus respectively on online forum discussions of a significant political event; a campaign of eco-protests; and the activism emerging from a website and forum dedicated to motherhood. The essay argues that new media have brought civic and political issues and the possibility to deliberate and act on them into the everyday lives of Bulgarians. As a result, the voice of Bulgarian civil society has grown stronger and has been able to penetrate the sphere of formal politics, sometimes with important consequences.

MUNDANE CITIZENSHIP IS A CONCEPT I WOULD like to introduce at the outset with the objective of capturing a set of forms of civic engagement that have arisen in a 'mediapolis' (Silverstone 2007) where citizens' competence at intermeshing reception and participation across a variety of media formats has grown significantly (Couldry 2009). These new forms have shown the potential to strengthen, enrich and expand 'the public connection' (Couldry *et al.* 2007), or the awareness of ordinary people about what goes on in their larger public world and their sense of enrolment and agency in that world. To use a somewhat graphic metaphor, new media have taken numerous people out of their deep and isolated private abodes and enabled them to respond, speak and act with regard to social and political issues, even if most of the time in elementary and inconspicuous gestures. Thus the two defining characteristics of mundane citizenship are: first, that it is intertwined with the routine activities and concerns of everyday living; and second, that it is crucially enabled by new media of communication. By introducing this category I do not mean to argue that empowering and activating citizenship is what new media do by design or necessity. The abundant literature concerning the effects of the internet on democracy has demonstrated that there are both bright and shadowy sides to this relationship. As a host of new communication technologies, the internet has been taken up by activists, organisations and politically engaged individuals in inventive and productive ways (Bentivegna 2006; Dahlberg 2007; Dahlgren 2009; Nielsen 2010). At the same time, it also lends itself to alienation and manipulation that can undermine the prospects of democratic civic and political participation among users (Chadwick 2006; Sunstein 2007). 'Mundane citizenship' is a working caption for a range of novel phenomena observed in my empirical research that I interpret as a set of new possibilities, as new ways of operating that place civic participation deep into the heart of everyday life.[1] I set out to investigate what it is possible to achieve with new media in the terrain of citizenship rather than to present a statistically measured account of what is real.[2] Elsewhere (Bakardjieva 2009) I have referred to this approach as the 'method of openings'. Through this method I will inquire into the circumstances of actual civic engagement facilitated by new media and what can be learned from them.

Citizenship is of course a broad category that has been the subject of a wide variety of conceptualisations and definitions.[3] The liberal tradition in political thought defines citizenship as a set of essential rights and freedoms that individuals possess as members of a liberal-democratic state. As such, individuals should be guaranteed the freedom to pursue their own understanding of the good life within a legislative framework that precludes harm to others. This tradition insists on clear and strictly imposed limits on the state's capacity to intervene in the private lives of citizens. The citizen thus emerges as a client of the state and consumer of the protective and enabling services that it has to offer. Against this prevailing background of understanding, the communitarian view of citizenship emphasises a cultural dimension of citizenship which represents a pillar of individual and group identity. Its premise is that community holds ontological primacy over the individual. Partaking in a moral and cultural order of shared values and meanings compels individuals to embrace the common good and to prioritise it over their private needs and interests. A third model—the republican conception of citizenship—emphasises the agency of citizens and their active participation in the political community. Republicanism promotes 'civic virtue', which includes the ability of the individual to set aside his or her personal interests in the name of the public good as well as the individual's active and devoted participation in public life. Classical republicanism has framed public life as the higher honourable ground on which human activity unfolds and puts down the private as far less worthy and significant. With a boundary so pointed and morally charged, the definition of the public as opposed to the private has become a sensitive matter and contested ground. Feminists, for example, have criticised republicanism for ascribing low status to women and their responsibility for sustaining the virtues of the private world. Another important move undertaken by feminist activists and scholars has been the one that calls for rethinking the very boundary between the public and the private as suggested by the formula 'the personal is political' (Hanisch 1970).

The theory of radical democracy, for its part, criticises republicanism for its elitist and overly elevated take on citizenship. Republicanism, this criticism goes, fails to recognise the realities of inequality and diversity in late modern capitalist society and

---

[1]This research comprises three case studies focused on new media use for civic engagement conducted in Bulgaria between 2007 and 2010. These cases constitute the empirical foundation of the discussion that will be presented in the following sections of this essay.

[2]See Lefebvre (1991, p. 9) on the critique of the real with the possible.

[3]For a detailed discussion see Isin and Turner (2002) and Dahlgren (2009).

the impossibility of all citizens being able to acquire civic competences and to actively involve themselves in the affairs of the *polis*. Moreover, the very idea of a public good is highly contested and close to impossible to agree upon among the members of such a complex political community. For that reason, the recognition of private and group rights and interests which may oppose and challenge the hegemonic understanding of the public good is a necessary condition for a contemporary pluralist democracy. For proponents of the radical-democratic view, citizenship is an aspect of individual identity borne through the numerous acts of positioning that individuals perform *vis-à-vis* their surrounding public world. Such acts gain a political edge when people identify themselves with collective entities along the friend–enemy axis (Mouffe 1993, 2005).

Mundane citizenship can be defined from the perspective of the radical-democratic model as subject-positioning and identity work within a dynamic context of intersecting and shifting public discourses and flexible collectives centred on social and political issues. Mundane citizenship is firmly rooted in private experiences, needs and concerns, but it sheds this shell through collective identification and movement from private to interpersonal, group and public discourse. Its mundane nature distinguishes it from the elevated standard put forward by republicanism both in terms of its practical manifestations, which can be unglamorous and even trivial, and its primary residence in the private sphere. Public discourses as they are received, oriented to and made personally meaningful within the private abode of their audiences are a major carrier of this private–public interaction. The multiple and subtle planes of meaning and negotiation at which mundane citizenship manifests itself lie in stark contrast to the formal institutional operation of liberal citizenship. Although it fundamentally relies on collective identities, mundane citizenship has a much more open, fluid and even dissentious nature than the communitarian ideal allows. At the same time it carries in itself elements that have been associated with all these models of citizenship and can look like each one of them in its various manifestations.

Mundane citizenship can be better understood when three interrelated levels of political engagement and action are acknowledged. First, this is the level of politics proper, or as Beck (1997) describes it, the officially recognised and institutionalised political and corporate system. Secondly, politics has a further mode of operation that Beck calls 'subpolitics' where agents coming from outside the official political institutions appear 'on the stage of social design', including different professional groups and organisations, citizens' issue-centred initiatives and social movements, and finally, individuals (Beck 1997, p. 103). As far as individuals are concerned, especially with respect to the developments unfolding within their private worlds, I have argued that a third layer of politics, and respectively citizenship, has to be taken into account, the level of subactivism (Bakardjieva 2009). Subactivism comprises small-scale, often individual and private decisions, discourses or actions that have either a political or ethical frame of reference and never appear on the stage of social design, but on the contrary, remain submerged in everyday life. It involves a variety of inconspicuous processes such as identity construction through subject positioning *vis-à-vis* social and political discourses and relations, the distinction between 'friend' and 'enemy' and identification with collective formations, the discursive re-enactment of debates and clashes with political frames of reference in the private sphere (everyday political talk), as well as practical actions and choices regarding matters of daily living that have wider social and political significance.

Subactivism is the hidden dimension of citizenship that provides a foundation for overt engagement at the levels of subpolitics and politics. It does not guarantee or necessarily transform into actions undertaken by citizens at those levels, but predetermines the possibility for such actions. That is why a central question for the investigation of citizenship as a complex phenomenon is what factors and conditions need to be in place for subactivism to break out of the confines of the private sphere and to percolate into the more visible and institutionalised spheres of activity characterising subpolitics and formal politics. The argument that follows will be woven around empirical evidence suggesting that a key factor for this leap to occur is the imaginative employment by citizens of new media and the novel communication practices growing and consolidating around these media. The main goal will be to identify the diverse 'modes of operation or the schemata for action' (de Certeau 1984, p. xii ) in which these practices organise themselves in a particular social and political context.

*New media and mundane citizenship in the Bulgarian context*

The research site for examining the relationship between new media and mundane citizenship in this project was Bulgaria, a relatively new democracy and a recent member of the European Union. While internet access and use in Bulgaria has grown at a steady rate in the past 10 years, due to the low average household income the country lags behind in European rankings (European Commission, Eurostat 2009). The percentage of the population that regularly uses the internet in Bulgaria was 30.7 in 2009 (National Statistical Institute 2009). It is a fact that large portions of the Bulgarian population have neither the means nor the knowledge or motivation to connect to the internet. At the same time, the young and the educated have eagerly embraced the medium and score highly on measures of intensity and innovativeness of internet use (National Statistical Institute 2009; Bakardjieva 2007). Already by 2006, the annually published e-Bulgaria Report (ARC Fund 2006) registered palpable growth in the number of Bulgarian blogs and the rising profile of several online communities and forums. In that period, the shift from use practices based mainly on email, browsing and interpersonal chat to Web 2.0 type activities began to take place. Thus, the potential of the new media to support civic engagement started to present itself to Bulgarian citizens. The following discussion focuses on concrete instances in recent Bulgarian history where this potential has been vividly demonstrated.

The research project on which this essay is based employed the method of a qualitative case study (Creswell 1998). It involved the discovery and delimitation of cases where mundane citizenship (as per the definition above) could be observed and its relationship with new media closely examined. The specific techniques of data collection included structured reading of media publications, discourse analysis of key texts and informal and formal interviews with informants who had participated in civic action. These informants were selected to represent diverse viewpoints and profiles so that a fuller picture of the events under consideration could be captured. Three individual cases were thus defined and examined in depth: the public discussion of the accession of Bulgaria to the European Union which occurred on 1 January 2007 in mainstream media and popular online forums; the campaign of civic protests that erupted around a 2007 court decision stripping Strandja Mountain of its status as a protected natural reserve; and a series of civic initiatives undertaken by the site bg-mamma over an extended period of time.[4]

In Bulgaria, examples of subactivsim facilitated by new media can be found in the context of online forums such as those offering news media audiences the opportunity to post comments on individual journalistic articles. Another type of discussion forum that attracts a mass following is hosted by larger internet portals such as dir.bg where participants themselves could open topics and subtopics with or without links to journalistic news items. The investigation of the role of internet-based media as tools for mundane citizenship can usefully start out with a close analysis of the nature and contribution of such forums. This is the objective of the analysis of the first case concerning Bulgaria's accession to the EU. Later on, Bulgarian civic society saw the employment of new media such as blogs and online forums in raising attention to particular social and political issues and the organisation of civic campaigns, sometimes with significant results. The case of the 'save Strandja' protests illustrates this different mode of operation of mundane citizenship. Strandja, a mountain in the south-east of Bulgaria, was about to lose its status as a protected natural territory by virtue of a decision by the Supreme Administrative Court. Environmentalists and young people took part in civic protests, which resulted in the reversal of the court's decision. Parallel to that, over several years the online community hosted on the site bg-mamma increased its popularity and public visibility. On a number of occasions it showed an ability to make its voice heard with respect to controversial social issues and to elicit responses and action from formal political institutions. In all three cases the citizenship of the participants involved remained fundamentally immersed in their everyday private lives and related to topical concerns arising in the course of these lives. Subactivism expressed in identifying with subject positions, echoing and reframing public discourses and building interpersonal allegiances remained the beginning and the end of the story in some of the cases. In others, however, subactivism spilled out of its 'private container' to grow into collective mobilisation and action that affected decision making at the highest stages of 'social design', those of government and parliament.

*Talking Bulgaria's way into the European Union*

The first case study is constructed around the lively debate marking the historic event of the official accession of Bulgaria to the European Union on 1 January 2007. It focuses on this 'critical discourse moment', as Gamson (1992) has recommended, in order to trace the commentaries that appeared in various public forums 'by sponsors of different frames, journalists and other observers' (p. 26). Within that colourful multitude of positions and alternative conceptual frames, the analysis seeks to sift out and weigh the contributions of ordinary Bulgarians. To what extent were the interpretations and reframings generated by such citizens in online forums visible and

---

[4]See http://www.bg-mamma.com, accessed 2 December 2010.

consequential for the public debate? What did they have to say about the nature of the forums themselves?

To address these questions, the main source of data was journalistic publications and forum contributions on the issue of Bulgaria's membership in the European Union during the period in which that issue was on the top of the public agenda— immediately preceding and following the date of the formal accession. The forums selected for the study were hosted respectively by the daily newspaper *Sega*; the online publication *Mediapool*; and the web portal dir.bg.[5] All of them were among the most popular of such forums with long-standing communities of users. Among the journalistic publications addressing EU accession on these news sites, those with the largest number of comments by forum participants were selected for examination. In this manner, smaller and more concrete, but lively and thematically and stylistically rich discourse moments and events were isolated within the larger critical discourse moment marked by the accession as a political event. These textual units (publications and the 'tails' of comments that followed them or threads organised around accession-related topics) were subjected to close reading with several objectives in mind: to identify and classify the diverse interpretative frames that different participants propose, or in other words, the different contexts and meanings associated with Bulgaria's acceptance into the European Union; to trace the interplay between these frames and identify the different discursive acts that they were subjected to; to identify the discursive repertoires or styles adopted by participants in the online forums and the rules and norms, if any, that organise discourse production; and to identify roles adopted by participants in the forums and types of relationships arising among them.

One of the most striking observations that emerged from this examination was the resemblance between the online forums and carnival, a cultural institution famously analysed by Bakhtin (1984). Without a doubt, performance in the forums proved to be much closer to the tradition of popular festive forms represented by the carnival than to any models of rational–critical political debate along Habermasian lines.[6] The observed online discussion forums reproduced many of the defining features of the carnival such as suspension of hierarchical differences, not only among participants, but also with respect to the authoritative institutions and voices dominating the accession discourse. Authorities were characteristically dethroned and symbolically kicked around, mocked and castigated. As is typical of the carnival square, prevailing behavioural norms and rules were disregarded, which resulted in a plethora of mock fights, clashes, insults, vulgarity and earthiness. Participants performed discursive actions hidden, or rather defined, by symbolic masks that often were chosen to suggest their strongly stereotyped positions and attitudes. In true carnival style, opposing views clashed to produce eruptions of debate and firm differentiation of participants along friend–enemy lines. Particularly visible were the lines drawn between forum members on the one hand, and journalists considered biased and 'bought' by political

---

[5]See http://www.segabg.com/; http://www.mediapool.bg/; and http://clubs.dir.bg, all accessed 2 December 2010.

[6]See Jankowski and Van Selm (2000) and Kaposi (2006) regarding online deliberation, and Kuntsman (2009) for a view on the carnivalesque properties of online forums; Gardiner (2004) discusses Bakhtin's concepts of dialogue and carnival as an alternative to the Habermasian rational–critical public sphere.

interests on the other. The following comment targets the critical views on the accession expressed by a prominent political commentator who enjoyed high status and recognition in socialist times as well as later, in the era of the new democratic press.

*Buffoon*: Jimmy-boy [refers to the well-known journalist whose article the discussion is trailing] has poured out his proletarian grief. Big tragedy for the Russian pipers [disciples] and the reddies [refers to the red colour, symbol of communism]—Bulgaria joined Europe instead of joining Russia! A-ah, it is so hard to resist pulling the *Makarovets* [a Russian hand gun] from under the pillow and firing out of grief . . .[7]

As much as this comment may voice its author's true feelings and beliefs, and undoubtedly employs deeply insulting labels, it also carries a devious playfulness that is probably evident to all forum participants regardless of their political colour ('reddies' included). This playfulness is revealed to the outside observer only later on in the string of exchanges when Buffoon turns around to explain to another 'forumist', as regular participants call themselves:

*Buffoon:* Peycho, my dear, how can I resist pulling the tails of the reddies? Just to cheer them up a bit, I thought. Look at the Euro-pessimism that has overcome them. It is worse than a hangover 😁😁.[8]

Along another firm line of distinction, that between forum participants as 'ordinary people' and the economic and political elites pushing and extolling the EU accession, arises another authority-challenging practice: the inveterate grounding of interpretations in the earthly conditions of everyday life of the majority of Bulgarians.

*UF1*: And now what? Same old . . . Even if they remove all borders, the hard-working Bulgarian like me won't have enough dough to afford a trip. If he ever does it, it will be with great sacrifices. How does this EU thing benefit me? It does me harm, if anything. Gas is more expensive, and from there the prices of all consumer goods go up, electricity is at European prices, but what about my salary . . .? Officially, the average salary in Bulgaria is 360 leva [about $240 per month] . . . How can you balance a family budget with this amount of money, even if both spouses work? . . . What borders are they talking to me about? What the heck?[9]

Along with the categorical and even vicious friend–enemy identification that animates most of the discussions, some of the voices of the forum clearly favoured a more tolerant and rational–critical debate. Such voices periodically attempted to calm the waters by introducing a wider and more reflexive perspective:

*Archbishop Nikiphor:* There should be joy [at Bulgaria's acceptance into the European Union], but we should also be realistic: the EU is still not inside Bulgarians—as an appreciation of [social] order and respect for society. Until this internal EU emerges, we will

---

[7]Posted on *Sega* forum, available at: http://www.segabg.com/online/new/articlenew.asp?issueid = 2519&sectionid = 5&id = 0001101, accessed 27 June 2012.

[8]Posted on *Sega* forum, available at: http://www.segabg.com/online/new/articlenew.asp?issueid = 2519&sectionid = 5&id = 0001101, accessed 27 June 2012.

[9]Posted on *Sega* forum, available at: http://forum.segabg.com/topic.asp?whichpage = 2&topic_id = 112695, accessed 27 June 2012.

be a different kind of Europeans. A European is a state of the spirit, not a political *status quo*. Don't get too upset with Dmitri [the journalist] ... Everybody has the right to his or her own opinion, be it one or another. The good thing is that we have a forum in which to express it, whatever it is.[10]

While it can be justly argued that such statements and calls for balance and tolerance appear rarely in the forum discussions, the fact remains that they are often made by some of the most respected participants and that all forum members and visitors are exposed to them. Even if one can imagine that the hottest and most extreme discussants would not be moved by such appeals, what the forum does is to open their eyes to the existence of opposite and alternative views among other representatives of the breed of 'ordinary Bulgarians' that can be held and proclaimed with the same passion and conviction as their own. This rampant, loud and irreducible diversity of social languages and views turns the forums into provocative environments for making sense of political events and developments in an open and inclusive way. In the forums, consensus is rarely achieved, but instead conflict is clarified and sharpened, a necessary moment, as Mansbridge (1999) has argued, for achieving equal recognition of opposing interests and views. Too often in the pursuit of rational consensus through formal deliberation the hegemonic definitions of the common good are reaffirmed and oppositional demands silenced. In contrast, to paraphrase Bakhtin, the carnival environment allows thought and speech to be placed under such conditions that the world can expose its other side: the side that is hidden from view that nobody talks about, that does not fit the words and forms of prevailing philosophy. The role of the carnival is that it opens up a space for positions permitting a look at the other side of established values, so that new bearings can be found (Bakhtin 1984, pp. 271–72).

Such bearings were actively and even frantically sought in the major historic moment of Bulgaria's accession into the European Union, a moment when a tremendous load of existing schemas forming the cultural stock of knowledge had to be revised and reshaped. Citizens had to identify with new positions offered in public discourses that spanned the whole range of possibilities between seeing themselves as proud and equal new Europeans, or as cheated and exploited pariahs in an imperial relationship with foreign powers. The various framings of the accession offered in institutionally- and professionally-produced discourses constructed subject positions to be adopted by Bulgarians, but forum participants did not slot themselves into any of them quietly and uncritically. Their debate questioned the received media and political framings of the event, refined, expanded, challenged or undermined these frames. By doing this, forum members were injecting numerous alternatives into the public debate, alternatives stemming from the specific social vantage point occupied by each one of them and from their situated knowledges. The forum gave these citizens the chance to rise above the inaudible rumblings generated around their private kitchen tables (Bakardjieva 2008) and to give their views visibility and legitimacy, but most importantly, to find support, or for that matter, meet with open challenge. Thus ordinary citizens were temporarily and within a limited field assuming the roles and

[10]Posted on *Sega* forum, available at: http://www.segabg.com/online/new/articlenew.asp?issueid = 2519&sectionid = 5&id = 0001101, accessed 27 June 2012.

20

responsibilities of public figures, negotiators and opinion leaders. Additional investigation could show whether and to what extent the musings of forum participants were heeded by journalists, public relations specialists and policy makers and if the turbulent discursive activity of forums influenced these actors in changing anything in their thinking and operation. While forums provided a supportive common ground to ordinary Bulgarians to reflect and make sense of the new dimensions of their European citizenship, they did little more than making the working of subactivism overt and shareable. The kind of civic agency that they supported remained purely verbal, disorganised and largely inconsequential for the larger community. Despite their passion and volume, discussion forums were easy to ignore and hardly ever influenced institutional politics.[11]

## Saving Strandja: the triple helix of the mediapolis

The second case study to be presented here exemplifies a different mode of operation of mundane citizenship where the voices of citizens transmitted by a range of traditional and new media managed to come together to make a definitive and powerful collective statement and with that to effect actual changes on the 'stage of social design'. This case centres around a memorable sequence of events in the summer of 2007 when, by a decision of the Supreme Administrative Court, the biggest Bulgarian natural reserve, Strandja Mountain, located in the south-east of the country, was due to lose its protected territory status. Under pressure from entrepreneurs interested in developing resort sites in the strip of the mountain bordering the Black Sea, the administration of one of the municipalities located in the territory had challenged the legal act granting the mountain its natural reserve status. While the lawsuit unfolded and the court decision was still pending, numerous environmentalist NGOs had worked to put up expert counter-arguments intended to defend the mountain as a natural reserve and to draw public attention to the issue. According to the accounts of the NGO representatives and other activists interviewed,[12] they had employed a wide range of techniques to engage the Bulgarian public including petitions, letters by prominent intellectuals, folklore concerts and children's painting competitions. Environmentalists had put forward a draft legislation to amend the act, an amendment to the act regulating the protected natural territories that would change fundamentally the approach to creating and maintaining natural reserves in the country. Despite all these efforts, the issue did not make it onto the front pages and central news programmes and remained at the margins of public attention.

In this context, the decision by the Supreme Administrative Court to cancel Strandja Mountain's status, and thus protection, as a natural reserve was announced on 29 June 2007. This decision immediately found its way into online news sites, initially without much interpretation or discussion. It left environmental activists

---

[11]For a detailed discussion of this study and the relation between online forums and earlier forms of everyday political talk in Bulgaria, see Bakardjieva (2008).

[12]Interview with Jordanka Dineva, coordinator, Bulgarian Foundation Biodiversity, 27 October 2009, Sofia and interview with Yanina Taneva, PR consultant, 26 May 2010, Sofia–Calgary, via Skype.

numb and helpless in the face of the tremendous disaster they believed awaited Strandja. Their means to react to this turn of events however were quite limited and already exhausted. As one activist explained,[13] none of the officially registered NGOs wanted to take it upon themselves to step up the campaign and call for street protests because that would reflect poorly on their image, legitimacy and further ability to interact with the authorities. At that moment a few environmentally concerned citizens with only weak connections to the formal NGOs took the initiative into their own hands by creating a site named savestrandja.net which announced a time and place for a civic protest. The site also gave instructions regarding the signs and slogans that participants should carry in order to indicate their cause and position. In this way, a common symbolism quickly began to emerge and began to glue together numerous fragmented groups of concerned people. Crucially, the link to the site and its message spread around the already popular online discussion forums, social networks and chats. Attention was also drawn to the site by many of the popular blogs whose authors not only transmitted the appeal, but added their insight to the public understanding of the controversy translating the legal and ecological arguments into ordinary language.

When the first protest—small, but centrally placed—erupted in a city square (29 June 2007, see Figure 1) the dynamic of public attention changed significantly, in no small degree thanks to the reports in popular blogs.

The bloggers were the first to publicise detailed accounts of the event accompanied by pictures and commentary thus leading the way in conceptually framing what was going on. References to blog posts about the protest floated across the Bulgarian internet space, exponentially increasing their exposure. Traditional media followed, sometimes taking cues from the blogs and the forums that kept their sights focused on the events. Incidentally, a couple of journalists themselves wrote blogs and reported both there and in mainstream publications, thus bringing the new and the traditional media unprecedentedly close to one another.[14]

This virtuous circuit of increasing public visibility which, as I have argued elsewhere (Bakardjieva 2010), was fuelled by the energy of the emergent triple helix of online media, traditional media and city square, commanded the attention of political players and forced them into action. As the wave of protests and its reflection in the media of all kinds grew, three different parties represented in the national parliament embraced the amendment proposed by environmental NGOs earlier in the process and introduced it for discussion in the Parliamentary Commission on Environmental Protection and on Water Resources. Within two weeks the amendment had passed its first reading and was transformed into formal legislation by 19 July 2007.

A detailed political analysis of this set of developments is certainly necessary in order to answer the question of exactly what factors led to this phenomenal victory for civil society over commercial and administrative players. The extent to which it was

[13]Interview with Jordanka Dineva, coordinator, Bulgarian Foundation Biodiversity, 27 October 2009, Sofia.

[14]See for example the news item by Simeon Pateev published in *Dnevnik*, 3 July 2007, available at: http://www.dnevnik.bg/dnevnikplus/2007/07/03/355654_protesti_i_aresti/, and the postings on the blog http://nabludatel.blogspot.com written by the same author (accessed 2 December 2010).

FIGURE 1. SPONTANEOUS PROTEST IN SOFIA ON 29 JUNE 2007.
*Source*: Simion Pateev, available at: http://nabludatel.blogspot.com, accessed 2 December 2010
(used with the author's permission).

actually a victory with substantive repercussions for the preservation of the Strandja Black Sea coast as a natural reserve is still being questioned by some observers.[15] I leave the final verdict to political experts and prefer to focus my discussion on the practices of mundane citizenship that manifested themselves in the context of these events. The most important of these practices included the prominent contribution of bloggers who acted as translators, interpreters and amplifiers of the civic concerns, especially thanks to their ability to interact with readers and to be at the head of traditional media in reporting the action in the street.[16] The online discussion forums proved to be fertile soil for the calls for attention to the issue and for organising the response. Further on, social networks based on dedicated sites or simply organisational email lists and interpersonal contacts helped in spreading the announcements and with the consolidation of the demands. Of critical significance was the materialisation of citizens' bodies out of all these virtual enclaves into the physical spaces of the city from where their demands reached the television screens and

[15]Interview with Ivan Bedrov, editor of the newspaper *Pari*, 16 August 2010, Sofia.

[16]Two prominent blogs that closely followed the events and emerged as foci of attention, discussion and organisation were Optimiced.com (http://www.optimiced.com/bg/2007/06/30/civic-protest-to-protect-strandja-park/) and NABLUDATEL (http://nabludatel.blogspot.com/2007/07/20.html; both sites accessed 2 December 2010.

newspaper pages read by the multitude of non-networked Bulgarians. With the ensuing decisive turn of public opinion in favour of the legal protection of Strandja as a natural reserve, politicians were pushed into legislative action, which finally sealed the deal in a way favourable for environmentalists.

Apart from the emergence of bloggers as public opinion leaders, the 'save Strandja' protests demonstrated the key role of new media in forging a connection between environmental NGOs and the online citizenry. The blogs and forums, as one of my informants observed, put the masses of supporters behind the otherwise meagre and expertise-focused environmental organisations. At the same time, the knowledge and leadership provided by these NGOs, the capacity for which had been attained throughout their long-term organisation-building and project work, provided the ideas around which otherwise unengaged citizens could unite their efforts. This can be seen as a clear moment of bridging together the subactivism of ordinary people nurtured mainly within their private spheres and interpersonal networks with the subpolitical stratum inhabited by NGOs and dedicated activists. All in all, the Strandja protests serve as an example of the potential of mundane citizenship to pierce through the other layers of the political system in critical moments when vital issues of the *polis* are at stake. In Bulgaria, specifically, these protests remained in the public memory also as a manifestation of the capability of ordinary citizens to rise above complacency and to tell the political and economic elites 'enough is enough'. Some commentators in Bulgaria (Popov 2007) saw them as a sign of the arrival of a new generation, empowered by new media among other things, that refuses to put up with the corruption practices that had thrived in the country for many years. Others agree that the protests were inspired not so much by environmentalist but by anti-corruption sentiments.[17] Whatever the case may be, the materiality and efficiency of mundane citizenship as a political force was proven beyond doubt in the Strandja events.

## *Mothers in arms: the commanding voice of bg-mamma*

The third mode of operation of mundane citizenship discerned in the framework of this project transpires in the case of a prominent online community that has acquired significant visibility and weight as a factor in Bulgarian social and political life. The site bg-mamma was created in 2002 to host the useful information accumulated in the course of discussions occurring in a forum devoted to maternity within the Bulgarian online portal dir.bg.[18] Gradually, it provided a plethora of its own forums organised around different topics related to pregnancy, maternity, child-rearing, type of family experience and others.[19] The common bond created by the maternity experience provided a powerful glue for the community that arose around the site.

Over the years, the site and its forums have registered remarkable growth not only in the numbers of participants, postings, themes and hits, but also in their internal

---

[17]Interview with Ivan Bedrov, editor of the newspaper *Pari*, 16 August 2010, Sofia. Many in the Bulgarian public saw the decision of the Supreme Administrative Court as serving the interests of commercial developers and having been 'bought' by them.

[18]ARC Fund (2006), and interview with site owner Julian Kuzmanov, 26 October 2009, Sofia.

[19]See for example: Babies: http://www.bg-mamma.com/index.php?board=97.0; Future Mothers: http://www.bg-mamma.com/?board=1.0 (accessed 2 December 2010).

structural organisation. Technologically, the site was updated at a moderate pace in order to maintain its easy and intuitive interface as well as to slowly expand the kinds of functions and features that would make it a hospitable place for ever more internet-savvy young women. It pioneered a business model based on commercial advertising as a source of funding and yet, at the same time, sustained a firm commitment to a community-service value system.[20] Unique for the site's internal regulatory mechanism are a set of rules of conduct spelled out explicitly and which are mandatory for new users to accept before they can receive access to the site. Central among them are the commandments prohibiting unsolicited commercial and political promotion and advertising.[21]

Even more noteworthy is the way in which these rules are enforced in the daily operation of the site. A dedicated team of moderators closely monitor the content and tone of the discussions and intervene every time they see a deterioration of the exchanges into profanity, insult or gibberish. They have the authorisation to ban users from the site temporarily or even permanently when the offence has been severe. These moderators are volunteers who invest substantial amounts of their personal time in their role. Some of them construe it as an opportunity to pursue personally important causes (for example breastfeeding in the case of one interviewed moderator)[22] and as a kind of community service that gives them the respect of others and a sense of fulfilment. Moderators also admit that the performance of this function in the forum has allowed them to grow both in terms of interpersonal and communicative skills as well as with respect to knowledge and social capital. These returns, they point out,[23] have sometimes led to palpable professional and personal gains in their offline lives.

There is more to be learned from the system through which content production and discussion in the site are regulated, but one of the main outcomes to highlight is that users flock to the site with ever greater interest and devotion. Many of them participate for years, long after the initial problem or curiosity that first led them to the site has been forgotten. Thus the site has developed circles of topical discussion around more advanced stages of the motherly and even grand-motherly career. It also offers a range of forums with thematically open free content (e.g. the Gossip-shop) where mere socialising occurs and all kinds of current and long-standing social and political issues are tackled. In the course of this activity regular users have made friendships and established friendship-like relationships with others whom they sometimes know only by their nickname, but whom they deeply respect and trust. Overall, a spirit of community and solidarity has established itself on the site. Members' sense of belonging to a collective entity has often led them to turn to the bg-mamma forums for support for a cherished civic cause. In such instances, the subactivism of individual site members has managed to catalyse the emergence of collectivities of significant size and energy, ready and willing to mobilise for civic action. Consequently, bg-mamma has won a name and image for itself as a formidable force to be reckoned with by different

[20]Interviews with Julian Kuzmanov, site owner, 26 October 2009, Sofia; and with bg-mamma moderators Reza, 13 July 2010 and Vache, 14 August 2010 (pseudonyms used in forum).

[21]Forum rules, available at: http://www.bg-mamma.com/index.php?topic=23714.0, accessed 2 December 2010.

[22]Interview with Boo (pseudonym used in the forum), bg-mamma moderator, 16 July 2010, Plovdiv.

[23]Interviews with Boo, 16 July 2010, Plovdiv and Reza, 13 July 2010, Sofia, bg-mamma moderators.

types of economic and political players. On one occasion the site was blamed for almost crashing a Bulgarian bank by circulating rumours about its bankruptcy that immediately triggered the massive withdrawal of investments by bank clients. This story was never proven to be true, but it can still serve as an indication of the perceived power and potential influence of the bg-mamma community.

More factually correct and well recognised are the numerous civic initiatives of bg-mamma members, many of which produced important changes in the Bulgarian state's social policy. Several social movements, for example the Movement of Bulgarian Mothers (*Dvizhenie na balgarskite mayki*) and Civic Alliance Smile with Me (*Grazhdansko sdruzhenie 'Usmihni se s men'*), have emerged from the site and have established themselves as key representatives of civil society taking part in policy making in their specific areas.[24] Thanks to well staged protests by mothers with their baby carriages on the official Day of the Child (1 June) in 2005, the plodding parliamentary discussions regarding the increase of maternal benefits were sped up and eventually produced legislation ensuring better state support for maternity and children. A forceful initiative triggered by a BBC documentary on the living conditions in a home for abandoned children with disabilities in Northern Bulgaria was undertaken in 2006–2007. Like in the Strandja case, it was informed by the expertise accumulated by NGO representatives working in that area. The mothers from bg-mamma carried out not only charitable activities to supply that home and similar ones with necessary materials, but also exerted pressure on central and local administrations which brought about inspections and changes in the homes' staff and management. Most importantly, the bg-mamma members, many of whom were mothers with disabled children, headed a sustained campaign to turn around government policy concerning children with disability, to fundamentally change the philosophy on which this policy was built and to get public opinion to realise and support the need for change. Once again, as in the case of Strandja, not only petitions and protests, but a wide media engagement was achieved. Although, no instantaneous improvement of the situation or abrupt shift in the government's approach was possible in this instance, the conversation was opened and the respective administrative bodies set on the path of new policy development.

Most recently, mothers from bg-mamma spearheaded a protest against planned liberalisation of the legislation regulating the growth and consumption of genetically modified organisms (GMOs) in Bulgaria. Once information concerning the drafted changes in the law started to crop up in different news outlets, some participants in the forum fired up a mobilisation campaign determined to stop these changes and, as they themselves put it, to protect their children and Bulgarian nature from the threat of scientific innovations with insufficiently understood long-term effects (petition prepared by bg-mamma).[25] As a typical example of mundane citizenship, this struggle stemmed out of the private commitment some mothers had made to feed their children with natural and local produce as a way to secure for them a healthy lifestyle. From a

---

[24]Interview with Teodora Piralkova, bg-mamma activist, Chair of National Alliance Smile With Me, 6 July 2010, Sofia.

[25]See petition text, available at: http://www.grazhdani.eu/text_peticija_gmo.html, accessed 2 December 2010.

personal choice of the individual family and household, in the face of political decisions that reflected on the private lives of the whole nation, this commitment transformed into a fiercely contested political issue.

As often happens in the popular–political imagination (but also in actual practice), government representatives involved in the legislative process were suspected by the mothers of selling out the national interest to American corporations and playing by the tune of foreign political figures transmitting these corporate demands. In response, the full palette of techniques of popular resistance was employed: mothers stood in the January cold of 2010, petition and children in hand, waiting for the prime minister to present their demands directly to him; personal connections were sought with Bulgarian scientists and members of the parliamentary committee working on the amendment; and notably, journalists who regularly covered environmental issues were recruited into the movement and their detailed understanding of how the legislative process works proved essential to the staging of the right moves at the right time.[26] These journalists also played the role of the movement's Trojan horse in the parliament building and some of them blogged and posted information on the bg-mamma site live as they sat and observed the committee's deliberations.[27]

Some of the most engaged bg-mamma members grew into real activists for the period of the protests, writing and editing pamphlets and petitions, chasing politicians, sitting in parliamentary committee meetings as well as in meetings of the *ad-hoc* civic alliance with environmental and professional organisations (including the Bulgarian Association for Bio-products (*Bulgarska asotsiatsiya za bio-produkti*) and the Bulgarian Association of Professional Chefs (*Bulgarska asotsiatsiya na gotvachite*)). Although, by doing this, they had moved out of their daily routines and into a sphere of political activity, they remained based in their own kitchens. There, they composed texts late at night, on their own personal computer so that no other household members could touch it for days or weeks. Their civic zeal was intermingled with a sense of guilt in relation to their young children who had not eaten a home-made meal for a while.[28] Although not all bg-mamma members supporting the action were involved to the same degree, a majority of them acted, in the words of one informant, as bees collecting information, digging out phone numbers of politicians and specialists, searching their interpersonal networks for links to people to lobby, preparing postcards with their children that insisted on the preservation of the natural fruits of the Bulgarian land.[29]

Once again, valuable lessons can be learned from the organisation of that particular campaign which ended with a victory for the civic alliance led by bg-mamma's activists. The proposed text that was intended to open Bulgaria to GMOs was turned down by the parliamentary committee and never entered the parliament floor. Without taking a position on the disputed issue of whether this resolution was right or overly protectionist and to what extent it is enforceable and sustainable in the long

---

[26]Interview with Nadalina Aneva, reporter at the newspaper *Sega*, 13 August 2010, Sofia.

[27]Interview with Boycho Popov, reporter, *BNews*, 2 July 2010, Sofia.

[28]Interview with ПО (pseudonym chosen by respondent), bg-mamma member, 29 June 2010, Sofia.

[29]See a bg-mamma theme focused on organising one of the civic protests against GMOs, available at: http://archives.bg-mamma.com/archive/index.php?topic=482706.msg14518839#msg14518839, accessed 2 December 2010.

term, Bulgarian civil society indisputably witnessed the power of mundane citizenship to transform into a major force operating at all three levels of politics and making a real difference, even if only within a very particular scope.

What distinguishes the case of bg-mamma's civic initiatives from the ones described in the previous cases was their emergence, not in an amorphous and spontaneous wave of civic energy ('save Strandja', or in a multivocal, dissentious and inconclusive argument among 'wild publics' (Gardiner 2004), but in a firm base of a relatively stable online community. Certainly, in its day-to-day operation bg-mamma does not represent an idyllic community. It is not comprised of regular, long-term friendly relationships among members. It pulsates and changes; it hosts as many quarrels as agreements. Nevertheless, it subscribes to a set of ethical norms that help establish a reliable moral entity, the members of which have a reasonable expectation that their views will be valued and their personal worth and dignity respected, enough to believe that they can find support for not only solving their private problems with baby food and nappies, but also for their civic concerns and passions. The faith in this kind of fluid, but nevertheless reliable, ethical community helped elevate gestures of subactivism above their private base into the realm of subpolitics and politics proper.

*Conclusion*

Mundane citizenship enabled by new media manifests the power of ordinary people who are not political operators or dedicated members of formal NGOs and social movements, to engage, participate and sometimes change developments on the large political stage of social design. The internet has allowed users to navigate public discourses and to identify with positions constructed in them, to challenge, change and reframe these positions from the comfort of their own homes and working offices as a matter of course in their daily life. The internet, most typically through open online forums, has brought Bulgarian users together with like-minded individuals on issues of social and political controversy, but even more importantly, it has brought them together with people of radically different ideological backgrounds and convictions. In that process, forum participants have had the opportunity to practise 'everyday political talk' (Mansbridge 1999) in an open setting, which has led to a growing awareness of diversity, the overcoming of pluralistic ignorance, clarifying and sharpening of conflict and building collectivities with friends. The internet has also offered subjects access to remote and often anonymous institutions of power, the most obvious example being the mass media and their representatives who can now be challenged in their agenda-setting and issue-framing privileges. Holders of state and economic power have in turn become the object of criticism, derision and castigation, even if only symbolically, sometimes to an extent that has commanded their and their officers' attention. Taken a step further in terms of community identification and loyalty to common causes, internet discussion sites, forums and blogs have supported collective organisation of civic action in the real world that has led to actual social change.

The discovery of such a diverse set of practices of mundane citizenship in the Bulgarian context is good news for Bulgarian democracy. For years on end observers and analysts have complained that the heavy cultural and political inheritance of totalitarianism has skewed the value system of Bulgarians and has deprived them of the ability to look beyond their private interests and care for the well-being of the larger community (Howard 2003; Paunov 2009). This preoccupation with individual survival, often at the expense of the common good, has been seen as a barrier to the development of true civic engagement and activism. Civil society organisations and NGOs in Bulgaria have been typically donor driven, funded by external sponsors, elitist and largely disconnected from the general population (Tancau 2007; Andreeva *et al.* 2005). Yet at the same time, Bulgarians have learned some important lessons during the period of transition. They saw a new political class grow out of former colleagues and kin, that is, from a place very close to their everyday lives, and subsequently experienced the deep effects of the decisions made by these political elites in the course of the economic reform. In that sense, the world of formal political institutions in a new democracy like Bulgaria is conceptually much closer to people's daily thoughts and struggles than in long-standing democracies where the political establishment is far removed (through a long history of class selection, professionalisation and institutional differentiation) from the everyday world.

The new media arrived in this cultural context and helped people draw links between concrete daily concerns, on the one hand, and political discourses and decisions, on the other: between the lofty statements of political leaders and journalists regarding a major political move and the personal experiences and worries of people at home; between a court resolution and the evident destruction of Bulgarian natural treasures such as the Strandja Black Sea coast; between the regulation of GMO proliferation and the health and quality of life of future generations. The new technical networks and communication forms in the hands of Bulgarians afforded the proverbial 'doing something about it' that previous generations have missed. They allowed new social bonds with other ordinary people to be found online, as the recounted cases show. The new media, importantly, helped connect the otherwise isolated and expertise-focused NGOs working in different areas with the energy of a mass of people ready to take over the city squares or stand in the cold for hours to indicate their demands. Finally, one can hope that the recounted events may have sensitised political leaders to the fact that new-media equipped citizens represent a significant factor to be reckoned with when decisions are made. The will of numerous ordinary people to participate first-hand in the legislative process around issues of concern was also registered in the course of some of the cases.

While it may be still too early to draw grand conclusions about the revolutionary potential of new media in the hands of active citizens, there are certainly multiple practical lessons to be learned by examining the mechanics and dynamics of successful manifestations of civic agency such as these. The models of interaction between bodies occupying different levels of the *polis* that emerge from such cases could be usefully seen as best practices to be consolidated and spread around. These are also practices to be defended from cooptation and distortion on the part of corporate interests that are bound to notice the transformative power of mundane citizenship. Thus, it remains an imperative task for academics, activists and engaged citizens alike to carefully study the instances highlighting the arrival of new possibilities for civic agency and to reflexively incorporate them into their future work.

*University of Calgary*

## References

Andreeva, D., Doushkova, I., Petkova, D. & Mihailov, D. (2005) 'Civil Society without the Citizens: An Assessment of Bulgarian Civil Society (2003–2005)', *Civicus Civil Society Index Report For Bulgaria* (CIVICUS: World Alliance for Citizen Participation), available at: http://www.balkanassist.bg/vfs/7a8335e7501a9c31ba5c5936e212bcb2_1/146e4472ed0781b564d987e71735d4ab.pdf, accessed 28 August 2010.

ARC Fund (2006) *E-Bulgaria 2006*, available at: http://www.csd.bg/imgShow.php?id=3177&art_id=7920, accessed 28 August 2010.

Bakardjieva, M. (2007) 'The Mysterious East: Pluses and Minuses in the e-Europe Equation', in Anderson, B., Brynin, B. & Raban, Y. (eds) (2007) *Information and Communications Technologies in Society* (London & New York, Routledge), pp. 62–77.

Bakardjieva, M. (2008) 'Bulgarian Online Forums as Carnival: Popular Political Forms and New Media', in Sudweeks, F. & Ess, C. (eds) (2008) *Cultural Attitudes towards Technology and Communication* (Murdoch, WA, Australia School of Information Technology, Murdoch University).

Bakardjieva, M. (2009) 'Subactivism: Lifeworld and Politics in the Age of the Internet', *The Information Society*, 25, 2, pp. 91–104.

Bakardjieva, M. (2010) 'The Internet and Subactivism: Cultivating Young Citizenship in Everyday Life', in Dahlgren, P. & Olson, T. (eds) (2010) *Youth, Civic Participation and ICT* (Gothenburg, Nordicom).

Bakardjieva, M. (2012) 'Reconfiguring the Mediapolis: New Media and Civic Agency', *New Media and Society*, 14, 1, pp. 63–79.

Bakhtin, M. (1984) *Rabelais and his World* (Bloomington, IN, Indiana University Press).

Beck, U. (1997) *The Reinvention of Politics: Rethinking Modernity in the Global Social Order* (Cambridge, Polity Press).

Bentivegna, S. (2006) 'Rethinking Politics in the World of ICTs', *European Journal of Communication*, 21, 3, pp. 331–43.

Chadwick, A. (2006) *Internet Politics: States, Citizens and New Communication Technologies* (New York, Oxford University Press).

Couldry, N. (2009) 'Does "the Media" Have a Future?', *European Journal of Communication*, 24, 4, pp. 1–13.

Couldry, N., Livingstone, S. & Markham T. (2007) 'Connection or Disconnection? Tracking the Mediated Public Sphere in Everyday Life', in Butsch, R. (ed.) (2007) *Media and Public Spheres* (New York, Palgrave Macmillan), pp. 28–42.

Creswell, J. (1998) *Qualitative Inquiry and Research Design: Choosing Among Five Traditions* (London, Sage).

Dahlberg, L. (2007) 'The Internet and Discursive Exclusion: From Deliberative to Agonistic Public Sphere Theory', in Dahlberg, L. & Siapera, E. (eds) (2007), pp. 128–47.

Dahlberg, L. & Siapera, E. (eds) (2007) *Radical Democracy and the Internet: Interrogating Theory and Practice* (London, Palgrave).

Dahlgren, P. (2007) 'Civic Identity and Net Activism: The Frame of Radical Democracy', in Dahlberg, L. & Siapera, E. (eds) (2007), pp. 55–72.

Dahlgren, P. (2009) *Media and Political Engagement: Citizens, Communication and Democracy* (Cambridge, Cambridge University Press).

de Certeau, M. (1984) *The Practice of Everyday Life* (Berkley and Los Angeles, University of California Press).

European Commission, Eurostat (2009) *Information Society Statistics at Regional Level* [statistical report], available at: http://epp.eurostat.ec.europa.eu/statistics_explained/index.php/Information_society_statistics_at_regional_level, accessed 29 August 2010.

Gamson, W. (1992) *Talking Politics* (Cambridge, Cambridge University Press).

Gardiner, M. (2004) 'Wild Publics and Grotesque Symposiums: Habermas and Bakhtin on Dialogue, Everyday Life and the Public Sphere', in Crossley, N. & Roberts, J. (eds) (2004) *After Habermas: New Perspectives on the Public Sphere* (Oxford, Blackwell), pp. 28–48.

Hall, S. (1996) 'Introduction: Who Needs "Identity"?', in Hall, S. & du Gay, P. (eds) (1996) *Questions of Cultural Identity* (London, Sage Publications), pp. 1–17.

Hanisch, C. (1970) 'The Personal is Political', in Firestone, S. & Koedt, A. (eds) (1970) *Notes from the Second Year: Women's Liberation*, available at: http://scholar.alexanderstreet.com/download/attachments/2259/Personal+Is+Pol.pdf?version=1, accessed 19 August 2010.

Howard, M. M. (2003) *The Weakness of Civil Society in Post-Communist Europe* (Cambridge, Cambridge University Press).

Isin, E. & Turner, B. (2002) *Handbook of Citizenship Studies* (London, Sage).

30

Jankowski, N. W. & Van Selm, M. (2000) 'The Promise and Practice of Public Debate in Cyberspace', in Hacker, K. L. & van Dijk, J. (eds) (2000) *Digital Democracy: Issues of Theory and Practice* (London, Sage Publications), pp. 149–65.

Kaposi, I. (2006) *Virtual Deliberation An Ethnography of Online Political Discussion in Hungary*, doctoral dissertation, Central European University, Budapest, available at: http://web.ceu.hu/polsci/dissertations/Ildiko_Kaposi.doc, accessed 12 November 2010.

Kuntsman, A. (2009) *Figurations of Violence and Belonging: Queerness, Migranthood and Nationalism in Cyberspace and Beyond* (Bern, Peter Lang).

Lefebvre, H. (1991) *Critique of Everyday Life. Vol. 1: Introduction* (London & New York, Verso).

Mansbridge, J. (1999) 'Everyday Talk in the Deliberative System', in Macedo, S. (ed.) (1999) *Deliberative Politics: Essays on Democracy and Disagreement* (New York, Oxford University Press), pp. 211–39.

Mouffe, C. (1993) *The Return of the Political* (London, Verso).

Mouffe, C. (2005) *On the Political* (London, Routledge).

Natsionalen statisticheski institute (National Statistical Institute) (2009) *Informatsionno obshtestvo— Danni '1.1.4. Litsa, regulyarno izpolzvashti internet—Danni—2009*, available at: http://www.nsi.bg/otrasal.php?otr=17&a1=491&a2=492&a3=493#cont, accessed 29 August 2010.

Nielsen, K. R. (2010) 'Mundane Internet Tools, Mobilizing Practices, and the Coproduction of Citizenship in Political Campaigns', paper presented at *Internet, Politics, Policy 2010: An Impact Assessment*, OII, Oxford, available at: http://microsites.oii.ox.ac.uk/ipp2010/system/files/IPP 2010_Nielsen_Paper.pdf, accessed 29 October 2010.

Paunov, M. (2009) *Tsennostite na bulgarite: Suvremenen portret na evropeyski fon* (Sofia, Universitetso izdatelstvo stopanstvo).

Popov, Y. (2007) 'Nashestvieto na strandzhanskite blogeri', *Dnevnik*, 5 August.

Searle, J. (1969) *Speech Acts: An Essay in the Philosophy of Language* (Cambridge, Cambridge University Press).

Silverstone, R. (2007) *Media and Morality: On the Rise of the Mediapolis* (Cambridge, Polity).

Sunstein, C. (2007) *Republic.com 2.0* (Princeton, NJ, Princeton University Press).

Tancau, M. (2007) *Civil Society in Romania and Bulgaria* (CEDAG, European Council for Non-Profit Organizations), available at: http://www.cedag-eu.org/uploads/File/CEDAG%20-%20Civil%20 Society%20in%20Romania%20and%20Bulgaria%202007.pdf, accessed 29 August 2010.

# The Role of Social Networking Sites in Civic Activism in Russia and Finland

## BORIS GLADAREV & MARKKU LONKILA

*Abstract*

This essay compares the role of the social networking sites, Facebook in Finland and Vkontakte in Russia, in organising civic activism. It is based on data collected about two successful campaigns mobilised by local residents against urban building projects in St Petersburg and Helsinki in 2009. Though in both cities these sites were important channels for transmitting information and organising and coordinating the campaigns, their role was clearly limited in terms of impartial, democratic discussion of the issue of common concern: the sites were rather used to build and reinforce emerging collective identities and to create consensus within the movement. In contrast to the situation in Helsinki, Vkontakte also had a central role in creating and maintaining ties between formerly isolated campaigns against building projects elsewhere in St Petersburg, thereby helping to build a 'network of grassroots resistance' in the city.

THIS ESSAY COMPARES THE ROLE OF SOCIAL network sites (SNS), namely Facebook in Finland and Vkontakte (In Contact) in Russia, in civic activism. It draws on the data collected about two movements organised by local dwellers against new building projects at Komendantskii Prospekt 40 (KP40) in the Primorskii district of St Petersburg, and in the Kumpula district of Helsinki, in 2009. Both campaigns used social network sites in their actions and, most importantly, both managed to stop the building projects. In particular, the campaign by the dwellers at KP40 was one of the very few recent successful actions against building developers in St Petersburg and has thus served as an example for several similar movements in the city.

The data come from a detailed examination of the relevant segments of Finnish Facebook and Russian Vkontakte that were created by users to support these campaigns. Our examination of these social media sites aims to find out first, how and to what extent they function as particular kinds of public spheres offering an arena for debate and discussion of issues of common concern, and second, what their role has been in practical organising of the movement, notably in recruiting followers,

We thank Jeremy Morris, Natalia Rulyova, Vlad Strukov, Tuomas Ylä-Anttila, the anonymous reviewer of *Europe-Asia Studies* and the members of the Helsinki Research Group for Political Sociology seminar for useful comments on the essay. The research has been supported by the Academy of Finland.

coordinating action and creating and maintaining networks between actors. In addition, we utilise thematic interviews with local activists (conducted in 2010, two in Helsinki and eight in St Petersburg, each lasting on average 1.5 hours), of which one interview in each city focused especially on the use of social networking sites in organising the local protests; as well as data from media debates concerning the campaigns in the two cities; and the personal experiences of the Finnish campaign of one of the authors. All our informants were assured anonymity.

The next two sections briefly review existing scholarship of the relationship between social networking sites and social movements, describe the emergence of the local movements and depict the nature of the campaigns against the building projects in the two cities. The two subsequent sections investigate the role of social networking sites in these campaigns. In both of these sections we will first briefly describe how the local dwellers' action group sites—Vkontakte in St Petersburg and Facebook in Helsinki—fit into the overall repertoire of online and offline actions conducted by the local inhabitants. Then we will describe how these sites were founded and what their practical organisation and structure was like. Finally, based on examination of the sites, we will address two aspects of civic activism: the kind of debates that took place on the sites concerning the various campaign-related issues, and how the sites were used for practical organisation and mobilisation of the campaigns.

## Social networking sites and civic activism

### Studies of the role of social networking sites in civic activism

Although the role of information and communications technology (ICT) and the internet in political and civic activism has been the subject of a vast body of research literature, this literature has produced somewhat heterogeneous findings.[1] This state of affairs is partly due to the dynamic and continuously changing nature of ICT. In addition, ICT and the internet consist of a great number of various technologies (Häyhtiö 2008; Breindl 2010). Consequently, addressing the relationship of the internet to collective action runs the risk of lumping together very different applications with diverse uses and functions. In order to avoid these pitfalls, our analysis focuses on Vkontakte and Facebook, two specific social networking sites, and on their use by two local social movements in Russia and Finland. We believe that analysing empirical data of particular social networking sites at close range, instead of addressing 'the internet' in general, will strengthen the analytical power of our text.

Moreover, the huge and growing popularity of both Facebook and Vkontakte renders them interesting objects of analysis in their own right. Facebook, for example, reached 500 million users in July 2010 which made it one of the most popular internet applications worldwide (Wauters 2010). The global hegemony of Facebook conceals,

---

[1]For ICT and civic activism, see for example, van de Donk et al. (2004), McCaughey and Ayers (2003), Dahlgren (2005), Breindl (2010) and Oates et al. (2006). For a recent review of this literature, see Baumgartner and Morris (2010, pp. 25–26). For the Russian internet see Schmidt et al. (2006), Lonkila (2008), Fossato and Lloyd (2008), Goroshko and Zhigalina (2008) and Etling et al. (2010), as well as publications in Digital Icons: Studies in Russian, Eurasian and Central European New Media (available at: http://www.digitalicons.org, accessed 15 January 2011).

however, important exceptions. In addition to China, Japan, South Korea and Brazil, the Russian Federation is one of the few countries where Facebook is not the leading SNS but is clearly in second place, with one million users, in comparison with its local rivals (Barnett 2010; Sweeney 2010). The most popular Russian social networking sites are Vkontakte and odnoklassniki.ru, of which the former claimed in August 2010 to have over 86 million registered users.[2] Vkontakte was also an obvious choice for one of the activists and our key informant at KP40, who founded the Vkontakte group for the movement. Despite the popularity of social networking sites, until recently relatively few studies have explicitly addressed their role in promoting activism (Langlois *et al.* 2009; Bortree & Seltzer 2009; Smith *et al.* 2009; Baumgartner & Morris 2010; Smuts 2010; Zhang *et al.* 2010; Johnson *et al.* 2011),[3] and these studies have produced mixed results. Smuts (2010, pp. 80–86), who investigated the role of Facebook during the 2008 presidential campaign in the USA, concludes, for example, that it both facilitated the formation of public opinion and included new participants in the political process. However, Baumgartner and Morris (2010, p. 24), who studied the relationship between the use of social networking websites and political engagement of young adults during the early stages of the 2008 presidential primary season, found that despite the promise of these sites to increase political interest and participation 'among a chronically disengaged cohort', there was little evidence to suggest that the sites had facilitated significantly greater political knowledge, engagement or participation. Rather, their analysis proposes that many young adults used the sites to look for information that conformed to their pre-existing political opinions and they were no more inclined to participate in politics than are users of other media (Baumgartner & Morris 2010, pp. 24–25). Further, Neumayer and Raffl (2008) argue that social software has the potential to promote grassroots activism, political inclusion and community building but note that local and global inequalities may constrain this potential. Johnson *et al.* (2011) discovered that social networking sites predicted offline and online political activities but did not boost intention to vote.[4] Finally, Zhang *et al.* (2010) examined, among other things, reliance on social media and social networking sites such as YouTube, Facebook and MySpace in 'civic' and 'political' participation. By civic participation they referred to activities that addressed community concerns through non-governmental or non-electoral means, such as working on a community project, whereas political participation denoted activities aiming to influence the selection of elected officials or to develop or

[2]Due to huge commercial interests, the user statistics should be interpreted with care. According to one study in May 2010, 31% of Russian internet users visited Vkontakte daily while the corresponding figure for odnoklassniki.ru was 21%, and Facebook was 1% (Reiting populyarnosti 2010). Vkontakte seems thus to be the clear leader of the two, particularly in north-western Russia. For studies on Vkontakte, see Khveshchanka and Suter (2010) and Suleymanova (2009).

[3]Danah Boyd's bibliography on social network sites is comprised of 338 book chapters and articles (available at: http://www.danah.org/researchBibs/sns.php, accessed 13 August 2010), of which only slightly more than 1% of the 338 chapters and articles related to contentious action.

[4]However, one clear finding concerning civic involvement on social network sites and blogs in the USA is that it is the domain of young people. Smith *et al.* (2009) found in their 2008 survey that the respondents under 35 years of age represented 28% of all respondents, but made up 72% of those who made political use of social networking sites.

implement public policy. The authors found that reliance on social networking sites was positively related to civic participation but not to political participation.

These findings suggest that more research is needed to spell out the actual impact of social networking sites on enhancing civic activism and participation. On the one hand, these sites seem to be quick and easy means to bring up and debate issues of common concern and to mobilise joint action, even on the global scale (Neumayer & Raffl 2008). On the other hand, the ease of participation has raised questions about the depth of the commitment of 'cyberactivists'. Moreover, both linguistic barriers and social, political and global inequalities cast doubt on the image of the internet as a homogeneous global space where everyone can be connected with everyone regardless of the constraints of physical space (Rohozinski 1999; Lonkila 2008). Finally, though social networking sites may have the potential to enhance activism, other uses are actually much more popular. As Danah Boyd has commented (2008, pp. 114–15), on social networking sites 'exchanging gossip is far more common than voting'.

Many of the studies reviewed above draw from survey data and seek to establish causal, law-like relationships between SNS use and political engagement, often in relation to election results in the USA. Though important, they tell us little about the processes through which these sites benefit emerging grassroots level social movements in real life, which is the aim of the current study. When analysing actual examples of social media, such as Facebook and Vkontakte, one has to take into account that they not only imply a particular view of sociability but also constrain and mould interaction in specific ways. Facebook, for example, has been built upon the idea of a personal network formed around each Facebook user who may be linked both to other users as well as to other nodes in and outside the system. Mejias (2010) remarks that this 'nodocentrism' and the 'privatised sociality' may have contradictory effects: for example, the increased user freedom to navigate in the social networking sites is conditioned by the corporate determination of how to develop the system, and the diversity of voices is countered by the homogenisation of platforms.

*Social networking sites as arenas and tools for social movements*

Our theoretical approach draws on the discussions regarding the public sphere and social movements, which we consider as mutually constitutive: movements are, on the one hand, dependent on the public sphere to further their causes and win supporters; on the other, they are also central actors in bringing new issues to public discussion. This intrinsic connection is emphasised by, among other authors, Tuomas Ylä-Anttila (2005), according to whom social movements played a central role in the very formation of the national public spheres:

> Not only have social movements contributed to the processes where issues are raised to the public agenda, but the very ideas and institutions of public debate have evolved in the course of political conflicts and through collective action in social movements. (Ylä-Anttila 2005, p. 425)

In terms of our actual empirical data, the social networking sites founded by the movement activists in Russia and Finland may be considered as particular kinds of public spheres where people could debate issues of common concern. In addition, these sites were also important in terms of actually organising and mobilising the movements. Therefore, we define these two aspects, or in fact functions, by using the metaphors of arenas and tools for civic activism. The public sphere function considers the social networking sites as arenas for debating the goals and strategies of joint action, whereas the organising function emphasises the role of these sites as tools for mobilising emerging social movements. In terms of the first function, the preconditions for an ideal Habermasian public debate should disregard the status of discussants, address issues of common concern, and by definition should not exclude any person.[5] In our empirical analysis we focus on the first point by asking to what extent debates in the social networking sites in the two cities strove to reach consensus based on the best arguments instead of social status and to what extent they were conducted free of domination.

Research suggests, however, that in real situations, and particularly in internet debates, the abovementioned ideal preconditions rarely exist. Dahlgren (2005, pp. 156–57) remarks on how the procedures of open discussion may result in 'rationalist bias'. This bias neglects, for example, affective, poetic, humorous and ironic modes of communication, and downplays the power relations built in to the communicative situations. Moreover, it forgets that in cases of extra-parliamentary politics the political discussion may, instead of consensus, strive for political mobilisation and building of collective identity. Many of these 'biases' were indeed detected in our study of the debates conducted in Vkontakte and Facebook.

The second function focuses on the role of Vkontakte in St Petersburg and Facebook in Helsinki in organising the movement; in other words, in recruiting followers, coordinating activities and building networks, both within the movement and between the movement and the outside world. Though related and sometimes overlapping with the public sphere aspect of social networking sites, in the remaining part of the essay these organising functions will be distinguished analytically from the public debating about the aims and strategies of the movement. This two-fold division structures the empirical analysis of the sections investigating the role of Vkontakte and Facebook in organising protests in St Petersburg and Helsinki.

### Campaigns against building projects in St Petersburg and Helsinki

#### 'No to the building of a new block of flats in the park at Komendantskii Prospekt 40'

The St Petersburg dwellers' campaign dates back to 2004 when the city administration granted permission to the Russian building company Severnyi Gorod (SG, Northern City) to build a new block of flats in a small park already surrounded by several massive blocks of flats at the corner of Komendantskii Prospekt and Shavrova Street in the Primorskii city district.

This provoked strong and consistent resistance amongst local dwellers against *uplotnitel'nayazastroika* (fill-in construction),[6] which turned into open conflicts in

---

[5]For the ongoing debate on the Habermasian view of the public sphere see, for example, Calhoun (1992) and Smuts (2010).

[6]The term '*uplotnitel'nayazastroika*' refers to the practice of constructing new buildings in old quarters with already existing infrastructure, which is profitable but may lead to the overburdening and

2007, 2008 and 2009. This resistance—including physical confrontations between the builders and residents who were blocking the building machinery's access to the park—led to the failure of the two first attempts to start building at KP40. However, SG took the case to court and got permission to continue building on 27 March 2009.

The last and most violent phase of the conflict began at nine o'clock in the morning on 1 October 2009, when SG launched a military-style operation at KP40 in order to build a fence around the park and restart building. A total of 33 lorries loaded with concrete fence-plates drove to the small park from different directions, unloaded the plates, and the builders started to erect the fence. They were guarded by the staff of a private security company who kept the furious dwellers from stopping the operation. Once the fence was completed, the builders started cutting the trees, which had been planted by the local residents at their own initiative and cost.

The same evening the angry residents decided to establish a round-the-clock guard in the park. They also came up with the idea of naming the emerging movement *Komendantskii Pyatachok* (Komendant's Plot) after Nevskii pyatachok—a piece of land on the banks of the Neva river which had been heroically defended by the Red Army against Nazi invasion. The dwellers broke the lock on the fence gate, entered the park area, and decided that four women would stay in the park overnight in order to prevent the company from continuing construction. At four o'clock in the morning, the private guards started to remove the women from the area, but their yelling woke up nearby dwellers who started phoning the media, politicians and the police. A crowd gathered at the park, and media attention and politicians forced the company to suspend construction.

This invasion started a month-long campaign, during which the dwellers continued the round-the-clock guard in the park and contacted local and national politicians, city administration, online and offline media and various NGOs. Their action repertoire combined Soviet-era forms of resistance—such as writing *zhaloby* (complaints) and organising a *subbotnik* (a Soviet-era tradition of voluntary work in which good citizens were expected to participate) to plant new trees in place of the ones cut down by the builder—with the most modern forms of protest such as creating an action group in Vkontakte.

The movement at KP40 was supported by several NGOs and social movements in the city, such as *Dvizhenie Grazhdanskikh Initsiativ* (DGI, Movement of Citizens' Initiatives), the ecological expert organisation EKOM, *Zelenaya Volna* (Green Wave) and the movements *Doloi Uplotnitel'nuyuZastroiku* (Down with Fill-in Construction) and *Zhivoi Gorod* (Living City). The local Communists (*Kommunisticheskaya Partiya Rossiiskoi Federatsiya*) and Just Russia (*Spravedlivaya Rossiya*) politicians also started to intervene and the mass media—including the media allegedly under the influence of the city administration—turned against SG. Vice-governor Alexander Vakhmistrov interrupted construction and finally, on 3 November 2010, Governor Valentina Matvienko announced the cancellation of the building permit and ordered the fence to be removed.

---

break-up of the old infrastructure. The new buildings are often erected very close to the ones constructed earlier and the construction often breaks technical and ecological standards. This practice became especially popular in St Petersburg in the 1990s and 2000s.

*'No to the new street in Kumpula valley'*

In April 2009, 500–1,000 dwellers gathered in the green valley of the Helsinki city district Kumpula to protest against the city board's decision to build a new road through one of the biggest parks and recreation areas in eastern Helsinki. This event, called *Laakson valtaus* (the Occupation of the Valley), received a lot of publicity in the local and national media and was mobilised through, among other means, interaction on Facebook. City politicians and national and local media were invited to the occupation of the valley, and it was professionally hosted by two women activists, both residents of Kumpula, who during the demonstration interviewed the city government members in public in a humorous and non-confrontational manner. The new road was for a bus line that was meant to improve the public traffic connections between the Helsinki University Campus of Natural Sciences and the neighbouring Institute of Arts and Design with the Helsinki University of Technology. The project was supported by the leadership of the University and the Department of Meteorology at the Kumpula campus, and resisted by, among others, the inhabitants of the Kumpula city district.

When the news of the street construction plan reached the local residents, a meeting was held by the inhabitants in *Kylätila* (Village Space), an apartment rented to the local Village Space Association. This NGO was run by dwellers and it organised several activities in Kumpula, such as a children's day care club, art and hobby groups, and the yearly Village Carnival which each spring turned the whole city district into a huge music festival.

In addition to the Occupation of the Valley, the Kumpula campaign contained, much as at KP40, a wide action repertoire: the dwellers contacted and lobbied local politicians and media and city administration, proposed alternative routes for the planned bus line and wrote complaints. Because of the dwellers' resistance and growing publicity, the construction plan for the new street was interrupted by the decision of the city government on 14 April 2009.

## The role of Vkontakte in organising protests in St Petersburg

This and the following section will describe in detail the role of the Finnish segment of Facebook and the Russian Vkontakte in organising campaigns in the two cities. The first subsection describes briefly the overlapping forms of online and offline activism, the history of the founding and the structure of the local movements' websites. The second subsection investigates the nature of the debate conducted on these websites— in other words, the 'public sphere function' of the sites. The third subsection focuses on three aspects of the websites' roles in organising the campaigns, namely mobilising, coordinating and networking the movement.

### Founding and structure of the Vkontakte website

During the campaign at KP40, online and offline actions were combined in ways which question the division between the 'virtual' and 'real' worlds. In addition to and overlapping with the face-to-face gatherings and informal discussions in the courtyard or staircase, and over the phone, the local dwellers wrote hundreds of conventional letters to every possible authority including the presidential and city administration and local and national politicians, reminiscent of the Soviet-era tradition of writing official complaints. Simultaneously 'computer-savvy' activists searched for information at the city administration websites, information portals and news websites, sending emails and posting updates in Vkontakte about the events of the ongoing battle in the park.

The Vkontakte group *Net stroitel'stvu doma v skvere na Komendantskom prospekte* (No to the Building of a Block of Flats in the Park at KP40)[7] was founded at the beginning of the third and most severe wave of conflict in October 2009 by Galina, a local resident in her thirties who had been involved in the campaign from the outset. As an experienced user, she decided to found a group for the dwellers' movement on Vkontakte after having received an anxious phone call from a friend who witnessed the cutting of the trees in the park through her apartment window. While establishing the group Galina found out about an already existing Vkontakte group in defence of the park. She contacted the founders and the two groups were joined.

Prior to and simultaneously with the Vkontakte group, the events at KP40 had been discussed in local media and on websites, such as the local e-newspaper *Fontanka*[8] and the website called Little One[9] founded for communication among St Petersburg parents. As a mother of small children, Galina had participated in the discussions on the Little One site and informed the other discussants of KP40's newly established Vkontakte group.

In August 2010, another group focusing on the events at KP40 was also involved in Vkontakte. The group *Spasem skver vmeste!* (Let Us Save the Park Together!)[10] had 244 participants and its photo archive documented the invasion of SG at KP40. However, the founder of the group did not disclose any personal information about himself, and at a closer glance the group wall does not contain debate relevant to the movement but is rather filled with spam messages. Hence, in this essay we focus on the group 'No to the Building of a Block of Flats in the Park at KP40' founded by a local activist whom we interviewed in May 2010 (see subsequent sections of this essay).

On 22 August 2010, the Vkontakte group still contained 89 of the original participants who had numbered around 100. In addition to Galina, there were two other organisers, one of whom was also an activist in Movement of Citizens' Initiatives (DGI), which played an important role in supporting, advising and enabling networking among the dwellers during the entire campaign. In addition, the group's Vkontakte website contained permanent links to four 'friendly groups' within Vkontakte. Three of these links connected the campaign at KP40 with similar ones in Lopukhinskii Park,[11] the park on Ivan Fomin Street, and with a group protecting the historical heritage of St Petersburg. The photo archive of the website comprised several photos of the struggle against the builders and private guards in the park and two video clips from the local TV news about the new bill concerning green areas of the city. The discussion on the group

---

[7]See http://vkontakte.ru/club12101017, accessed 17 December 2011.

[8]See www.fontanka.ru, accessed 19 June 2012.

[9]See www.littleone.ru, accessed 19 June 2012.

[10]See http://vkontakte.ru/club12156583, accessed 17 December 2011.

[11]The URLs of these groups are http://vkontakte.ru/club15902308, http://vkontakte.ru/club8731757 and http://vkontakte.ru/club3924910, respectively; sites accessed 17 December 2011.

wall contained 94 postings. The first of them, dated 3 October 2009, informed the user of the founding of the Vkontakte group *Za sokhranenie istoricheskogo naslediia Pitera* (For the Conservation of the Historical Heritage of Piter).[12] Among the other postings were, for example, children's drawings of the park, announcements of meetings, supporting voices by activists from other similar groups and references to the media debate concerning KP40. The last relevant posting on the group wall, from 29 June 2010, reminded readers about the ongoing campaign against the construction in Lopukhinskii Park which was, similarly to KP40, a target for a Northern City building project. The discussion forum of the group contained 10 thematic 'threads' which will be dealt with at the end of the next subsection.

*The public sphere function: informing and debating*

In both the cities, the SNS sites were used, in addition to other online and offline means, to inform activists about the issues relevant to the campaign. The founder of the St Petersburg Vkontakte group emphasised the role of the internet in general to get quick access to information and news and transmit them to other activists:

> We monitored the sites of the city parliament (*zakonodatel'noe sobranie*), ecological organisation EKOM and our St Petersburg internet publication Fontanka.ru where information is being published about these kinds of issues. The same with television: we managed to watch all broadcasts immediately.[13]

Her words were confirmed by scrutiny of the Vkontakte group wall and discussion forums which, for example, were used to refer to the traditional and internet media news about the campaign in real time:

> Just a while ago on the Fifth Channel, the programme 'Peterburg hour' showed how the private guards of the building company are carrying tents and people out of the park ....[14]

> There are very useful comments on the Fontanka blog/46613.html on Komendantskii Prospekt 40, go and read it! We have to start a blog.[15]

In addition to reporting on the mass media coverage to the group members, the Vkontakte wall also informed readers about the acts and decisions of the construction company, city administration or other relevant actors in the conflict, including the stand taken by the 'environmental prosecution authority' (*prirodookhrannaya prokuratora*) in St Petersburg:

> The environmental prosecution authority of St Petersburg is against the building in the park. The employees of this monitoring organ reacted to the scandalous events which have emerged

---

[12]See http://vk.com/club3924910, accessed 19 June 2010.
[13]Author's interview with Galina, May 2010.
[14]5 October 2009, available at: http://vkontakte.ru/wall-12101017, accessed 17 December 2011.
[15]13 October 2009, available at: http://vkontakte.ru/wall-12101017, accessed 17 December 2011.

around the park at Komendantskii Prospekt. Its representatives visited yesterday [the park in the] Primorskii city district to find out on the spot if the cutting of the trees was legal ….[16]

Finally, on 15 October the site contained a cheerful posting:

Watch the news!!!! 20 minutes ago Matvienko cancelled the (building) permission.[17]

Despite the good news, the suspicious activists continued the round-the-clock guard in the area until the fence was finally torn down on 4 November 2009.

The postings above are illustrative of the links between (the increasingly converging) traditional and new media: relevant information found in traditional media was brought to the attention of the movement through Vkontakte. Conversely, the local journalists in traditional media could use Vkontakte as a source of information for their own work provided that they registered as members.[18]

The debate conducted at the Vkontakte site was nevertheless far from an ideal Habermasian debate where all viewpoints are reflected upon and given equal attention. Since the Vkontakte group was established at the height of the violent conflict, the local dwellers had neither time for nor interest in this kind of debate. Threatened by the common 'enemy', very few critical voices were raised and the discussion was rather aiming at building and strengthening the collective identity of the emerging movement. As often in web discussion forums, the tone of a lone disagreeing voice was more provocative than argumentative:

…. People, you and your 'meetings' are ridiculous. Don't you have other things to do? There is no doubt that they will build [in the park]. And they'll be right. Why should the park be populated by homeless people drinking cheap wine? ….[19]

This provocation caused furious replies and threats of exclusion from the forum. The strength of the reaction is probably explained by a combination of the rude tone of the posting, the violent nature of the conflict, suspicions of the postings having been 'paid for' by the construction company, as well as the anonymous nature of internet communication in general.

The movement's Vkontakte discussion forum was divided into 10 discussion 'threads'. They included, among other things, expression of support for the dwellers' campaign by DGI activists; organising the round-the-clock guard in the garden; informing about similar conflicts on Prospekt Koroleva, Lopukhinskii Park and Ivan Fomin Street; a posting about Matvienko's decision to cancel the building permission; and information about the new bill proposal concerning the green areas of the city. As is evident, these threads were mostly about informing or mobilising, not about rational

---

[16]4 October 2009, available at: http://vkontakte.ru/wall-12101017, accessed 17 December 2011.

[17]15 October 2009, available at: http://vkontakte.ru/wall-12101017, accessed 17 December 2011.

[18]However, the closed architecture of Vkontakte constrains the public of the activist. Unlike in LiveJournal, for example, Vkontakte pages cannot be accessed by non-members and are not indexed by the Russian search engines such as Yandex (private communication with Philip Torchinsky, St Petersburg, May 2010).

[19]14 October 2009, available at: http://vkontakte.ru/wall-12101017, accessed 17 December 2011.

debate striving for consensus. The number of the discussants was also limited to the most active local activists. Closest to ideal Habermasian debate, and also attended by a greater number of discussants, were the two threads where the renovation of the park was being discussed after the victorious battle.

### Mobilising and coordinating action

Prior to the conflict, many of the residents in the massive block of flats consisting of between 500 and 1,000 apartments hardly knew each other. It was the heat of the struggle that brought the dwellers together, also creating long-lasting ties of friendship. In addition to all other means (telephone, face-to-face contacts), Vkontakte had a central role in organising the campaign. It was used, for example, to organise taking turns in the round-the-clock guard in the park, to which a separate discussion thread was devoted on the Vkontakte site:

> Dear neighbours!!!
> I would like to draw your attention to a topic which is important for all, that is, the guarding of the park. Our most important task is to hang on, not to give up, not to leave the park! In this group there are basically young people, could you please replace the grandmothers! They are on duty day and night and God forbid what might happen. Enrol for the duty, come to the tent!!! Or phone me at [telephone number provided].[20]

The site was similarly used to encourage dwellers to join various meetings and events organised by the movement, such as the one designed to create pressure on the city administration:

> Today 12 October a new meeting will be arranged at 18 because tomorrow the government of St Petersburg will discuss the issue of the building at the 'Komendant's plot'. The city councillors said that the people must declare their wish to have a PARK instead of a BLOCK OF FLATS in this place. All you who are free from fear and prejudices, C O M E![21]

The word 'fear' illustrates vividly how Russian civic activists have to take into account the very real fear of being physically assaulted. The threshold for engaging in this kind of high-risk activism is therefore very likely more elevated than, say, in Finland, and to step over this threshold, the support from and close ties with other activists were very important. Vkontakte was an important avenue in building and maintaining these ties, as the following subsection will reveal.

### Creating networks

It is one of the basic findings of social movement research that recruiting and mobilising often happen along the lines of already existing structures and networks (Diani & McAdam 2003). Though the initial group of activists indeed included four

---

[20]14 October 2009, available at: http://vkontakte.ru/topic-12101017_21715785, accessed 17 December 2011.

[21]2 October 2009, available at: http://vkontakte.ru/wall-12101017, accessed 17 December 2011.

families of Afghanistan war veterans—some of the blocks of flats at KP40 were allocated for these veterans—it was rather the conflict and the ensuing campaign itself which enlarged the movement and created new ties between the activists, as in the case of Galina and her neighbour Irina:

> We did not know each other [before the campaign]. We lived in this house for 20 years and did not know anyone ... but now we became friends, as did our kids, and we go to the gym together .... We all got to know each other, young and old.[22]

In addition to creating and maintaining ties among the local dwellers and activists, Vkontakte helped to build ties between similar, but formerly unconnected local campaigns in various parts of the city. The first way to create these connections was to post a request for help and solidarity on the wall of another movement, such as the request by the KP40 group founder 'to support flash-mob action on the Mariinskii yard'[23] or the following posting by a dweller from the neighbouring Ivan Fomin Street sent to the KP40 wall:

> Dear neighbours! We have similar situation. Our beloved park at Ivan Fomin and Prospekt Prosveshcheniya is being threatened by builders. Share your experience with us!!!! Let us not concede our favourite places of recreation to arbitrariness. Thank you in advance![24]

These and other similar postings illustrate the exceptional role of the campaign at KP40 as one of the first successful attempts by the dwellers to stop 'fill-in construction projects' in the city.[25] KP40 therefore had exemplary value for the activists of the later campaigns, such as the one at Ivan Fomin Street cited above or the one at Lopukhinskii Park, whose activists also posted a message at the KP40 Vkontakte site.

The second way to build connections between KP40 and other campaigns was to create a permanent hyperlink to another campaign on KP40's Vkontakte website under the title 'friendly groups'. A Vkontakte user interested in locating similar campaigns in the city could then easily move from one group site to another following the hyperlinks of this emerging virtual network of resistance. A quick search in Vkontakte and other internet portals and local media revealed the existence of dozens of such campaigns in other city districts of St Petersburg. This network created unexpected new connections both online and offline between otherwise unconnected activists:

> ... people came to us from organisations which we did not even know existed. Suddenly one evening when we were conducting the round-the-clock guard in the tent, young people from some organisation showed up. I do not even know if they were from 'Green Wave' or some other organisation. They were about thirty youngsters, around 18–20 years of age, who told us that they knew our history, and wanted to help us with the guarding of the park. And they

---

[22]Author's interview with Galina, May 2010.

[23]25 June 2010, available at: http://vk.com/club12101017, accessed 19 June 2012.

[24]2 November 2009, available at: http://vkontakte.ru/wall-12101017, accessed 17 December 2011.

[25]In addition to the persistent resistance by the movement, there were certainly other factors working in favour of the dwellers, such as the tensions between SG and the city administration. In this essay our focus is, however, on the campaign.

were sitting with us in the tent during the nights ... they learned about us on the internet, but I did not see them ever in Vkontakte. Most probably they found out about us through Fontanka.ru. ... who they were, I don't know to this day. We did not even get to know each other well. In this way the internet is interesting: you will meet unknown people who learned about you on the net![26]

Finally, individual activists could simply join as members of the other protest groups in Vkontakte in order to support and keep track of the unfolding of similar events elsewhere. Galina, for example, was still following the evolution of the conflict at Ivan Fomin Street after the campaign at KP40 was over:

> Our contact with them [Ivan Fomin] kind of faded, but I still belong to their group [in Vkontakte] .... We write on their wall, and we may exchange phone numbers: they phone or I phone them, we will exchange the news.[27]

In all, Galina's own personal development testifies to the gradual widening of her range of interests which were originally only focused on the defence of her 'own' park. Prior to the campaign at KP40 she had been 'afraid to get involved' in public protests and only did so because 'they started building in front of our window'.[28] Now, with the help of her Vkontakte network, she was actively keeping in touch with and supporting activists involved in similar struggles in other parts of the city.

## The role of Facebook in organising action in Helsinki

### Founding and structure of the Facebook website

As in St Petersburg, the Helsinki activists combined online and offline means in their campaign. They contacted media, lobbied the city administration and politicians, recruited followers and kept in touch with fellow activists through face-to-face meetings, via mobile phones, on social networking sites and through email distribution lists.

The fact that the activists of the small village-like city district Kumpula knew each other beforehand and could rely on the existing resources of the Village Space Association as well as of the Kumpula Society, another local NGO, made communicating and organising resistance much easier. Our informant Anna, a Finnish woman in her thirties, who was an employee of the Village Space Association had, for example, three different email distribution lists at her disposal. The first was used for keeping in touch with the core activist group of the movement. The second one—compiled beforehand for the use of the Village Space Association—contained the email addresses of 350–400 Kumpula dwellers, while the third was comprised of roughly 100 email addresses of various media outlets. Among these media contacts were *Helsingin Sanomat* (News of Helsinki), the biggest national and Scandinavian daily, all other Helsinki area newspapers and journals, the newspapers of all Finnish political parties, and the news desks of all the TV channels with the personal email

[26]Author's interview with Galina, May 2010.
[27]Author's interview with Galina, May 2010.
[28]Author's interview with Galina, May 2010.

addresses of their journalists. Thus, any important news about the movement could reach all relevant national and local media in an instant.

As in St Petersburg, Finnish activists wrote letters and complaints to the city administration. Compared to the hundreds of letters written by individual Russian dwellers, the Finnish complaints in the form of conventional letters were fewer, addressed only to the responsible official city organs, and officially undersigned by the local Kumpula Society. However, the dwellers were also encouraged to send personal emails to the city board and city planning board members, whose addresses were published on Facebook. As in St Petersburg, the Facebook site *Ei katua Kumpulan laaksoon* (No to the Street in Kumpula Valley) and the event *Laakson valtaus* (Occupation of the Valley) were founded after a worried phone call from a friend of our key informant Anna, who had found out about the building plans and suggested that something had to be done. Since the friend had small children and little time at her disposal, Anna founded the group on a Thursday. Over the weekend the number of participants exceeded 600 people, and it reached over 2,500 people at the height of the campaign. As at KP40, the Finnish Facebook website was used to, among other things, organise and coordinate practical tasks, such as distributing paper flyer ads for the Occupation of the Valley. On 22 August 2010, the group still consisted of 1,985 members.

While the photos at KP40's Vkontakte website were mostly of the violent scenes of demolishing the park and of the confrontations between the dwellers and the builders, the Finnish Facebook group, on the contrary, contained idyllic pictures of the park and the valley during all seasons. The photos from the Occupation of the Valley demonstration pictured smiling people and families with small children having a nice time in the park. There were no policemen or displays of any kind of violence or confrontation to be seen. The group's Facebook site contained seven permanent links. In addition to the two links to newspaper articles about the campaign, they included a link to a personal blog titled *Laakson henki* (Spirit of the Valley) founded in defence of the valley by one of the activists;[29] to a Facebook site in defence of the neighbouring Vallila valley; to an exhibition of the Helsinki City museum with historical pictures of the Kumpula valley; and to another Facebook site defending a part of the central park in Helsinki.

The first posting on the group wall is from Thursday, 29 January 2009, when the dwellers had just learned of the building plans. The decisive city board meeting on the issue was to be held the following Monday, and the first postings, in addition to expressing their support for the movement in general, also encouraged people to write directly to the city board members and vice-members. The postings on the wall informed readers about the media debate concerning the campaign, discussed alternative options for the planned bus line, and coordinated and mobilised individual activities.

*The public sphere function: informing and debating*

As in St Petersburg, the Finnish activists were feeding relevant information 'from outside' the movement (for example from the media, the city administration and

---

[29]See http://laaksonhenki.wordpress.com/, accessed 17 December 2011.

politicians) into a Facebook site. One local dweller published a long email she had received from a Green League politician and member of the city board. The politician regretted that the reporting of the city board session's voting about the street project in the media was too short and laconic and gave a detailed account of the voting and stances of various parties. Another local dweller was exceptionally active in informing about the unfolding of the events and had posted altogether dozens of messages on the Facebook site. Often these postings contained a reference to his Spirit of the Valley blog. Thus, the Facebook group, among other things, helped dwellers to closely follow the processing of the case in the city administration and political apparatus. This was greatly aided by the close personal ties between some Kumpula activists and local politicians.

As in St Petersburg, much of the debate in the Finnish Facebook group focused on arguments against the building project, with only a few dissenting voices being raised. The one who did was much more argumentative than his provocative Russian counterpart presented above in this essay.

> Though I like the area of the Kumpula botanical garden and the Kumpula mansion, I don't really understand ... how a street meant precisely for public traffic will cause such a strong protest. You just cannot develop Helsinki with a NIMBY mentality dictated by the dwellers living in [picturesque and expensive detached] wooden houses ... if you want to create a functioning and modern city for all dwellers. The street for public traffic would benefit the locals, the work commuters, the university staff and students.

The reference to the NIMBY (Not In My Back Yard) mentality may be illustrative of the difference between Finnish and Russian political culture and deserves to be dealt with in more detail.[30] In organising the Occupation of the Valley the activists took pains to distinguish the campaign from NIMBY movements—in other words from the forms of activism which would only defend the residents' narrow, local interests. In the Finnish political culture, a NIMBY label attached to a movement could be used as a powerful rhetorical weapon against the movement, whereas in St Petersburg the label was used in neither SNS nor the media debate about the campaign. Thus, in Helsinki the NIMBY accusation was also taken seriously and replied to by one of the Kumpula activists:

> This isn't a Nimby project at all. You will realise this if you just analyse the issue a bit more closely. First, dropping the building project [in Kumpula] would not relocate the problem to another city district as it typically would in Nimby protests. The building project would not

---

[30]Generalisation based on two cases can naturally be indicative only. The differences observed may be due to differences between the cases—e.g. targeting the protest against the city administration in Helsinki but against a private construction company in St Petersburg, the socio-economic composition of the inhabitants of the two city districts, and the very nature of the parks (a large and traditional recreation area in Helsinki compared to a small park created by the dwellers themselves only recently). All these factors cannot be dealt with in this essay. On the other hand, the two cases also shared important similarities. Both were mobilised in 2009 by the local inhabitants against building construction projects threatening their neighbourhood. They used all means at their disposal to stop these projects and, as our analysis suggests, the repertoire of their actions were in many ways similar.

make any sense, even if it were implemented in some other part of the city. I will deal with the Nimby problems more closely in my blog.[31]

Except for the NIMBY debate in Helsinki, the SNS groups in the two cities contained little discussion concerning the tactics and strategies of the movement. This was probably due to the public status of the groups' websites which could be scrutinised at any time by adversaries. Therefore, it is plausible to think that many of the tactics and strategies were rather discussed in face-to-face meetings, phone calls or email exchanges between the dwellers.

### The organising function: mobilising, coordinating, networking

#### Mobilising and coordinating action

Though the Kumpula activists founded—in addition to the action group's website—a separate Facebook 'event site' for the Occupation of the Valley[32] the original group site was also used for mobilising and coordinating action. Local dwellers were, for example, recruited to distribute paper flyers in the city district, and they reminded members of the group about the 'Occupation' through the site:

> Please remember the Occupation of the Valley action on Tuesday, 10 March, between five and six o'clock. We have invited leadership of the University and the Department of Meteorology, the members of the city board and city planning board, the chairmen of all parties at the city council and media. There will be some news broadcasting prior to the event at least in a few radio programs and in the Helsinki News on Sunday. Bring with you your own signs and banners: 'For the valley' and 'No to the street in Kumpula Valley'. See you on the barricades! P.S. If someone wants to distribute flyers on Monday directly to the people e.g. in the Arabia shopping mall, please contact me![33]

Moreover, people were encouraged to write emails to the city board members in order to create political pressure. In the city board and city planning board the street project divided the parties, of which the Leftist Union and Green League were against the construction. Particularly for the Green League this was a natural stand given that the Kumpula area was probably the greenest Helsinki city district in terms of its dwellers' voting in municipal elections.

> It is great that this group has grown so fast! Will you all also write to the city board members until Monday morning? If you don't have time for a long and well-argued message, even a few lines will do. The main thing is to send a lot of mail since the politician is afraid of the movement of the masses ☺. And to repeat: all the email addresses of the city board members and their deputy members will be found here.[34]

---

[31]Posted on 14 February 2009.
[32]We did not separately analyse this event.
[33]Posted on 6 March 2009.
[34]Posted on 1 February 2009.

Even a small percentage of the 2,000 group members could thus generate reasonable pressure on the politicians by email. One of the authors of this essay, for example, sent a personal email to all city board members, many of whom replied within a few days.

*Creating networks*

Unlike in St Petersburg, in Helsinki there were few efforts to use the Facebook group to create links with other similar campaigns (though such links could have certainly been created otherwise). One of them was the Facebook group consisting of the proponents of the neighbouring Vallila valley. In addition, one of the discussants took up a similar conflict in Tampere, one of the biggest Finnish cities located 100 miles north of Helsinki:

> Hi, would it be useful in this campaign to hear from the experiences of others who have formerly had similar experiences? In Tampere, for example, they managed to protect a local quarter of [idyllic, detached] wooden houses but it required years of persistent work. There are certainly still a lot of means which have not been used. This fb-page is also a brilliant idea.[35]

As a whole, however, it was clear that creating connections with other groups in similar situations through social networking sites was not as important in Helsinki as in St Petersburg. This was not because of the lack of conflicts in Helsinki: one of the Finnish website discussants noted that similar conflicts between city planners and local dwellers were ongoing constantly in other parts of the city. Rather, the Finnish activists could both utilise a less restrictive media environment and lean on the already existing network of city district associations through which their board members were invited to participate in the Occupation of the Valley and support the Kumpula campaign.

*Conclusions*

The role of social networking sites in protest movements in St Petersburg and Helsinki was similar in many respects. Though in both cities these sites were important channels for transmitting information and organising and coordinating the campaigns, their role was clearly limited in terms of impartial, democratic discussion of the issue of common concern. The sites were rather used to build and reinforce the emerging collective identity and to create consensus within the movement than to encourage Habermasian public debate. This does not, however, prove the ultimate unsuitability of SNSs as 'new public spheres' (Smuts 2010). Mobilising for an ongoing conflict in general leaves little room for differing voices, and this tendency is even more underlined under the conditions of high-risk activism, such as in our Russian case. When confronting physical violence, discussion is destined to remain secondary. Therefore, our comparative analysis rather suggests further research on these sites under different conditions.

[35]Posted on 2 February 2009.

Regardless of the nature of the debate on the sites, mere involvement in it may have lasting effects. Research suggests that interpersonal discussion about politics leads to an increase in political involvement and empowerment (Zhang *et al.* 2010, pp. 78–79). Though we do not have follow-up data for the discussants on the Russian and Finnish sites, the personal history of our Russian key informant Galina showed how participating in the campaign and the creation, maintenance and debate of the Vkontakte page evoked a continued interest in civic activism which exceeded the narrow confines of her own neighbourhood. This observation was confirmed by another of our respondents—an activist of the Movement of Citizens' Initiatives (DGI)—according to whom each civic campaign, regardless of its outcome, pushes one or two persons towards activism.

A clear difference between the two cities emerged in the comparison of the role of SNSs in creating networks both within the movement and between the similar campaigns elsewhere in the city. Unlike in Helsinki, the campaign at KP40 was not mainly built on already existing civil society structures or social ties in the neighbourhood, but rather created these ties. Even though the movement participants were living in the same block of flats around the park, many did not know their neighbours prior to the campaign. Though the strength of the emerging ties was probably in large part due to the actual experiences of the hours spent in the round-the-clock guard in the tent during the cold October nights, in the creation and maintenance of these ties Vkontakte seem to have had an essential role.

Also in contrast to the situation in Helsinki, Vkontakte had a central role in creating and maintaining external ties or 'bridges' between formerly isolated campaigns against building projects elsewhere in St Petersburg. These bridges consisted first, of permanent links between 'friendly groups' created by the movements' website administrators; second, of the membership of activists in two or more contentious groups simultaneously; and third, of the personal friendship ties made through Vkontakte between two activists from different groups.

The strength of these personal bridges between campaigns may well be related to the experience of the threat of violence, which was probably the most striking difference for the outside observer between the Russian and Finnish cases. When joining the movement at KP40, the local dwellers in St Petersburg, unlike those in Helsinki, had to take into account the very concrete risk of being exposed to physical violence. While this was not a factor conducive to mass recruitment, those who joined were highly devoted to the common and concrete cause of defending their own neighbourhood and park.

The importance of the personal bridges in Vkontakte might have to do with the sharing of this feeling of devotion and risking one's physical integrity. In addition it may also be related to the implicit modelling of sociability in Facebook and Vkontakte around the notion of the personal network. This notion describes social life as a web of relations anchored around a focal person (ego) and containing different types of network members (alters) such as family, kin, friends and colleagues. This way of looking at social life fits especially well in post-Soviet Russian daily life, where personal networks play a particularly important role (Lonkila 2011).

We propose that the nature of these networks—indeed, probably the nature of the social tie itself—may be different in the two countries. The duties and obligations of

Vkontakte 'friends' in Russia and Facebook 'friends' in Finland may differ and consequently the personal 'bridges' connecting various campaigns through these ties of friendship might support a different amount of weight and prove vital for the emerging Russian civil society.

*University of Tampere*
*Centre for Independent Social Research, St Petersburg*

## References

Barnett, E. (2010) 'Mark Zuckerberg Confident that Facebook will Reach One Billion Users', *Telegraph.co.uk*, 23 June, available at: http://www.telegraph.co.uk/technology/facebook/7849912/Mark-Zuckerberg-confident-that-Facebook-will-reach-one-billion-users.html, accessed 17 December 2011.

Baumgartner, J. & Morris, J. (2010) 'MyFaceTube Politics: Social Networking Web Sites and Political Engagement of Young Adults', *Social Science Computer Review*, 28, 1, pp. 23–44.

Bortree, D. & Seltzer, T. (2009) 'Dialogic Strategies and Outcomes: An Analysis of Environmental Advocacy Groups' Facebook Profiles', *Public Relations Review*, 35, 3, pp. 317–19.

Boyd, D. (2008) 'Can Social Network Sites Enable Political Action?', in Fine, A., Sifry, M., Rasiej, A. & Levy, J. (eds) (2008) *Rebooting America* (New York, Personal Democracy Press), pp. 112–16.

Breindl, Y. (2010) 'Critique of the Democratic Potentialities of the Internet: A Review of Current Theory and Practice', *triplC*, 8, 1, available at: http://triple-c.at/index.php/tripleC/article/view/159/165, accessed 11 August 2010.

Calhoun, C. (1992) 'Introduction: Habermas and the Public Sphere', in Calhoun, C. (ed.) (1992) *Habermas and the Public Sphere* (Cambridge, MA, The MIT Press).

Dahlgren, P. (2005) 'The Internet, Public Spheres, and Political Communication: Dispersion and Deliberation', *Political Communication*, 22, 2, pp. 146–62.

Diani, M. & McAdam, D. (2003) *Social Movements and Networks. Relational Approaches to Collective Action* (Oxford, Oxford University Press).

Etling, B., Alexanyan, K., Kelly, J., Faris, R., Palfrey, J. & Gasser, U. (2010) *Public Discourse in the Russian Blogosphere: Mapping RuNet Politics and Mobilization*, Berkman Center Research Publication no. 201-11, 19 October, available at: http://cyber.law.harvard.edu/sites/cyber.law.harvard.edu/files/Public_Discourse_in_the_Russian_Blogosphere_2010.pdf, accessed 2 January 2011.

Fossato, F. & Lloyd, J. (2008) *The Web that Failed. How Opposition Politics and Independent Initiatives are Failing on the Internet in Russia* (Oxford, Reuters Institute for the Study of Journalism).

Goroshko, O. & Zhigalina, E. (2008) 'Political Interactions in Blogs', *The Russian Cyberspace Journal*, 1, available at: http://www.digitalicons.org/issue01/issue1/goroshko_and_zhigalina.php?lng=English, accessed 19 June 2012.

Häyhtiö, T. (2008) 'Teemana internet ja kansalaisaktivismi', *Politiikka*, 50, 1, pp. 1–4.

Johnson, T., Zhang, W., Bichard, S. & Seltzer, T. (2011) 'United We Stand? Online Social Network Sites and Civic Engagement', in Papacharissi, Z. (ed.) (2011) *A Networked Self: Identity, Community, and Culture on Social Network Sites* (New York, Routledge), pp. 185–207.

Khveshchanka, S. & Suter, L. (2010) 'Verchleichende Analyse von profilbasierten sozialen Netzwerken aus Russland (Vkontakte), Deuthschland (StudiVZ) und den USA (Facebook)', *Information, Wissenschaft & Praxis*, 61, 2, pp. 71–76.

Langlois, G., Elmer, G., McKelvey, F. & Devereaux, Z. (2009) 'Networked Publics: The Double Articulation of Code and Politics on Facebook', *Canadian Journal of Communication*, 34, 3, pp. 415–34.

Lonkila, M. (2008) 'The Internet and Anti-military Activism in Russia', *Europe-Asia Studies*, 60, 7, pp. 1125–49.

Lonkila, M. (2011) *Networks in the Russian Market Economy* (Houndmills, Basingstoke, Palgrave Macmillan).

McCaughey, M. & Ayers, M. (eds) (2003) *Cyberactivism: Online Activism in Theory and Practice* (New York & London, Routledge).

Mejias, U. (2010) 'The Limits of Networks as Models for Organizing the Social', *New Media & Society*, 12, 4, pp. 603–17.

Neumayer, C. & Raffl, C. (2008) 'Facebook for Global Protest: The Potential and Limits of Social Software for Grassroots Activism', in Stillman, L. & Johanson, G. (eds) (2008) *Proceedings of the 5th Prato Community Informatics & Development Informatics Conference 2008: ICTs for Social Inclusion: What is the Reality?*, available at: http://cirn.infotech.monash.edu.au/assets/docs/prato2008papers/raffl.pdf, accessed 19 June 2012.

Oates, S., Owen, D. & Gibson, R. (eds) (2006) *The Internet and Politics. Citizens, Voters and Activists* (London & New York, Routledge).

Reiting populyarnosti (2010) 'Reiting populyarnosti sotsial'nykh internet-setei', *VTSIOM Press-vypusk*, 1501, 25 May, available at: http://wciom.ru/index.php?id=268&uid=13526, accessed 19 June 2012.

Rohozinski, R. (1999) *Mapping Russian Cyberspace: Perspective on Democracy and the Net*, UNRISD Discussion Paper 115 (Geneva, United Nations Research Institute for Social Development), available at: http://unpan1.un.org/intradoc/groups/public/documents/untc/unpan 015092.pdf, accessed 17 December 2011.

Schmidt, H., Teubener, K. & Konradova, N. (eds) (2006) *Control + Shift. Public and Private Usages of the Russian Internet* (Norderstedt, Books on Demand GmbH).

Smith, A., Lehman Schlozman, K., Verba, S. & Brady, H. (2009) 'The Internet and Civic Engagement', *Pew Internet & American Life Project*, available at: http://pewinternet.org/Reports/2009/15–The-Internet-and-Civic-Engagement.aspx, accessed 29 December 2010.

Smuts, L.-M. (2010) *Social Networking Sites as a New Public Sphere: Facebook and its Potential to Facilitate Public Opinion as the Function of Public Discourse—A Case Study of the 2008 Obama Campaign*, PhD thesis, Stellenbosch University, available at: https://scholar.sun.ac.za/bitstream/handle/10019.1/4209/Smuts,%20L.M.pdf?sequence=1, accessed 28 December 2010.

Suleymanova, D. (2009) 'Tatar Groups in Vkontakte: The Interplay between Ethnic and Virtual Identities on Social Networking Sites', *Digital Icons: Studies in Russian, Eurasian and Central European New Media*, 1, 2, pp. 37–55, available at: http://www.digitalicons.org/wp-content/uploads/2009/12/Dilyara-Suleymanova-DI-2.4.pdf, accessed 17 December 2011.

Sweeney, M. (2010) 'Mark Zuckerberg: Facebook "Almost Guaranteed" to Reach 1 Billion Users', *The Guardian*, 23 July, available at: http://www.guardian.co.uk/media/2010/jun/23/mark- zuckerberg-facebook-cannes-lions, accessed 29 August 2010.

Van de Donk, W., Loader, B., Nixon, P. & Rucht, D. (eds) (2004) *Cyberprotest: New Media, Citizens and Social Movements* (London & New York, Routledge).

Wauters, R. (2010) 'Zuckerberg Makes It Official: Facebook Hits 500 Million Members', *TechCrunch*, 21 July, available at: http://techcrunch.com/2010/07/21/facebook-500-million/, accessed 29 August 2010.

Ylä-Anttila, T. (2009) 'The Globalization Debate: A Public Justifications Analysis', paper presented at the *9th Conference of the European Sociological Association*, 2–5 September, Lisbon, Portugal, available at: http://www.mv.helsinki.fi/home/ylaantti/PJA.pdf, accessed 19 June 2012.

Zhang, W., Johnson, T., Seltzer, T. & Bichard, S. (2010) 'The Revolution Will Be Networked: The Influence of Social Networking Sites on Political Attitudes and Behavior', *Social Science Computer Review*, 28, 1, pp. 75–92.

# From Blogging Central Asia to Citizen Media: A Practitioners' Perspective on the Evolution of the *neweurasia* Blog Project

CAI WILKINSON & YELENA JETPYSPAYEVA

*Abstract*

This essay examines the development of one regional blogosphere, the Central Asian 'Stanosphere', through a focus on the *neweurasia* blog project. The *neweurasia* project began in 2005 as an English-language volunteer-run blog project about the former Soviet republics of Central Asia and the Caucasus, rapidly becoming one of the most visited blogs about the region. Following this auspicious start, over the next five years *neweurasia* developed into a multi-language locally driven project with more than 80,000 unique page views on average per month. Despite its indisputable successes, the project was often a steep learning curve for all involved. In this essay, we examine *neweurasia*'s evolution from 'blogging Central Asia' towards a citizen media project, and reflect on some of the issues and challenges encountered. On the basis of our discussion, we reflect upon how *neweurasia*, and citizen media in general, can maximise its impact on the nascent Stanosphere, in the process helping to give Central Asia a voice in the global blogosphere.

IN 2007, RECALLS A JOURNALIST FROM KAZAKHSTAN who was at that time writing about online media for a Kazakh-language newspaper, 'the editors corrected the word "blog" to "blok", and that's how it went to print'.[1] It was only when the Kazakh Prime Minister opened his own blog in January 2009 that the word began to be used more widely and to be understood in the republic.[2] Anecdotes such as this are a useful reminder that while on a global level declarations that blogging was 'past its prime as a cultural, business and political phenomenon' were already being made in 2006 (Drezner & Farrell 2008, p. 1), there are considerable disparities in regional and linguistic blogging trends that benefit from a closer examination.

The internet has frequently been touted as a medium with the potential to facilitate 'democratisation' or even affect political change (Mcglinchey & Johnson 2007, p. 273; Kulikova & Perlmutter 2007; Srinivasan & Fish 2009, p. 559), particularly in relation to countries classified as 'not free' or 'partially free', such as Russia and the five post-Soviet republics of Central Asia (Freedom House 2010). However, in practice low

---

[1] Email questionnaire, Kazakh-language bridge blogger, 31 January 2010.
[2] http://primeminister.government.kz/blogs/masimov_k, accessed 30 January 2012.

internet penetration, the relatively high cost of internet access and especially external traffic, as well as government regulation and interference, have all impacted on the development of so-called new media in Central Asia (Driesbach *et al.* 2009). The impact of these circumstances can clearly be seen in relation to the popularity of blogs, one of the earliest and most popular Web 2.0 phenomena, in the region.[3] While the global blogosphere contains around 133 million blogs (Winn 2009), and the Russian-speaking blogosphere comprises an estimated 7.4 million blogs (Yandex 2009, p. 2), blogging in Central Asia remains a minority and mainly elite activity, with 'Central Asia's future leaders... the segment of the population most frequently exploring the web' (Pickett 2006, p. 3).

Nevertheless, since the early 2000s, a distinct 'Stanosphere' has formed,[4] reflecting Zuckerman's notion that, as the global blogosphere continues to expand, an increasing number of regional and non-English, language-specific blogospheres will emerge (Zuckerman 2008). This regional blogosphere is comprised of blogs and bloggers in or from Kazakhstan, Kyrgyzstan, Uzbekistan, Tajikistan and Turkmenistan. It is multilingual, with the main language being Russian, but with a growing number of blogs in Kazakh and, to a lesser extent, Tajik, Kyrgyz and Uzbek, as well as in English. Overall, in the absence of any reliable statistics, a rough estimate would be that as of 2010 the Stanosphere comprises no more than a few thousand blogs, with the overwhelming majority in Kazakhstan and/or in Russian.[5]

The first Kazakhstani blogs appeared in the early 2000s, and Kazakhstan has remained the regional leader for blogging with several local blog platforms, including Yvision.kz, Blogos.kz and On.kz, as well as Kazakh-language localisation of WordPress underway.[6] Due to the dominance of Russian as the language of choice for blogging there is a large overlap with the *Runet* blogosphere, especially in the case of platforms such as LiveJournal. However, over the last few years, a fledgling Kazakh-language blogosphere has developed thanks to the concerted efforts of bloggers such as Urimtal, who lists more than 80 Kazakh-language blogs in his blogroll.[7]

In keeping with infrastructural and political limitations, development has been slower in the other republics. In Kyrgyzstan, blogging remained limited to 'a few hundred' people until the late 2000s,[8] when the Hivos-funded Bishkek-based NGO Kloop Media Foundation founded a local blog platform with servers in-country.[9] While this made blogging more accessible to people in the republic by reducing data traffic costs and increasing server reliability and speed, as in Kazakhstan blogging remains far from a mass activity; as of March 2009 there were only 550 blogs on the

---

[3]While there are a growing number of definitions for what a blog is, the central characteristics are that it is a webpage or site maintained by an individual or group on which entries, which may be text-based or feature audio and/or visual content (or any combination of these), are displayed in reverse chronological order. Posts will frequently feature hyperlinks to other online content and readers are normally able to comment on posts, creating opportunities for asynchronous online interaction.

[4]The word was coined by Ian (2008) on the blog *Beyond the River*.

[5]Personal correspondence with a leading Central Asian blogger, 24 May 2010.

[6]http://qazaqblogshilar.wordpress.com, accessed 2 November 2011.

[7]http://urimtal.wordpress.com, accessed 2 November 2011.

[8]Email questionnaire, Kyrgyz-language bridge blogger, 15 January 2010.

[9]http://www.kloop.kg/blogs, accessed 2 November 2011.

Kloop platform, which had then been open for 18 months (Journalist News 2009). Tajikistan's and Uzbekistan's blogospheres have grown still more slowly, hindered both by a lack of access to the internet and also government controls on access to information, as well as fragmentation caused by bloggers choosing variously to write in Roman, Cyrillic and Arabic scripts.[10] Finally, the proportion of the population with regular internet access is so low in Turkmenistan, at under 1%, that talk of a Turkmenistani blogosphere is all but meaningless (Pickett 2006, p. 6).

## Introducing the neweurasia project

In this essay, we focus on one particular blog in the Stanosphere, *neweurasia*.[11] We chart the project's development between 2005 and late 2009 and reflect on what it suggests about the evolution of blogging and citizen media in the Central Asian republics. In the following section we provide a detailed history of *neweurasia*, charting its progression from a volunteer-run English-language group blog to a donor-funded multilingual citizen media project, before turning our attention to exploring a number of the key issues that have emerged over the project's lifetime in the latter part of this essay. Specifically, we reflect upon the effect of the Central Asian context on *neweurasia*'s evolutionary trajectory and some of the logistical and organisational challenges encountered by the project.

We conclude by considering what *neweurasia*'s evolution suggests about the development of online citizen media and journalism and the implications for the future growth of blogging and online media in Central Asia. We suggest that *neweurasia*'s experiences indicate that citizen media projects can maximise their sustainability in two main ways. Firstly, the development of a flexible 'amalgamation model' that combines elements of social and conversational media with conventional Western journalistic standards is likely to be most responsive to the demands of all stakeholders, from readers to the bloggers themselves, as well as donors, for example. We believe that such responsiveness is critical for online media projects to survive, despite the challenges in managing competing demands and changing interests and agendas.

Secondly, *neweurasia*'s success in building up a sizeable readership and expanding from a volunteer blog undertaking to being a donor-funded citizen media project appears to be in no small part due to its ability to facilitate the exchange of information and opinion between different culturo-linguistic blogospheres both inside and beyond Central Asia. In this respect, we argue that citizen media projects would do well to incorporate a 'bridging' function into their operation models. By doing so, such projects facilitate the development of as wide a public sphere as possible online by acting as info-cultural bridges in the global blogosphere. This is of particular significance in regions of the world such as Central Asia that are often perceived as being remote and inaccessible to outsiders for historical, cultural and linguistic reasons. In addition, adopting this function offers a way for citizen media projects to more

---

[10]This appears to affect Tajikistan to a greater extent than Uzbekistan, and in the case of blogs in Arabic script creates an overlap between the Stanosphere and the Persian-language blogosphere, or 'weblogestan' (Doostdar 2004) that is centred on Iran.

[11]http://www.neweurasia.net, accessed 2 November 2011.

clearly differentiate themselves from 'traditional' media providers, albeit potentially entailing more explicit alignment with the normative agendas of donors as a way to access funding—a topic we shall return to in the penultimate section of this essay.

In writing this essay, we have drawn on the various funding proposals and reports produced over the course of the project between 2006 and 2009,[12] as well as posts from *neweurasia* itself. A questionnaire was also distributed to the *neweurasia* email list in English and Russian in early 2010, with recipients invited to reply to as many or as few questions as they wished. A total of eight neweurasians responded, almost all at considerable length. Some of their comments are included verbatim in our discussion and their input contributed greatly to the overall shape of the essay, for which we are very grateful.

We also draw on our own experiences of new media, blogging in general, and being involved with *neweurasia* in various capacities. Yelena, a journalism graduate, joined *neweurasia* as Russian-language managing editor in January 2009, having previously worked on a variety of online media projects in Kazakhstan including Gazeta.kz,[13] as well as organising the 2009 BarCamp Central Asia.[14] Cai has been involved with *neweurasia* in a variety of ways since its founding in 2005, first as a blogger and editor on the English-language Kyrgyzstan blog and later as a board member, and also as managing editor between January and April 2008. We have sought to combine these sources with our experiential knowledge of the project to produce a reflexive account of *neweurasia*'s evolution that considers its relationship not only to the Stanosphere but also to the ongoing development of blogging and other 'conversational media' (Fitzgerald 2007, p. S4).

As a final note, we would like to emphasise that the views expressed in this essay are the personal interpretations of the authors, and should not be taken as the official opinion of *neweurasia*, nor any person or organisation associated with the project. Ethically, our involvement in the project has meant treading a relatively fine line between aiming to provide as full a discussion of the project as possible, but also ensuring that confidentiality is respected and the interests of all involved in *neweurasia* as an ongoing and active project are unharmed. In balancing these interests, as well as accommodating the limits of a single essay, we have inevitably left some issues unexplored and criticisms unaddressed. While this is likely to be to the frustration of some readers, we would emphasise that this empirical account is very much a first step in exploring part of the Stanosphere and, as such, arguably inevitably raises far more questions than answers. At the same time, we hope these questions may help to develop a further research agenda for new media studies both in relation to Central Asia and, potentially, more widely.

### *The history of* neweurasia

The *neweurasia* project was founded in 2005 by Ben Paarmann, building on the foundation laid by the student-run *Thinking East*, based at the School of Oriental and

---

[12]We would like to thank Ben Paarman for permitting us to use these documents.
[13]http://gazeta.kz, accessed 2 November 2011.
[14]http://www.barcampkz.net, accessed 2 November 2011.

African Studies, University of London.[15] The initial aim was to create a forum for young people, particularly university students, from Central Asia and the Caucasus to talk and exchange their experiences of life in their countries with their peers from outside the region. Specifically, as *neweurasia*'s original mission statement put it, the aim was to 'effect positive changes in our troubled era', by 'bringing together young academics from around the world... [and] offering these young thinkers space to publish their ideas on Central Asia and the Caucasus, unrestrained from the agendas of established publishers' (neweurasia 2005). Blogging was the logical medium to facilitate such exchanges, given the project's target demographic, who were likely to have access to computers with an internet connection, as well as possessing the prerequisite writing and language skills and, most importantly, a desire to write about a wide range of topics and share their opinions.

The project went live in the final quarter of 2005 with blogs focusing on the five post-Soviet Central Asian republics as well as Armenia, Azerbaijan and Georgia. These 'country blogs' were united around a 'homebase', which provided a single landing page for visitors, quick links to each blog and recent posts on them, as well as posts about the region more generally. Initially, *neweurasia*'s team was made up of around 10 students based mainly in the USA, UK and Germany who all contributed on a voluntary basis. Significantly from the perspective of ensuring a smooth launch, everyone involved had prior experience of blogging, either due to owning a personal blog, or contributing to other group blogs or online projects such as *neweurasia*'s predecessor *Thinking East* or *Registan*.[16] At this early stage *neweurasia* was formally a solely English-language project, despite the fact that almost all of the bloggers had a working knowledge of at least one of the region's languages.

Over the first seven months, contributors posted about 500 posts to *neweurasia*'s blogs (neweurasia 2006). While calls for contributors were posted on the homebase and distributed via email lists, and efforts were made to recruit contributors through personal contacts and local networks, these had only a very limited impact and the vast majority of the posts were made by a core of nine or ten contributors. Perhaps inevitably given the scope of the project and the voluntary nature of contributions, the individual country blogs developed at very different rates; while the five Central Asian blogs and the homebase were all updated 'regularly' or 'very regularly', the three Caucasian blogs were 'practically idle' by March 2006 (neweurasia 2006).

Despite the uneven development of the constituent blogs, *neweurasia* made a strong start, garnering an average of 1,500 visitors per day at the end of the first seven months and attracting attention from academics and policy makers working on the region (neweurasia 2006). Coverage of the Azerbaijani parliamentary elections in November 2005[17] and the Kazakhstani presidential election in December 2005 (Ben 2006) were seen as particular highlights, as was the first 'country blogs universal post' in February 2006, which saw simultaneous posts about HIV and AIDS in all of the countries covered by *neweurasia* as well as Mongolia (Jessica 2006).

---

[15]http://thinking-east.net, accessed 2 November 2011.

[16]http://www.registan.net was founded in 2003 by Nathan Hamm.

[17]Full coverage available at: http://web.archive.org/web/20060215010002/azerbaijan.neweurasia.net/?cat=2m, accessed 22 December 2010.

Building on these initial successes, May 2006 saw the start of a new stage in *neweurasia*'s development. A partnership with the Czech-based news media non-profit organisation, Transitions Online (TOL),[18] was established in order 'to promote blogging as a tool of free speech in Central Asia' as well as help realise *neweurasia*'s ambition to be 'the prime regional blog and student network'.[19] The joint venture saw the project's scope and approach to blogging change considerably. Key changes included phasing out the Caucasian blogs and starting to offer blog content in Russian. Funding came from the Dutch NGO Hivos and the Open Society Institute-funded Eurasianet.[20]

Central to this expansion and diversification was embedding the concepts of bridge blogging and paid blogging into the project. Bridge blogging involves actively creating and sustaining connections, or 'bridges' between blogs in order to 'reach across gaps of language, culture and nationality to enable interpersonal communication' (Zuckerman 2008, p. 48), and provide a counter to the 'echo-chamber' effect of online activity, whereby bloggers are more likely to link to other blogs that hold similar views to their own (Reese *et al.* 2007, p. 252). Zuckerman goes on to explain the defining characteristic of bridge blogs: 'They are distinguished from the vast majority of blogs by their intended audience: while most blogs are targeted to friend, family or countrymen, bridgeblogs are intended to be read by an audience from a different nation, religion or culture' (Zuckerman 2008, p. 48). It was felt that bridge blogging would improve the coherence of *neweurasia*'s content, as well as helping to create a unique identity for the project that would help to differentiate it from other blogs in or about Central Asia.

It was decided to try and recruit five bridge bloggers to the project, each of whom would be paid a salary and be responsible for a particular Russian-language country blog, acting as a bridge to nascent country-specific blogospheres, as well as translating posts from Russian to English and *vice versa*. The bridge bloggers would also be responsible for recruiting other contributors for their blogs, who would be paid a fee for each blog post of suitable length (about 300 words) and quality. The bridge bloggers' other responsibility in terms of actual blogging was to post regular round-ups of the Stanosphere to the homebase, which were also published on the English version of *Global Voices*.[21] Given the new financial arrangements, targets were also set for how many posts per week the bridge bloggers were expected to write.

In the event, four bridge bloggers were employed, with a vacancy for a Turkmenistani bridge blogger remaining. The results of this strategy for *neweurasia*'s continued development were impressive. By the end of the year-long grant period in April 2007, internet traffic had grown to between 110,000 and 118,000 unique visitors per month and there were between 60 and 100 posts on the site every month.[22] In addition, this phase of development marked the start of *neweurasia* becoming a source of news for people in Central Asia, rather than mainly being a news platform for

---

[18]http://www.tol.org, accessed 2 November 2011.
[19]'Neweurasia Strategy Paper #1', April, internal document for *neweurasia*'s management team.
[20]http://www.eurasianet.org, accessed 2 November 2011.
[21]http://globalvoicesonline.org/, accessed 2 November 2011.
[22]'TOL/neweurasia Proposal', February, internal document for *neweurasia*'s management team.

Westerners: Bishkek and Almaty were the two largest sources of visitors to the site, with visitors from Kazakhstan accounting for 20% of overall traffic, in third place after the USA (29%) and the UK (24%).[23] In July 2006 *neweurasia* was blocked in Uzbekistan, which was seen as a positive indication of the project's success as a source of alternative reportage about Uzbekistan.[24]

Following on from this success, the TOL and *neweurasia* partnership was continued for a further year, with the scope of the project once more being expanded. As reflected in the title of the proposed project, 'Building Blogging in Central Asia', the aim was no longer simply to promote blogging in general, but to try and proactively develop the Stanosphere and the blogospheres of the individual countries by targeting and supporting specific groups who were marginalised or under-represented in the mainstream media. This revision of *neweurasia*'s central aim was brought about not only to increase explicit congruence with the agendas of the project's donors concerning media freedom and broader democratisation but also due to growing awareness that if *neweurasia* were to continue to develop and achieve maximum impact, there was a need to develop individual country strategies rather than using a one-size-fits-all approach.

The role of the bridge bloggers expanded considerably in the second year of partnership with TOL. In addition to their blogging duties, the bridge bloggers were expected to undertake outreach activities including blogging trainings in the country for which they were responsible.[25] Over the course of the year-long project cycle a total of 28 training sessions were held in Kazakhstan, Kyrgyzstan and Tajikistan, reaching 344 people (neweurasia 2008, p. 5), as well as a variety of other outreach activities including presentations at a number of conferences, media interviews and running a fully funded three-day new media conference in Bishkek in November 2007 (neweurasia 2008, pp. 6–9, 24). The number of bridge bloggers was also expanded by the creation of local-language blogs for Kazakhstan, Kyrgyzstan, Tajikistan and Uzbekistan, as well as the re-establishment of the Turkmenistan blog, which had been dormant for the previous year. Unlike the rest of the bridge blogger posts, which continued to be funded by Hivos, the Open Society Institute provided funds to support the Turkmenistan blog and a bridge blogger,[26] who joined *neweurasia* in May 2007.

The second phase of the TOL and *neweurasia* partnership also included a number of 'spin-off' activities to promote blogging in the region. One of the main undertakings in this respect was the proposal to offer financial support for a limited number of so-called thematic blogs. Applications were invited from people in Central Asia wishing to start and maintain a blog on a particular topic separately from *neweurasia*. In return for regular posting, these bloggers would be paid a monthly stipend. A call for

---

[23]'History', *neweurasia*, no date, available at: http://www.neweurasia.net/history/, accessed 10 December 2010.

[24]'Neweurasia Strategy Paper #1', April, internal document for *neweurasia*'s management team.

[25]In the case of the Kazakhstani, Kyrgyzstani and Tajikistani bridge bloggers only. It was not deemed viable to hold trainings in Uzbekistan or Turkmenistan for safety reasons. However, some Uzbek-language trainings were later run in Osh, Kyrgyzstan.

[26]'Building Blogging in Central Asia: Interim Report May 1, 2007–December 15', December, internal document for *neweurasia*'s management team.

proposals was released in April 2007 and eventually five projects were selected on a variety of topics including gender and women,[27] borders in Central Asia,[28] life in Uzbekistan[29] and the Tajik economy.[30] The *neweurasia* project also launched a Best Central Asian Blog competition in December 2007 with the aim of raising the profile of existing bloggers in the region and encouraging them to build up their online networks.[31]

A further and very welcome aspect of this project cycle was the opportunity for *neweurasia*'s members to meet in person. In July 2007 TOL hosted a New Media Essentials training course in Prague that was attended by eight of *neweurasia*'s bloggers. The course was followed by a two-day *neweurasia* workshop with inputs from the project's managing editors and board members, which offered the chance for additional training on how to run training sessions and to exchange experiences and ideas. A number of neweurasians also attended the Bishkek New Media Conference in November 2007, and *neweurasia* was also represented at the February 2008 Riga BarCamp by four bloggers and the management team (neweurasia 2008, pp. 24–26).[32]

Towards the end of the grant cycle, it became apparent that despite a three-month extension to the project's funding, there was likely to be a hiatus until further funding could be secured. Overall the project was felt to have been successful, with approximately 2,616 posts made over 15 months and monthly average page views being 100,000 (neweurasia 2008, pp. 13–14). Furthermore, posting was relatively evenly distributed between the different language blogs, with 800–900 posts being made on each blog (neweurasia 2008). Particular successes were seen as the rapid expansion of the Kazakh-language blogging community, which was spearheaded by *neweurasia*'s Kazakh-language bridge blogger, the introduction of podcasts, and increasing attention from mainstream media, including the reposting of materials.[33]

Nevertheless, there were some concerns about the efficacy of the training programme and the amount of time it demanded from the bridge bloggers, and that the use of posting targets had resulted in a 'quantity over quality' mentality amongst some bloggers. Although the Kazakhstan and, to a lesser extent, the Kyrgyzstan blogs had developed well, there were continuing problems with the Uzbekistan and Tajikistan blogs for a number of reasons, including electricity shortages in Tajikistan and the Uzbekistan bridge blogger being based in Kyrgyzstan.

[27]http://genderstan.wordpress.com/; http://community.livejournal.com/uzbek_woman/, accessed 2 November 2011.

[28]http://bordersca.wordpress.com/, accessed 2 November 2011.

[29]http://realuzbekistan.kloop.kg/, accessed 2 November 2011.

[30]http://tajeconomy.wordpress.com/, accessed 2 November 2011.

[31]http://www.neweurasia.net/cross-regional-and-blogosphere/best-central-asia-blog-2007-2/. The winners are listed at http://www.neweurasia.net/cross-regional-and-blogosphere/best-central-asian-blog-2007-results/, accessed 2 November 2011.

[32]http://www.neweurasia.net/media-and-internet/long-way-to-barcamp/, accessed 2 November 2011.

[33]In some instances there were problems with other Central Asian news sites reposting without attribution. In one case in December 2007 an article by a contributor to the Uzbekistan blogs about Islam in the Fergana Valley was reposted to the website of *Hizb-ut-Tahrir* (see the section 'Issues in *neweurasia*'s development').

In addition, the Turkmenistan blog had been seen as problematic from the donors' perspective due to a relative lack of analytical posts on explicitly political issues.[34] This issue, which represented if not an outright clash between donors and bloggers then certainly their differing expectations of blog content, was resolved by employing an additional English-language Turkmenistan bridge blogger with a formal academic background and hence greater familiarity with the style of post desired by donors. This solution was viewed very favourably by readers, as a post from *Beyond the River* made clear:

> Between the two of them [the Turkmenistan bridge bloggers], they provide a great picture of everyday life in a country few know anything about, and in the past year their wing of neweurasia has been one of the best sources for political analysis on Turkmenistan that I've seen, period. Great photo posts, too.
>
> Turkmenistan has no journalism to speak of, and the exile press is at best tendentious. These two bloggers are making themselves heard in public (albeit anonymously) in a country where such a thing doesn't exist and it's hard for American bloggers to appreciate the real danger of doing it there. One can only hope that the duo can increase their ranks and bring more young Turkmen into the *neweurasia* fold, as it gradually gets easier to obtain access to the internet. I wish them the best of luck for 2008 and look forward to reading more from them. (Beyond the River 2008)

The delay in securing further funding meant that the latter half of 2008 saw *neweurasia*'s development stalled, content generation declining and the departure of a number of bloggers and editorial staff. While this was not a welcome development, it did present a logical time for *neweurasia* to undergo a 'global shakeup' in response to concerns that 'the network had become too dependent upon initiatives from the top, its bloggers too quota-minded, and its goals too diffuse' (neweurasia 2009). The solution proposed by the founder, Ben Paarman, was a synthesis of 'new structure, new accountability and new blood' (neweurasia 2009). Management of the project was streamlined with a new editorial team at the heart of the project and four local language co-ordinators (all former bridge bloggers) for the Kazakhstan and Kyrgyzstan blogs 'orbiting' the management team in the words of Christopher Schwartz, the English-language managing editor. In addition, a site translator was hired to help strengthen *neweurasia*'s 'bridging' function between different language-based blogospheres in the region, and *neweurasia* was rebranded, with the new logo having the tag line 'citizen media' instead of the previous 'blogging Central Asia'.

These organisational and administrative changes had largely been completed by January 2009, when funding was secured from Hivos for a project entitled 'Strengthening and Expanding Citizen Journalism in Kazakhstan and Kyrgyzstan'. While the primary focus was on the two above-mentioned republics, it was agreed that a small proportion of the funding could be used to support continued coverage of Tajikistan, Uzbekistan and Turkmenistan, as well as of the region in general (neweurasia 2009). The project's aims represented a further revision of *neweurasia*'s aims, with greater focus now being on 'professionalising' blogging by fostering the

---

[34]'Interim Report #1: May 15–October 31', November, internal document for *neweurasia*'s management team.

adoption of Western journalistic skills and standards amongst bloggers, as well as developing spin-offs of the project's local-language blogs and promoting 'the use of New Media for advocacy purposes among local NGOs' (neweurasia 2009).

This revised focus, while arguably a natural progression from *neweurasia*'s original aim of promoting intercultural dialogue, nonetheless meant that the project acquired a far more explicit normative agenda that was underpinned by Western notions of citizen journalism's form and function in Central Asia. Although in practice the benefits to both *neweurasia* and its bloggers have been considerable in terms of professional development and sustainability, there are broader implications that bear consideration. In particular, as we consider in the penultimate section, the shift towards a Western-led 'digital capitalism' model for electronic media in the Central Asian republics risks, at least in the short term, limiting the Stanosphere's potential to become self-sustaining due to the need to court external funding at the expense of developing local and, most crucially, prosuming audiences.[35]

The decision was also made to redesign and relaunch the website in order to combine content from the Russian and English versions of the country blogs into a central homebase available in either English or Russian (Schwartz 2009). The local-language blogs for Kyrgyzstan and Kazakhstan kept their own separate sites, accessible from the new homebase, as did the dormant Uzbekistan and Tajikistan blogs. The redesigned versions of the Kazakh- and Kyrgyz-language sites were finally launched in January 2010 (Schwartz 2010), although the still-dormant Tajik- and Uzbek-language blogs have not been redesigned.[36]

Despite these changes, a number of issues were identified as requiring action if *neweurasia* was to reach its full potential. Even though as of May 2009 posting levels had not recovered to those prior to the funding hiatus, *neweurasia* was nonetheless still 'the single largest source of citizen-generated content in Central Asia' (neweurasia 2009). However, reflecting back on the history and development of the project through its various cycles, *neweurasia*'s management team were beginning, in their words, 'to re-think our model as a blogging venue' and moving increasingly towards actively developing into a high-quality and trustworthy source of citizen media (neweurasia 2009). This new model has emphasised 'quality over quantity' in terms of content, and represents a convergence between *neweurasia*'s initial aim in 2005 and the need to ensure the project's longer-term sustainability by becoming more attractive to potential sources of funding, whether non-commercial or commercial.

While this development almost certainly represents *neweurasia*'s best chance for long-term sustainability, it still remains to be seen where exactly *neweurasia* will best fit in the Stanosphere and indeed, the wider blogosphere, in the future. As we discuss in the next section, there is a range of factors that will affect the project's further evolution, including technological, personnel and logistical issues; donor, sponsor, partner and/or advertiser agendas; and the perceived importance and viability of bridge blogging as a feature of the blogosphere. Perhaps the most crucial factor,

[35]Please see Sokolova, in this collection.
[36]http://neweurasia.net/tajikistan/ and http://neweurasia.net/uzbekistan/, accessed 2 November 2011.

however, will be managing to respond promptly to the changing demands of an increasingly sophisticated readership both outside and within the region.

## Issues in neweurasia's development

### Contextual issues

The regional scope of the project was viewed extremely positively by *neweurasia*'s bloggers and readers alike, and continues to be a unique selling point for the project in comparison with almost all other blogs in the Stanosphere (neweurasia 2007, p. 40). However, maintaining the regional scope in any meaningful way has been contingent upon successful individual country strategies and, as *neweurasia* expanded to include more languages, bridge blogging.

As previously noted, it was only after the first 18 months that *neweurasia* had accumulated sufficient knowledge about the specific conditions of each country to begin tailoring strategies and goals to individual countries. Kazakhstan and Kyrgyzstan proved least problematic, since in Bishkek and Almaty internet penetration was sufficient to sustain local blogging communities. In addition, the project was fortunate to recruit bridge bloggers who were mindful of the local context and willing to take the initiative in localising their presentation of *neweurasia*. One very tangible example of this was the Kazakh-language bridge blogger's decision to substitute the name *neweurasia* for '*jangaeuraziya*' (a direct translation into Kazakh), on the grounds that 'a name in English would put people off straight away'.[37]

Political conditions in the other three Central Asian republics, meanwhile, presented particular challenges. The chief concern was ensuring the personal safety of bloggers covering these countries and ensuring that they did not fall foul of government restrictions on the media. In the case of both Uzbekistan and Turkmenistan, despite initial efforts to find in-country bloggers, in the end the Uzbekistan blogs were co-ordinated by an Uzbek blogger from Osh in southern Kyrgyzstan, while the first Turkmenistan bridge blogger, a Turkmenistani national, was based in Western Europe. Nevertheless, security concerns remained and ensuring that all neweurasians knew how to blog anonymously was vital, resulting in the distribution of a 'Safe Blogging and Online Privacy Made Easy' guide in English, Russian and Uzbek.[38] The importance of such measures was made clear in December 2007 when an article about Islam in the Ferghana Valley was reposted to the *Hizb-ut-Tahrir* website, causing much alarm for one of the bridge bloggers on the grounds that the author, if identified, could potentially be accused of 'cooperating with a terrorist organisation' by the Uzbekistani or Kyrgyzstani authorities.[39]

Local conditions also impacted on the recruitment of writers to the project. 'My biggest problem was the recruitment of new authors. Often, people were suspicious

---

[37]Email questionnaire, Kazakh-language bridge blogger and country co-ordinator, 31 January 2010.

[38]http://www.neweurasia.net/resources/, accessed 2 November 2011.

[39]Email correspondence, 3–5 December 2008. The original article was posted at http://www.neweurasia.net/politics-and-society/resurgence-of-islam-in-fergana-valley/, accessed 22 December 2010.

about *neweurasia* since it is funded from abroad', commented the Tajikistan bridge blogger (neweurasia 2008, p. 19). Interestingly, people's caution impacted primarily on their willingness to blog for the project, rather than on their perceptions of *neweurasia*'s reliability and trustworthiness as a source of information. The Uzbekistan bridge blogger also acknowledged that he had encountered problems recruiting contributors, especially for the Uzbek-language blog. However, in his opinion 'this was due to the fact that most people who want to write on our blog want to do so in Russian', rather than to concerns about personal safety (neweurasia 2008, p. 20). More generally, recruitment was made harder by the fact that many people were not familiar with what a blog was and did not necessarily see the point of blogging.

In order to help remedy people's unfamiliarity with blogging and how it can be used, training sessions were run as part of the outreach programme. This response reflected the view that it is a lack of applied knowledge about how to blog (or indeed how to use the internet more generally), rather than technological or economic conditions, that acts as a brake on the Stanosphere's development. While the cost of internet access is undoubtedly an issue for some users, it is important not to overstate the impact of economic factors; data from a 2008 survey on reasons for internet non-use reported that only 1% of Kyrgyzstani non-users cited cost as the main obstacle to gaining access, while 31% explained their non-use as a simple lack of interest (Driesbach *et al.* 2009). This finding, as well as *neweurasia*'s experiences, suggests that, in Central Asia at least, until people see blogging or citizen media as relevant to their own lives and actively seek opportunities to engage with it, it will remain an activity limited to certain cultural elites such as higher education students, civil society activists and journalists. Until this shift occurs, it is difficult to see the Stanosphere evolving much beyond the top-down initiatives of international donors that remain primarily focused on local digital elites.

*Logistical and organisational issues*

'Management' was the one-word answer offered by one neweurasian who worked as a country editor between 2007 and 2008 when asked what the worst aspect of *neweurasia* was.[40] It would be fair to say that the sheer logistical challenges involved in keeping the project functioning presented the management team with an extremely steep learning curve. Prior to funding being secured for the first time in 2006, the organisational structure was extremely loose, but no less effective for this. Looking back, one of the key reasons for this was that the number of people involved was still relatively small at around 10 and everyone had previous experience of blogging independently or on group blogs. In addition, *neweurasia* was first and foremost a labour of love for those involved, with people's motivations being distinctly idealistic, if not outrightly utopian, in terms of a shared belief in the potential of blogging to facilitate the exchange of ideas and create communities of interest.

Acquiring funding, while greatly enhancing *neweurasia*'s capacity and potential, required far more active and structured management. Formal management arrangements, with *neweurasia* having a managing editor based in Prague at TOL, were put in

[40]Email questionnaire, former Kyrgyzstan country co-ordinator, 9 January 2010.

place rapidly. However, the task of maintaining communication with the network of bloggers in Central Asia proved more problematic, not least due to sporadic internet access and the fact that the bridge bloggers were only contracted to work part time. In addition, the lack of opportunities for regular face-to-face meetings meant that it was not always possible to offer the bridge bloggers as much support and mentoring as would have been ideal. While allocating funding for the training in Prague in July 2007 and a number of events in the region to enhance blogging skills helped to improve the situation, it highlighted the importance of finding individuals with previous experience of blogging and, ideally, journalism, as well as appropriate language skills, in order to maximise impact and ensure quality to create a virtuous circle of solid readership numbers.

A further organisational challenge was posed by the expansion of outreach activities from simply recruiting contributors into the provision of in-country trainings demanded by the second year of the partnership with TOL. This represented a considerable change to *neweurasia*'s scope that had not previously been envisaged. While additional remuneration, as well as editorial support, advice and training was in principle available to the bridge bloggers, over the course of the 2007–2008 grant cycle it became apparent that there was a growing tension between the required training activities and blogging responsibilities for some of the bridge bloggers, who understandably found it difficult to fit all their responsibilities within the constraints of their contracted hours. Frustration was heightened on several occasions by delays in payment caused by the additional time required for the transfer of funds from the Czech Republic and official invoicing procedures. Although most people were understanding about the reasons for the delays, the inconvenience caused was obviously undesirable, not least due to the potential to erode *neweurasia*'s reputation (and by extension that of its donors) and people's desire to participate.

Furthermore, concerns were voiced by several of the bridge bloggers that running training sessions was not an effective way to either recruit bloggers to *neweurasia* nor to 'seed' the Stanosphere. This concern was subsequently borne out by the fact that no participant in any Russian-language training session became a regular blogger, either for *neweurasia* or independently.[41] To a large extent the very limited impact of the training sessions was a further reflection of the need to ensure that the trainings were suitable for the target audiences. One of the key problems was the huge variation in the knowledge levels of participants, as well as with the quality and reliability of internet connections. For example, at one training seminar in Tajikistan, some participants had never used email before, which necessitated an impromptu revision of the planned seminar to teach this skill, rather than the more 'advanced' tasks of how to set up and post to a blog. The situation was further complicated by the fact that while there were 15 computers available, only two or three could be used simultaneously for internet access (neweurasia 2008, p. 8). In light of these issues, over the most recent project cycle, *neweurasia* revised its training focus to target existing bloggers and raise awareness of journalistic principles and skills (neweurasia 2009), further moving the project's focus towards citizen media rather than the promotion of 'grass roots' blogging.

---

[41]'Interim Report #1: May 15–October 31', November, internal document for *neweurasia*'s management team, p. 7.

*Personnel issues*

As noted above, finding suitable people was vital for *neweurasia*'s continued development and growth. In the event, this meant recruiting people who could not only blog, but who were also passionate about blogging and new media. The importance of this was made starkly clear in the second half of 2009 during the funding hiatus. With the departure of a number of long-term neweurasians, it became increasingly evident that 'the success of the project thus far had been dependent on a small number of very passionate people'.[42] Furthermore, while paying people to blog may initially encourage people to start their blog or to contribute to an existing one, unless it is a particularly lucrative activity, financial incentives alone are rarely enough, as *neweurasia*'s experiment with thematic blogs demonstrated; once funding ceased, in the majority of cases the blogs became dormant within a matter of months.[43]

While the availability of funding meant that *neweurasia* was able to provide financial incentives, several bloggers felt strongly that this was not a positive approach and undermined the 'civic' nature of the project by reducing blogging to being 'just about the money'.[44] Certainly, by 2009 it was recognised that the paid blogging model had negatively impacted on the quality of the site's content on occasion, with the new management team prioritising a 'ruthless' refocusing on the production of 'quality content' in the latter half of 2009 (neweurasia 2009). At the same time, paid blogging assignments continue to feature, reflecting the apparent widespread belief that financial benefit is vital for continued interest in blogging or participation in the creation of content. This perspective was clearly evident in comments by the author of Transitions Online's review of *neweurasia*: 'I also believe that unless the bloggers can build up a (partial) career in writing or photography, i.e. leading to some paid assignments, the majority will tire of blogging and look for something else within a year or two'.[45]

However, others feel that such assertions are misguided. Arguing strongly against the paid blogging model, one of the bridge bloggers from Kazakhstan asserted that 'someone who has recognised that blogging is a commitment and who has the financial means to run a blog is not in so much need of payment as the project's initiators thought, regardless of which country one is talking about'.[46] Certainly, for several of the bloggers, the benefits of working on *neweurasia* were not seen primarily in financial terms. Rather, respondents talked about the opportunity to develop new skills, blog in English, meet like-minded people and exchange ideas. Significantly, both of these activities took place as much offline as online, reflecting a key characteristic of the Stanosphere's community in that virtual and 'IRL' social networks have significant amounts of overlap. In the longer term, it seems likely that it will be incentives such as opportunities for personal and professional development that sustain people's involvement with citizen media.

---

[42]'neweurasia Citizen Media Project', Final Mid Term Review Report, September 2009, internal document for *neweurasia*'s management team, p. 26.

[43]As of December 2010, even *Real Uzbekistan*, which was updated until July 2009 and the *Uzbek Woman Blog*, which was active until July 2010, appear to have become inactive.

[44]Email questionnaire, thematic blogger, 5 May 2010.

[45]'neweurasia Citizen Media Project', Final Mid Term Review Report, September 2009, internal document for *neweurasia*'s management team, p. 31.

[46]Email questionnaire, Kazakhstan bridge blogger, 12 January 2010.

The debate surrounding the perceived efficacy of paying people to blog reflects divergent views about how to conceptualise the role of prosumers such as bloggers. Sokolova (pp. 1565–83) identifies two main perspectives: on the one hand, bloggers' co-participation in the creation of material can be seen as an example of 'gift culture' in which gifting, rather than purchasing power or the ability to command a fee for work, confers status. Proponents of digital capitalism, meanwhile, would argue that bloggers are providing labour and should therefore be remunerated to avoid (or at least mitigate) their exploitation. Interestingly, it appears that while *neweurasia*'s bridge bloggers espoused views that emphasised the gifting aspect of involvement, the project's donors appear to have assumed that a digital capitalist perspective would predominate, which led to the notion of pay per post blogging.

This divide arguably reflects the different norms of these two occupational communities: online activists such as *neweurasia*'s bridge bloggers or open source developers are socialised into a gift economy that emphasises reciprocity in terms of information and/or knowledge sharing (Bergquist & Ljungberg 2001, pp. 308–9). In contrast, Western donors bring with them the capitalist assumption that personal financial reward is the key motivation for engagement in a particular activity. Yet, as discussed above, the financial benefits of blogging appear to exert at best a limited incentivising effect in isolation from social and personal development. As such, *neweurasia*'s stated desire 'to impress upon the next generation of Central Asian bloggers an image of *neweurasia* as less a network than a cause [sic]' is, encouragingly, perhaps less idealistic than it may appear at first glance, suggesting an awareness of needing to take into account both perspectives to maximise long-term sustainability and help navigate the tensions between the two

More broadly, these tensions raise the question of the ability of donors or other funding providers to influence the development of projects such as *neweurasia* and the implications of such dynamics. While it may be tempting to see donor influence primarily in terms of distorting or overriding local agendas, this is an over-simplification that assumes not only that donor's interests and local interests are in conflict, but also that local actors are unable to resist, contest or subvert donor agendas. In practice there are multiple agendas and interests at play, resulting in an ongoing and dynamic process of negotiation between those involved. In *neweurasia*'s case, this process was to an extent made smoother by the fact that participation in the project was on a voluntary basis, meaning that there was a shared overall aim, and negotiations focused primarily on how best to reach it. Even so, as discussed in relation to the outreach programme above and more broadly, increasing agenda localisation and being responsive to local contexts and interests has been a vital factor in the project's continued development.

### *Convergence, amalgamation and bridge blogging: towards a sustainable new media model in Central Asia?*

'At the start, I didn't like the idea [of *neweurasia*]. I called the project a *stengazeta* in discussions with friends', acknowledged one journalist who subsequently became involved with the project as a bridge blogger. 'My thinking was that authors from Kyrgyzstan were writing into nowhere, just for the sake of there being posts, there was no commentary or discussion'.[47] Such comments reflect what Tilley and Cokley describe as an 'ongoing hegemonic struggle' over definitions of 'journalism' between traditional, professional journalists and the new breed of 'citizen journalists' (Tilley & Cokley 2008). Yet despite the division between 'professional' media and 'more informal, citizen-based, non-traditional forms' that forms 'the most important conceptual boundary highlighted in blogosphere' (Reese et al. 2007, p. 238), projects such as *neweurasia* suggest that there is increasing convergence and complementarity between the aims of many bloggers and professional journalists—namely to produce trustworthy, timely, high-quality content that meets the demands of their target audiences, while avoiding 'churnalism' (Tilley & Cokley 2008, p. 102), in which the speed of delivery and volume of content negatively impacts on quality.

The question, therefore, is how this process is occurring and what is the impact on the media involved. Since its founding, *neweurasia* has moved from being a traditional group blog to a blog-based generator of citizen media. As such, the project serves as an ongoing example of the increasing importance of media convergence, defined as 'an ongoing *process*, occurring at various intersections of media technologies, industries, content and audiences' (Jenkins 2001, p. 93). On the level of the Stanosphere, *neweurasia* reflects the expansion of the blogosphere's boundaries to embrace other Web 2.0 phenomena such as social networking, podcasts and video clip sharing sites such as YouTube (Drezner & Farrell 2008, p. 2), as well as new forms of blogging such as microblogs (for example Twitter) and short-form blogs (for example Tumblr) in an example of what Jenkins calls 'cultural convergence' in that content is generated across multiple channels. In conjunction with greater exploitation by content producers of 'social or organic convergence' (defined as 'consumers' multitasking strategies for navigating the new information environment'), Jenkins posits that 'storytellers will use each channel to communicate different kinds and levels of narrative information, using each medium to do what it does best' (2001, p. 93).

Secondly, *neweurasia* has been able to create an occupational 'bridge' between blogging as primarily a personal and highly subjective activity and professional online journalism. The result is a project that is currently unique in the Stanosphere as an example of a model of new media development whereby the dominant dynamic is not the amateurisation of journalism, but rather the professionalisation of blogging, in a process of 'amalgamation' by which journalism absorbs blogging into its occupational identity and practices (Lowrey 2006, p. 419). However, in contrast to the somewhat zero-sum view of the relationship between blogging and journalism presented by Lowrey, which casts amalgamation as a hostile response from journalism to the 'blogging challenge' (2006, p. 493), we suggest that *neweurasia* represents an example of 'benevolent' amalgamation between professional citizen media whereby the benefits of blogging—immediacy, range of voices, audience interactivity—are successfully married to the recognised benefits of traditional media, namely editorial control, accuracy and reliability. While this trajectory owes much to circumstance and serendipity, it also highlights how the blogosphere—or in this case, the Stanosphere—can been seen to be 'supplementing and interconnecting the work of professional journalists', rather than threatening it (Reese et al. 2007, p. 239).

---

[47]Email questionnaire, former Kyrgyzstan bridge blogger and country editor, 9 January 2010.

In addition bridging the occupational spheres of professional journalism and blogging, the project also fulfilled a second 'bridging' function, this time between the global blogosphere and the Stanosphere. This concept was central to the project from the very start, putting *neweurasia* firmly in the sub-category of 'bridge blogs'. Although predating Zuckerman's coining, *neweurasia* was fundamentally a bridge blog from its very inception insofar as the motivation for its creation was to facilitate the exchange of views between young people from Central Asia and the rest of the world. At the end of the first funded project cycle in 2007, the role of the bridge bloggers was viewed as central to the success of *neweurasia*, as the proposal for continued funding made clear:

> The decision to invest in bridge bloggers has definitely paid off. Compared to other blogs on Central Asia, the *neweurasia*/TOL project has the advantage of representing both local and international voices. By linking to blogs in local languages and doing regular roundups, the bridge bloggers create a gateway through which English-language readers can enter the local blogospheres. Bridge bloggers also are able to get local bloggers more visibility in the outside world, thus making their voices more prominent among English-language conversations that dominate the global blogosphere.[48]

Significantly, bridge bloggers also operate on the intra-regional level, creating links between Russian-language blogs and those in Kazakh, Kyrgyz, Uzbek, Tajik and Turkmen. This bridge function and ability to facilitate exchanges between different geographical and linguistic blogospheres is arguably one of *neweurasia*'s greatest strengths and selling points. Crucially, the presence of bridge blogs can help reduce the 'echo chamber' effect that can occur as self-selecting online communities form and only interact with other communities with which there is a commonality (for example, language or socio-political views), isolated from the wider blogosphere (Reese *et al.* 2007, p. 236). However, fulfilling this function requires either bloggers with strong language skills in at least one non-native tongue, or the use of translators. While *neweurasia* employs translators, which helps to ensure that posts get translated between Russian and English, maintaining bridges between English and Russian and the local languages has been more problematic, and it is hoped that this situation will improve in the future as the Tajik, Kyrgyz, Uzbek and possibly even Turkmen blogospheres develop.

The presence of these processes of convergence and amalgamation in the *neweurasia* project point to the increasing presence of transmedial practices—that is, practices of using multiple media formats and platforms—in Central Asia amongst a number of different communities. In the case of occupational communities, these practices are central in blurring the boundaries between the categories between journalist and blogger, resulting in the formation of the still vaguely defined category of 'citizen journalist' as a hybrid of the two. Significantly, movement can be from blogger to citizen journalist or from 'traditional' journalist to citizen journalist; several of *neweurasia*'s country bloggers, for example, worked as journalists prior to becoming involved with the project as a way to engage with a new format, while others started out blogging individually before joining *neweurasia* and developing their knowledge of

---

[48]'TOL/neweurasia Proposal', February, internal document for *neweurasia*'s management team, p. 11.

journalistic practices. In either case, we suggest, in agreement with Bowman and Willis, that the defining characteristic of a citizen journalist is that she 'is not out to cover something, but to share it... they want to tell everyone about their passion' (2005, p. 8). The experience of *neweurasia* to date suggests that the categories of citizen journalist, blogger and mainstream media journalist can be contextual and constituted by practice. Viewed this way, neweurasians gained access to a channel of communication and accompanying socio-cultural benefits that is in addition to those offered by stand-alone blogging and mainstream journalism, highlighting the increasing relevance of transmedial practices not just for the production and distribution of content, but also identity.

Looking to the future of *neweurasia*, the 2009 interim project report concluded that 'Perhaps such is the nature of blogging, but we are actively envisioning a *neweurasia* future that will sooner or later be without us (i.e., the current management team); it would be foolhardy to do otherwise' (neweurasia 2009). This does not necessarily mean, however, that *neweurasia* will become self-sustaining without further investment of time and, for better or worse, money. Indeed, as Bowman and Willis caution, online 'communities will not survive on the "Build it and they will come" ethos. They require constant attention, involved leadership, and, most important [sic], nurturing' (Bowman & Willis 2005, p. 8).

This is the greatest challenge for *neweurasia*, and Central Asian citizen media in general. Currently the vast majority of internet users in the region are passive consumers rather than 'prosumers' who both consume and generate online content. For example, Driesbach *et al.*'s interviews with 21 internet users in Kyrgyzstan suggests that online engagement is focused primarily on consumption rather than production, with interviewees estimating they spend approximately 80% of their online time reading and only 20% posting (2009). The result is a Stanosphere that is at present dominated by a relatively limited number of 'blog stars'.[49] As Adil Nurmakov, *neweurasia*'s co-ordinator in Kazakhstan put it, 'Blogs emerge when the level of internet usage grows and the user him/herself evolves into a prosumer' (neweurasia 2008, p. 7). The evolution of *neweurasia* thus far suggests that this process is still in its early stages amongst Central Asia's internet users. However, with the continued efforts of blogging enthusiasts and the further development of citizen media projects, it seems likely that the Stanosphere will continue to develop and expand, taking on forms that meet the demands of internet users from both Central Asia and other parts of the globe.

*Deakin University*
*neweurasia*

*References*

Ben (2006) 'The Elections on *neweurasia*', *neweurasia*, 12 March, available at: http://www.neweurasia. net/cross-regional-and-blogosphere/the-elections-on-neweurasia/, accessed 19 November 2010.
Bergquist, M. & Ljungberg, J. (2001) 'The Power of Gifts: Organizing Social Relationships in Open Source Communities', *Information Systems Journal*, 11, 4, pp. 305–20.

---

[49]Email questionnaire, Kyrgyz-language bridge blogger, 15 January 2010.

*Beyond the River* (2008) 'Looking Ahead to 2008 in the Stanosphere', 1 January, available at: http://web.archive.org/web/20080212104650/http://beyond-the-river.com, accessed 22 December 2010.

Bolton, T. (2007) 'News on the Net: A Critical Analysis of the Potential of Online Alternative Journalism to Challenge the Dominance of Mainstream News Media', *Scan Journal*, 4, 2, available at: http://scan.net.au, accessed 17 November 2010.

Bowman, S. & Willis, C. (2005) 'The Future is Here, but do News Media Companies See it?', *Nieman Reports*, 59, 4, pp. 6–10.

Doostdar, A. (2004) '"The Vulgar Spirit of Blogging": On Language, Culture, and Power in Persian Weblogestan', *American Anthropologist*, 106, 4, pp. 651–62.

Drezner, D. W. & Farrell, H. (2008) 'Introduction: Blogs, Politics and Power: A Special Issue of *Public Choice*', *Public Choice*, 132, 1–2, pp. 1–13.

Driesbach, C., Walton, R., Kolko, B. E. & Seidakmatova, A. (2009) 'Asking Internet Users to Explain Non-use in Kyrgyzstan', *Proceedings from Professional Communication Conference*, IPCC 2009, IEEE International, pp. 1–6.

Fitzgerald, K. (2007) 'Blogs Fascinate, Frighten Marketers Eager to Tap Loyalists', *Advertising Age*, 5 March, p. S-4, available at: http://adage.com/article/special-report-digitalreport030507/blogs-fascinate-frighten-marketers-eager-tap-loyalists/115322/, last accessed 29 June 2012.

Freedom House (2010) *Freedom in the World, 2010 edition*, available at: http://www.freedomhouse.org/template.cfm?page = 15&year = 2010, accessed 16 November 2010.

Ian (2008) 'Looking Ahead to 2008 in the Stanosphere', *Beyond the River*, 1 January, available at: http://web.archive.org/web/20080212104650/http://beyond-the-river.com/, accessed 17 November 2010.

Jenkins, H. (2001) 'Convergence? I Diverge', *Technology Review*, June, available at: http://www.technologyreview.com/business/12434/, accessed 9 January 2012.

Jessica (2006) 'HIV/AIDS in Eurasia: Donor Politics and Priorities', *neweurasia*, 26 February, available at: http://web.archive.org/web/20061110235015/neweurasia.net/?p = 231, accessed 22 December 2010.

*Journalist News* (2009) 'Kyrgyzstan stolknetsya s bumom internet-populyarnosti v blizhaishie paru let' ['Kyrgyzstan Will See a Boom in the Popularity of the Internet in the Next Couple of Years'], 12 March, available at: http://www.journalist.kg/?pid = 173&nid = 311, accessed 16 November 2010.

Kulikova, S. V. & Perlmutter, D. D. (2007) 'Blogging Down the Dictator? The Kyrgyz Revolution and Samizdat Websites', *International Communication Gazette*, 69, 1, pp. 29–50.

Lowrey, W. (2006) 'Mapping the Journalism-Blogging Relationship', *Journalism*, 7, 4, pp. 477–500.

Mcglinchey, E. & Johnson, E. (2007) 'Aiding the Internet in Central Asia', *Democratization*, 14, 2, pp. 273–88.

*neweurasia* (2005) 'What is neweurasia? Our Mission', 18 December, available at: http://web.archive.org/web/20051218202307/http://www.neweurasia.net/, accessed 19 November 2010.

*neweurasia* (2006) 'Press Release: Neweurasia.net Blocked by Uzbekistan Government', 26 July, available at: http://www.neweurasia.net/cross-regional-and-blogosphere/press-release-neweurasia-net-blocked-by-uzbekistan-government, accessed 10 December 2010.

*neweurasia* (2007) 'Response Summary Report #2: *neweurasia* Readership Survey', 7 November.

*neweurasia* (2008) 'Building Blogging in Central Asia: Final Project Report 01/05/2007–31/07/2008', September.

*neweurasia* (2009) 'Strengthening and Expanding Citizen Journalism in Kazakhstan and Kyrgyzstan: Interim Report January–May 2009', July.

Pickett, J. (2006) 'Citizen Journalism in Central Asia: Challenges and Opportunities of the Growing Online Community', *Eurasia21*, available at: http://eurasia21.com/cgi-data/document/files/eurasia21_-_citizen_journalism_in_central_asia_-_Pickett.pdf, accessed 16 May 2010.

Reese, S. D., Rutigliano, L., Hyun, K. & Jeong, J. (2007) 'Mapping the Blogosphere: Professional and Citizen-based Media in the Global News Arena', *Journalism*, 8, 3, pp. 254–80.

Schwartz (2009) '*neweurasia* has relaunched!', *neweurasia*, 24 May, available at: http://www.neweurasia.net/cross-regional-and-blogosphere/neweurasia-welcome-back, accessed 10 December 2010.

Schwartz (2010) 'Deer tek guy ov nu-urazia: wee haz ur webzyt', *neweurasia*, 27 January, available at: http://www.neweurasia.net/cross-regional-and-blogosphere/deer-tek-guy-ov-nu-urazia-wee-haz-ur-webzyt/, accessed 10 December 2010.

Srinivasan, R. & Fish, A. (2009) 'Internet Authorship: Social and Political Implications Within Kyrgyzstan', *Journal of Computer-Mediated Communication*, 14, 3, pp. 559–80.

Tilley, E. & Cokley, J. (2008) 'Deconstructing the Discourse of Citizen Journalism: Who Says What and Why It Matters', *Pacific Journalism Review*, 14, 1, pp. 94–144.

Winn, P. (2009) 'State of the Blogosphere 2008', *Technorati*, 21 August, available at: http://technorati.com/blogging/article/state-of-the-blogosphere-introduction/, accessed 17 March 2010.

Yandex (2009) 'Trends in the Russian Blogosphere', *Spring*, available at: http://download.yandex.ru/company/ya_blogosphere_report_eng.pdf, accessed 17 March 2010.

Zuckerman, E. (2008) 'Meet the Bridgebloggers', *Public Choice*, 134, 1–2, pp. 47–65.

# Blog Medvedev: Aiming for Public Consent

## DMITRY YAGODIN

### Abstract

This essay investigates the blog of Dmitrii Medvedev whose presidency marked the move of the Russian political field towards online communication. The case is an example of a hybrid model of blogging positioned between the official press service and informality of participatory media. I use the concept of symbolic power to explore the blog's search for public consent. The study also draws on a theoretical distinction between broadcast and post-broadcast forms of communication. In a generally hostile environment of the Russian blogosphere, the blog succeeds due to its discourse of effectiveness and connections with media networks both traditional and online.

DMITRII MEDVEDEV'S FIRST PRESIDENTIAL VIDEO MESSAGE was uploaded onto the official Kremlin webpage in October 2008, during the first year of his presidency.[1] In April 2009 the service was duplicated on the LiveJournal blog platform (Blog Dmitriya Medvedeva 2009) and became available for commenting. It was at this point that the online community proclaimed Medvedev a blogger. The decision of the Russian President to start a personal blog propelled the phenomenon of blogging to a new level of public attention. Medvedev's new social role brought him closer to bloggers themselves; some of them challenged the president's status as a blogger while others legitimated his presence through participation and consent. It is the investigation of this support and acceptance, rather than a critique, that makes up the essence of the essay. Medvedev's presidency does not mark any significant turns in Russian official policies. After two consecutive presidential terms, Vladimir Putin moved down to the position of prime minister, but remained *de facto* leader of the country.[2] Thus, not being completely independent in his decisions, Medvedev played a role of moderniser, soothing the growing liberal criticisms, and the blog as a symbolic tool helped him with that. I argue that the state armed with blogging is an example of a new logic in the Russian blogosphere and Russian political history.

To make sense of the Russian presidential blog I use a theoretical distinction between broadcast and post-broadcast forms of communication (Holmes 2005). The former implies a one-way flow of media messages to an audience most typical of radio and television. The latter, on the other hand, allows for a two-way interaction

---

[1] See http://blog.kremlin.ru, accessed 12 December 2009.
[2] This text was written before September 2011 when Putin announced his return as Russian president.

characteristic of internet media. I claim that, drawing on the internet and television, Medvedev's blog accumulates symbolic power of public consent and, thus, establishes its voice in the Russian blogosphere and generally among the Russian public.

The empirical materials include two types of data. Firstly, the results of a web-based survey that was conducted in July 2009 with the help of the web service http://try.surveymonkey.com. The service provides tools to create online questionnaires, to collect responses and to export data for further analysis. I compared participatory practices in three political blog communities: Blog-Medvedev (BM);[3] a political protest movement known as Namarsh-ru;[4] and general political discussions in Ru-Politics.[5] A random sample of bloggers registered in these communities was invited to respond to a set of questions about their motivations and modes of participation. The web survey was open for one month so that even rare visitors to the communities had a chance to respond. Despite a relatively low response rate (17.6%) and other potential biases, the total of 808 completed questionnaires provides some insight. In my research the sample frame was retrieved from friendship (membership) lists of the above mentioned blog communities. Naturally, internet access is a default attribute of blogging. However, such sampling excludes people who follow blogs without signing up as blog 'friends'. Therefore, the results of the survey are limited to registered users. A very low response rate may have several reasons and implications. As I tried to reach bloggers individually via LiveJournal's internal messaging system, many might have taken the request from an unknown sender as irrelevant junk mail. Perhaps timing also played its role, and some potential respondents were on summer vacation at the time of the survey. As the 'possibility of nonrespondents being different from respondents is likely to be greater when the response rate is lower' (Dillman 2007, p. 209), the results based on the survey do not infer any general claims. Instead, the survey provides useful background information for this study. The BM content is used for further analysis in the essay. Examples are drawn from BM entries and comments posted between January 2009, when the blog became actively discussed in the media, and October 2010, when my research was concluded.

The research is mainly based on discourse analysis of the media features of the blog and draws upon Pierre Bourdieu's distinction between the journalistic field and the field of politics (2005). Firstly, I put the subject of politics and blogs into theoretical and empirical contexts; secondly, I outline the state policies in relation to the idea of an information society and the growing role of blogs in Russia; thirdly, I show the distinctive features of Medvedev's blog and compare it with other prominent blogs and blog communities; the final part of the essay addresses the discourse of effectiveness produced by Medvedev's blog.

## *Politics in blogging*

Sometimes blogs appear as separate and self-sufficient media. Such projects seem to be well positioned in segments of specialised knowledge; for example, Netnewz.ru[6]

---

[3]See http://blog-medvedev.livejournal.com, accessed 12 December 2009.
[4]See http://namarsh-ru.livejournal.com, accessed 12 December 2009.
[5]See http://ru-politics.livejournal.com, accessed 12 December 2009.
[6]See http://netnewz.ru, accessed 15 January 2010.

publishes news about the IT industry; Ruconomics.com[7] works as an interactive encyclopaedia and provides an analysis of current developments in economics. Yet, more often blogs are part of networks, as in the case of a platform-based LiveJournal where all blogs are easily connected via mutual comments, friend lists and community building. What is more important, bloggers may have quick access to commenting tools of other online media. For example, LiveJournal users in Russia regularly comment on the journalistic texts of Gazeta.ru and Kommersant.ru, popular online news sites.[8] While Gazeta.ru is an online medium, Kommersant.ru is a web version of a daily newspaper widely read by the Russian political and financial elite. In terms of the integration with blog services, journalists from these media outlets remain accountable for what they publish in the face of critical assessments by bloggers.

Politicians have always sought publicity and positive media representations to gain or maintain symbolic power, for as shown by Hannah Arendt (1972), the key constitutive element of power lies in the realm of public opinion: 'It is the people's support that lends power to the institutions of a country, and this support is but the continuation of the consent that brought the laws into existence' (Arendt 1972, p. 140). 'Support' and 'consent' are therefore needed in order to challenge the hegemony of an existing order or to reinforce one's own position within the discourse of political thought. Tacit consent is also an asset, since only a sheer disagreement might undermine the powerful initiative or opinion, which is powerful as long as it is supported by others. In 'obtaining passive consent' Antonio Gramsci saw the way toward the realisation of certain hegemonic or counterhegemonic projects (2001, p. 531). Arendt's idea of political action as 'finding the right words at the right moment' (1958, p. 26) and Gramsci's concept of hegemony as consent caused by prestige and confidence (1971, p. 12) inform and supplement each other in my analysis. I show how Medvedev's blog needed to act politically in order to establish its own voice via social consent. A positive reputation is precisely what the state, or any other interested actor, may try to establish in the blogosphere.

Politicians are perhaps more interested than anyone else in being favourably represented in online media. Apart from a presence on their official websites, Russian political leaders have long ignored online audiences, carefully avoiding online social networks and blogs and relying mostly on old forms of media. This is how Vladimir Putin's control over television channels guaranteed him the support of citizens and a 'greater leverage with Russia's elites' during his presidency (Burrett 2011, p. 8). With several exceptions, the most prominent of which are Nikita Belykh, the governor of Kirovskaya *Oblast'*, Sergei Mironov, the speaker of the upper house of the Russian parliament, and Vladimir Zhirinovskii, the leader of Liberal Democratic Party of Russia (*Liberal'no-Demokraticheskaya Partiya Rossii*, LDPR), there were no top-ranking political actors present in the Russian blogosphere more or less until 2009.

Meanwhile, blogs have considerable influence on public opinion as well as on policy makers. For instance, Farrell and Drezner (2008) have shown that in spite of the relatively small audience, the blogosphere normally attracts elite social actors.

[7]See http://ruconomics.com, accessed 15 January 2010.
[8]See http://www.gazeta.ru and http://kommersant.ru, both accessed 15 January 2010.

Journalists and other intellectuals, politicians and civil activists, all those shaping the nation's understanding of politics constitute a core of such an audience. There are, however, numerous factors involved in the elite's interpretations of the blogging narratives. Similarly, it is hard to evaluate empirically the effects of a separate blog entry in that it is always part of a broader discourse. In other words, blogs facilitate ongoing political debates in society and do not redefine them radically. Richard Davis outlined emerging forms of electronic discussions in *Politics Online* (2005) and several years later he provided an updated overview of 'the role of blogs in American politics' (2009). He is particularly careful about making predictions that online communication would revolutionise politics. According to Davis, what online communication may enable is setting media and public agendas, and creating new visions of existing agendas (2009, pp. 178–81). I discuss the rising power of the blogosphere in detail below.

Research perspectives vary according to national contexts. Recent studies on Russian political blogging tend to cover themes of political resistance and identities. For example, Floriana Fossato considers bloggers as a force that has been failing to challenge the political regime (Fossato *et al.* 2008). She continues her critical inquiry by arguing that the Russian cybercommunity has become a common place for state propaganda and manipulations (Fossato 2009). Eugene Gorny has shown that the Russian-language blogosphere is an arena where cultural identities are being articulated (2006). Alexanyan and Koltsova have focused on the questions of national and cultural borders (2009).

The concept of the blogosphere in Russia is often connected to LiveJournal. The web service was created in the USA in 1999 and rapidly became popular in other countries. The first ever Russian-language blogs also appeared on LiveJournal and the ensuing development of the national blogosphere has been closely connected with this service. As a result, the Russian initials '*Zhe-zhe*' from '*Zhivoi zhurnal*' (Live Journal) are widely used as a synonym for blogging. Despite the rapidly growing presence of other blog platforms—http://www.liveinternet.ru, http://blogs.mail.ru, http://www.diary.ru—*LiveJournal* remains the key online resource for the Russian blogging community (Alexanyan & Koltsova 2009, p. 66). Andrei Podshibyakin, an active blogger himself and the author of the book on the history of blogging in Russia (2010), defines LiveJournal as more politicised than any segment of the Russian society (Larina 2010).[9]

## Towards the network state

Before joining the blogosphere, and even more insistently after, Medvedev had been publicly emphasising the importance of internet communications for politics. A reputation as a technically savvy president had become part of his high media profile. For example, at the end of 2009, Medvedev took part in the Forum of European and Asian Media (FEAM-2009) in Moscow where he underscored the positive role of online interactions with citizens and called on other officials to start blogging

[9]For his blog, under the nickname 'Evil-ninja', see http://evil-ninja.livejournal.com, accessed 10 January 2010.

(Gazeta.ru 2009 ).[10] This event marked a notable turn in the state's media strategies. The signal was sent to inert officials to become involved in what was seen before as something marginal and irrelevant. The blogosphere, in turn, received a message that a new type of relation between the state and its citizens was being established. Was it a readiness for a dialogue with new audiences or an attempt to merely strengthen the symbolic power of the state in the blogosphere? Perhaps both, as each of the two strategies would inevitably benefit from one another.

In the early 2000s, the Russian state began considering possible ways of modernising the economy and politics in line with the rapidly developing realities of digital technologies. A federal programme called 'Electronic Russia' (*Elektronnaya Rossiya*) was meant to improve the communication processes within state institutions and to provide connections between the authorities and the citizenry (Trifonov 2004). In other words, these were clear incentives to set up the foundations of a 'network state' which, compared to the old forms of state organisation, would in principle function more efficiently. This is because the network state can maintain a stronger hold over public opinion when it involves social actors in the field of communication within the realm of political power (Castells 2009, pp. 38–42).

However, a disappointment followed. In February 2009, after almost a year in office, Medvedev acknowledged the failure of the federal programme by calling the planned shift to electronic government 'a chimera' (Kuz'min 2009). The president made it clear that there was nothing to boast of, even after his personal involvement in the issue over the period of seven years prior to his presidency. Launched in January 2002, 'Electronic Russia' went through a series of revisions and, yet, eventually remained unfulfilled. Internet penetration in the country had barely reached 32.3% by September 2009 (Europe Internet Statistics 2009). According to Medvedev, public offices, although having been equipped with computers by that time, still processed all the records on paper (Kuz'min 2009). For example, having sent a document to officials, citizens could not keep track of its moves from one office to another. Therefore, the state's aspirations for building an information society in the near future started to fade away.

As a result, the government revisited the objectives of 'Electronic Russia'; the outcome of the revision was the introduction of a new federal programme called 'Information Society' (*Informatsionnoe obshchestvo*) to be implemented in 2011–2018. I believe that this new framework will determine the future government's media and communication policies. Otherwise, the above mentioned backwardness in the long term may result in a decrease of symbolic power accumulated by the government. In my view, at times when the Russian citizenry has instant access to alternative sources of information via social media and the internet in general, the obsolete and rigid state structures risk losing their control over people's minds. While the modernisation of the state institutions has made no substantial progress, some social groups, namely bloggers, have achieved noticeable influence on *Runet*.

In this context BM seems to strengthen the influence of the government and function as an event of symbolic value. However, it appeared in a very hostile environment, for the Russian blogosphere is known for its widespread criticism of the

---

[10]See also Toepfl, in this collection.

current political regime and predominantly negative evaluations of the government's policies. The scepticism of some experts raised expectations that the intervention of the government in the blogosphere would neither affect the balance of power inside online communities nor bring serious changes to the state of affairs offline.[11] Similar attitudes turned out to be common among many bloggers. However, I aim to prove the opposite by revealing the mechanisms of BM empowerment.

## The rising power of the blogosphere

John B. Thompson distinguishes four types of power—economic, political, coercive and symbolic (1995). Thompson sees the origins of the last in the actions of production, transmission and receiving of meaningful symbolic forms: 'Individuals are constantly engaged in the activity of expressing themselves in symbolic forms and in interpreting the expressions of others' (1995, p. 16). Following Bourdieu's field theory, the struggle for resources of symbolic power—trust, consent and support—is to be seen in the interactions between the field of politics and the journalistic field, with each of them claiming 'the imposition of the legitimate vision of the social world' (2005, p. 36). The field of journalism broadly implies cultural production of meaning, thus including the practices of blogging. The two fields conflate in BM, making the clash of 'legitimate visions' inevitable.

Manuel Castells (2009) extensively elaborated on the two main mechanisms of power formation in a networked environment, namely, violence and discourse. He concluded, for instance, that 'constitutional access to coercive capacity and communicative resources that enable the co-production of meaning complement each other in establishing power relationships' (2009, p. 13). Accumulation of power in the blogosphere follows the same pattern. Symbolic violence in blogs may be realised, for example, through the usage of abusive language and moderation based on a system of internal rules. While blog comments may be a common place for insults and provocations, they can also be limited to a certain community of users and accepted 'friends'. Also offensive commentators might be easily banned from further participation.

The network theory emphasises the importance of a network's outreach which 'largely determines its power and usability' (van Dijk 2006, p. 97). Therefore, it is logical to assume that the opposing network communities would compete with each other for a broader outreach, social support and consent. The political agenda set by BM has faced a lot of adversarial views. Political blog communities, two of which I refer to below, radically challenge the official agenda of BM and thus present competing nodes of networking. It is, therefore, the ability of a community to establish and maintain a viable network that matters. Castells calls such relationships 'the network-making power', which, according to him, is the paramount form of power in the network society (2009, p. 47).

One of the reasons for the growing power of blogging is its media nature. The mediatisation of society in general and politics in particular (Hjarvard 2008; Lundby 2009) is evident in many instances. The examples include electoral campaigning, the

---

[11]See for example Samigullina and Bocharova (2010).

increasing role of press secretaries, political talk shows and debates on television. All these media forms manifest a 'celebrification' of politics (Street 2004). In this respect, the ability of the internet to produce celebrities, as happens on the Russian internet (Strukov 2010), presents new opportunities for politicians. Nick Couldry, one of the theorists of mediatisation of the social, emphasises that 'the argument is not just about the forms of political performance or message transmission, but about the incorporation of media-based logics and norms into political action' (2008, p. 377). Also there is much evidence of the impact that the new media technologies have on the strategies of political campaigning and image making. For example, a study of the German blogosphere has shown how blogs enrich political actors with new forms of communication diversifying their campaigns (Albrecht *et al.* 2007).

Currently the digital media environment requires more complex mechanisms of persuasion and political public relations (PR). In this respect the state representatives, members of parliaments and national political leaders are currently facing both challenges and opportunities. The spread of new media can affect the established relationships between political institutions and the citizenry. Two aspects of such transformation are the liberation of communicative processes from 'the arbitration of journalism exercised in traditional media circles' and 'innovative forms of direct interaction' (Bentivegna 2002, p. 50). The increasing speed of media interactions multiplies the challenges when it becomes harder to have a fully controlled and predictable pattern of communication.

Media vary in their capacity to store symbolic meanings that they produce in time and space. Durability or fixation of the symbolic forms could last from momentary forms (radio and television broadcasting) to centuries (print), both being confined to distinct locations. The internet, in contrast, redefines these positions. 'Timeless time' and 'space of flows' (Castells 2009, pp. 33–35) are two defining characteristics of contemporary digital social networking. The space of flows refers to the 'technological and organizational possibility of practicing simultaneity without contiguity' (Castells 2009, p. 34).

'Space of flows' transforms our understanding of national media systems. People may comment and publish their stories in online media despite geographical distances and state borders. Conversely, many national media expand their reach through the web by translating materials into other languages. Hence Russian blogging should be seen in terms of the common language and culture, not so much in terms of its geographical location. Given the information from LiveJournal profiles one may conclude that bloggers from post-Soviet states, the USA, Israel and Germany are active participants in Russian blog communities. Therefore it is more appropriate to discuss the Russian-language blogosphere consisting of various nodes and networks. In the Russian blogosphere the authoritative bloggers and communities—nodes—generate respective networks of 'friends' and 'members'. A common virtual location on the blog platform LiveJournal facilitates the network-building capacity, making collective initiatives more effective.

The second characteristic of digital social networks—'timeless time'—is based on the negation of sequence by compressing time, blurring the sequence of social practices, mixing of present, past and future within the same text (Castells 2009, p. 35). It has been noted that the internet differs from other media in its treatment of content over time (Danowski & Park 2009, pp. 341–42), for the texts produced in the past are as quickly retrievable as the newest ones. In the Russian blogosphere, seemingly forgotten arguments re-emerge as if there were not a time span at all. Often bloggers link to their old entries putting them into the context of newer stories. Similarly web tags allow the indexing of blog entries according to specific categories or themes and to provide quick access to past content.

## A blog without a blogger

The controversy over Medvedev's bloggership results from a simple fact: there is no such thing as the personal blog of the president. There is only a LiveJournal blog community where a team from Medvedev's administration posts video clips and moderates discussions. It is more a form of social network of bloggers where Medvedev plays the role of a non-interactive subject. Therefore, BM is at once a symbolic presentation of the president and a discursive node within the blogosphere. Anyone registered on the LiveJournal platform can set up a community where discussion groups are formed according to a chosen theme. Indeed, thousands of such communities happen to constitute the core of the Russian blogosphere, which is often identified with LiveJournal. So, Medvedev functions as simply a unifying theme for one of these communities. Thus, despite the common view of the president as having a blog, Medvedev formally has no blog in its standard form. The blog rather looks like an imitation where no real author is visible and where other bloggers have limited access. Medvedev as a named blog holder neither posts anything himself, nor uses blogging tools to interact with others. Only video messages, often not meant for the blog directly, constitute the essential contents of his blogging. In fact anyone could have created a similar blog account to upload videos of the president. Medvedev as a blogger, however, seems more authentic on his Twitter account, where he is likely to post messages himself. However, this practice by the president (as well as Medvedev's participation in other social networks) and its symbolic implications are not part of this study.

The first Russian-language blogs appeared on LiveJournal in November 1999; after years of rapid growth the number of blogs reached 2 million in July 2009 (Podshibyakin 2010, pp. 7–10). In this respect BM is relatively young but it is already quite popular: there were more than 14,000 members in spring 2010. Along with the members there are also readers who do not even have to be registered on LiveJournal in order to have access to the information in the blog. By 2009, an estimated readership of the community stood at roughly 90,000 people (Beluza 2009). For many users, BM is attractive because it is about the president. This distinguishes Medvedev's blog from others that had to demonstrate interesting content to gain popularity. For example, over 4,500 users joined the community during the first two days of its existence. However, this was just a measure of formal participation based on the list of subscribers as in reality many bloggers were disappointed with censorship and moderation and the lack of interaction in the community. For instance, one blogger criticised the moderators for the removal of comments with links to non-Russian resources (Kurdakov 2009). In another case a blogger was banned from further participation for 'criticism of the presidential administration' (Cehutq91 2009). Still thousands of comments appeared in BM. This raises the question of whether this would have been possible if the blog community were not about the president.

In BM, before being published a comment first goes through a gatekeeper's hands— a team of moderators. The procedure does not only create a serious bias in the representation of bloggers' opinions, it also slows down communication as comments may stay unpublished for several days. Still, many presidential videos receive hundreds, sometimes even thousands of comments, and discussion threads continue to expand long after the initial message is published. At the same time, the qualitative aspect of participation is rather low. Services of blog ratings place BM far below many other communities with a similar number of members. According to the Yandex chart for communities in the Russian blogosphere, it was not even in the top 500 communities in August 2010 (Yandex 2010). This means that those who happen to post their comments in the community are rarely active and known bloggers.

On the one hand, BM follows the model of a forum open for everyone; on the other hand, it does not follow the logic of post-broadcast communication. Whereas post-broadcast communication indicates higher interactivity within the online environment (Holmes 2005), BM tends to be a predominantly a one-way broadcast medium. Firstly, most videos in the community are not original blog messages; they are often excerpts from television broadcasts. Secondly, neither the president nor the community moderators interact with bloggers regarding comments on the stories, even if they are explicitly asked to do so. Nor do they in any way develop already existing topics by updating the information in the subsequent blog entries. A common example of where the president looks isolated from the blogging audience appears in the video in which Medvedev talks at the Global Policy Forum that took place in Yaroslavl in September 2010. His speech was about the prospects for democracy-development in Russia. In a comment to the video, a blogger asks the president if specific amendments to the federal law will be made. The question receives no reply and there is no continuation of the subject in other comments (Teisejas 2010).

During the first year of its existence, BM had on average about seven video messages per month. This is a very small number compared with the hundreds of stories circulating in other communities. No doubt the press service could have published many more videos of the president, but perhaps more stories would have attracted less attention to each of them. The analysis of the community's dynamics derived from the number of comments made by the bloggers shows that the first several months were the most productive in terms of both published presidential messages and reactions to them. As for the content, there were serious changes throughout the first year; for example, in the beginning the original video messages by Medvedev were created for publication in the blog.

In the early stages BM's content looked more blog specific. In a way, the president used to talk directly to the online audience more often in 2009 than in the following year. While the blog has remained an important medium for Medvedev's most recent statements, it looked as if the president and his team receded from active engagement with the bloggers. The initial position in constructing an imaginary dialogue was substituted by videos taken from various press conferences and official meetings of the president with governors. As a result of the increased formality and the declarative tone of the content, the blog further diverged from post-broadcast ideals of interaction.

## Symbolic empowerment

To assess the specifics of BM empowerment as a community, I compare it with two other well-known political communities, deliberative Ru-Politics (http://ru-politics.livejournal.com) and protest-driven Namarsh-ru (http://namarsh-ru.livejournal.com). Ru-Politics, created in July 2003, is the largest political community in the Russian segment of LiveJournal with more than 10,000 registered members. It is an example of a forum for individual postings and comments about them on various topics related to political life. Namarsh-ru was launched in 2006 to unite political activists involved in protest movements throughout the Russian cities. There are nearly 1,500 registered members in that community.

BM may be also compared to other personality-based blogs. As we saw, BM distances itself from the audience at the level of Medvedev's personal representation (scarce interactivity) and establishes a hierarchical structure of power relations through a chain of intermediaries. This chain includes the members of the press office: speech writers, the video production and editing team, and moderators of the blog. By contrast, some prominent bloggers like Rustem Adagamov[12] or Artemii Lebedev[13] create their content themselves. The former is known for his photo works taken from around the world, the latter attracts users with his original columns and ability to induce readers' participation (Podshibyakin 2010, p. 192). Moreover, both bloggers often react to remarks in commenting threads and actively quote other bloggers. These two aspects—the density of interaction and individualised content— are absent in BM. In its turn BM follows another pattern, a different logic of communication.

The way BM imposes the logic of its presence on the blogosphere is closely connected with the establishment of symbolic power. The basic strategy in exercising power that is common to network structures implies the practices of inclusion in and, most importantly exclusion from, a network (Castells 2009, p. 50). The results of my web survey show that the majority of the respondents from BM are residents of Moscow (44%) and St Petersburg (13%), i.e. 57% reside in these Russian cities. With over 14% of respondents living abroad, only one third were from provincial Russia. Such a distinctive dominance of the two cities is true for the other two political communities. The most cosmopolitan was Namarsh-ru with 69% from Moscow and St Petersburg, the cities where most protest movements take place. The number for Ru-Politics was a little more than 58%, and the share of respondents from abroad was the highest (16%). What do these numbers tell us about the category of inclusion? They indicate that BM has more or less the same geographical outreach as other communities, and still includes a few more participants from different regions of Russia.

---

[12]See Drugoi, http://drugoi.livejournal.com, accessed 10 January 2010.
[13]See Tema, http://tema.livejournal.com, accessed 10 January 2010.

One of the survey questions was about the length of their blogging experience. The results show that half of the BM participants started individual blogging in 2005–2006, almost all of them before the end of 2007. Therefore, the core users have been involved in blogging for at least three to five years. This means they were not just driven to use the blogosphere by the opportunity to comment on BM entries. The survey was also meant to reveal the level of activity in the three blog communities. I found that the most active among the three communities were the members of Ru-Politics (only 30% of respondents do not participate in discussions); Medvedev's blog demonstrates the lowest level of engagement (60%). One third of the respondents from Ru-Politics and Namarsh-ru actively defend their positions, unlike the BM participants (12%). Another question in the survey concerned the issue of civic activities outside virtual space. This aspect is predictably high for protest bloggers with 54% being regularly involved in street actions. Ru-Politics demonstrated a significantly lower level of activity (13.5%), whereas the result for BM (1%) indicated presumably a almost zero interest in active protests. Similarly, volunteering is a rare type of activity for participants in Medvedev's community (5.1%) and it is highest among protesters (11.1%). The results support the argument for the low level of interaction in BM. Various factors might have affected inclusion in the blog, and a certain degree of loyalty is one of them.

An important source of symbolic empowerment lies in the ability of BM to stimulate public debate. The most discussed video of the president has been 'On the development of the internet in Russia' (Blog_d_medvedev 2009a). It received more than 4,500 comments and threads in the blog kept on growing for months thus demonstrating the concept of timeless time at its best. For example, some of the earlier video addresses by the president received responses for more than a year. However, the time dimension in the presidential blog appears less flexible, since there are structural constraints as to the lag between an actual response and the placement of comments online. In individual blog entries one can find evidence of the process of pre-moderation that sometimes takes up to several days. In the earlier stage of Medvedev's blog moderators had been rejecting up to 70% of comments (Webplanet.ru 2009). No doubt many users lost patience and interest in the service waiting for their comment to be published. The inconvenience of the time lag along with strict guidelines for participation, which is very unusual for the blogosphere but rather typical for bureaucratic structures, has resulted in the formation of an audience, in a way, disciplined by such conditions.

Topics originating from BM sometimes become widely discussed outside the community. For example, a video clip devoted to the Great Patriotic War (Blog_d_medvedev 2009b) received over 900 comments and was frequently referred to later in public discussions about the meaning of the historical event for modern generations. The television channel Rossiia used the video from the blog as part of its news report (Kozhevin 2009). So, commenting threads on BM are also extensions of public discourses. Danowski and Park consider the length of a comment thread to be a measure of the 'discussion-generating power' of a topic—extending the outreach of the network (2009, p. 340). The actual impact of such discussions throughout the blogosphere is hard to estimate. However, those constructive elements employed by BM may stimulate the necessary consent. For example, in one of the comments a blogger reacts approvingly to a video in which the president visits the Kaspersky laboratory, a Russian computer security company. The title of the comment, 'It is high time' (*Davno pora*), encourages a discussion about the state's innovation policy (Emperators 2009). The thread attracted various views on the issue, including critical remarks about insufficient budgeting for science.

Important outcomes of such thread discussions are constructive dialogue and reasonable arguments. Such results would hardly be achievable in nonregulated disputes that are so common for the blogosphere. Besides triggering online debates BM draws up some of the appropriated external flows for its communicative logic and rejects others, thus strengthening its symbolic power. The moderators' gatekeeping role, similar to that in the mass media, defines the overall direction of discussions within the community. Apart from a simple agenda setting there may be other benefits in that work. The moderators know more about the mood of the blogosphere than others are able to sense based on the threads of the community. In addition, the press office of the president acquires a convenient tool to pick out some newsworthy events for extensive media attention.

## Coping with pressures

As has been established, for example by Deibert and Rohozinski (2010) and Strukov (2009), the Russian internet is relatively free of political and legal pressures. According to Google's transparency report, Russian officials and courts passed only one request for content removal from the company's web search results during the period from July to December 2010. This can be compared to the countries with the highest numbers: Brazil—263, South Korea—139, Germany—118 (Google 2011). However, in reality the blogosphere's autonomy has been an issue on several occasions. For example, the acquisition of LiveJournal by the Russian company SUP Media in August 2006 was seen as the first step to controlling the site. Eventually, the change had no noticeable effect on bloggers as no restrictive actions have been reported. The removal of blogging ratings from Yandex's website, the national search engine, in December 2009 and the public discontent that followed was another alarming example. Again the change had only slowed down the process of further celebrification of bloggers. Indeed, there were dozens of other blog rating services, and the one on Yandex had just become more habitual over the course of time. Much more valid concerns were caused by several hacking attacks on LiveJournal in April 2011. The so-called distributed denial-of-service attacks (DDoS) disrupted the service for several days. It is always very difficult to establish the real force behind such actions, and in this case experts only agreed the motives had most likely been political, in other words that the attacks were induced by Kremlin ideologists. SUP representative Ilya Dronov contended that the ultimate goal of the attacks was to manipulate the audience of LiveJournal and to weaken it by diverting it to other social networks (Igrick 2011). Although, it could also be interpreted as competitive struggles between online companies, such an argument would have little support, especially considering the growing political agenda of the blogosphere.

While cyber attacks are perhaps most irritating for active internet users, the frequent ideological struggles in the Russian blogosphere start rumours about the special team of bloggers hired by the Kremlin to manipulate public opinion. The political analyst Mikhail Tulskii claimed in an interview with *Novaya Gazeta*, that after the Ukraine's Orange Revolution in 2005, the Kremlin allocated 'up to 1,000 dollars a month' to each member of the '*brigada*'—a group consisting of dozens of bloggers paid to fight the opposition on the internet (Balashova 2009). In addition to rumours, there was a big scandal that included the popular blog by Aleksei Navalny, who is known for his investigations into embezzlements in the major Russian state companies. After the scandals, Navalny's blog was spammed by offensive comments (Navalny 2011). In a less conspicuous manner, bloggers might face irrelevant or inappropriate messages as a reaction to the politically critical content published on their blogs, but such offensive forms of interaction are not typically Russian and can be found elsewhere. A common strategy to cope with such disturbances is to forbid comments from non-registered users or to ban the offenders. As for Medvedev's blog, this strategy is realised in the form of pre-moderation which, as mentioned already, slows down the interaction.

Some of the key rules of BM are: first, a comment will not be published in the case that it contains elements of discrimination, offence or a threat or any other violation of the Federal law; second, no foul language can be used, and the moderators monitor for its open or hidden use, for example, through transliterations; they also ban references or links to commercial information; third, comments must be understandable and must be written according to the norms of the Russian language; fourth, attempts to file a petition or anything of a similar character, a personal complaint or a request (when a response is required) are not allowed;[14] and fifth, a comment should not be part of a virtual flashmob (a large number of identical or similar comments from different people).[15] The rules do not exercise political censorship as such, but rather they establish a mode of participation in the community, which considering the limitations turns out to be relatively low.

Because of the work of moderators, extreme dissenters are not visible in BM. In one example the moderators refused to publish comments about the involvement of some high-ranking officials in illegal hunting. The accusations appeared after a helicopter crash in the Altai region in January 2009. Allegedly, officials were hunting straight from the helicopter. In the accident seven passengers died and four survived. BM did not allow any comments about the topic on the pretext that it was too sensitive for the victims' families. It is an open question whether to treat moderators' decision as relating to ethical issues or as an example of political censorship. Nevertheless, a visible drawback in this respect has to do with the strict regulations that do not stimulate participation.

However, mild criticism and a small number of complaints do appear in BM comments and seem to add authenticity to the process and eventually become part of the larger consent and support strategy. Selection mechanisms construct an audience that is ready to follow the guidelines set out in the blog. These guidelines constitute the

---

[14]This rule in a way denies an old custom of appealing to national leaders, whether Tsarist or Soviet, personally; in the Soviet period it was legitimised in the discourse of the so-called '*khodoki k Leninu*' also presented in the Socialist realist painting 'Peasant Representatives Visiting Lenin' by Vladimir Serov (1950, Moscow, State Historical Museum).

[15]'Blog-Medvedev—Community Profile', available at: http://community.livejournal.com/blog_medvedev/profile, accessed 5 December 2010.

resources of 'symbolic violence' (Bourdieu 2005). The voice of the state, therefore, does not follow typical patterns of blogging culture. Rather, it has imposed the same hierarchical strategies and relations characteristic of less interactive broadcasting strategies.

## *Visual effectiveness*

A distinguishing feature of BM is its video content and its general perception as a visual blog. As it happens, video blogging is not so common on the Russian blogosphere as yet. This is because it requires extra resources and skills to maintain a blog and depends on the availability of broadband connections. As a result of the lack of such resources, in Russia text-oriented blogs and blog communities are still prevalent. At the same time, some use of visuals and videos is complementary in many blogs. LiveJournal provides easy-to-use templates and simple publishing tools, video postings being just one of the options, but BM uses the blog space predominantly for video postings (see Figure 1).

I argue that the video nature of BM is only partially dependent on the broadcast form of communication. The first rupture with television traditions occurs at the level of genre. For example, video entries bear no sign of any journalistic involvement (i.e.

FIGURE 1. SCREENSHOT OF MEDVEDEV'S BLOG, 10 FEBRUARY 2010.
*Source*: Blog_d_medvedev, 10 February 2010, available at:
http://blog-medvedev.livejournal.com/45329.html, accessed 17 September 2010.

there is no voice-over, no presence of a reporter asking questions). Moreover, the videos achieve a high degree of authenticity by revealing their production cues, as for example, the presence of the camera crew is often revealed. In fact, some videos (Blog_d_medvedev 2009c) use several cameras and occasionally members of the production team are seen on-screen. Also many clips in the blog are too long for traditional TV programmes and instead they look more like well-made home videos. For example, one of the blog stories tells of the president's visit to the school where he studied as a child. In the video, he talks to his teacher and plays football on the school playground (Blog_d_medvedev 2009c). Although it is noticeable that the video was carefully staged, many comments expressed positive attitudes because the bloggers enjoyed the openness of the president, him being 'a very ordinary person with an ordinary childhood in an ordinary school' (Be_joy 2009). In a comment regarding another video entry, Medvedev was called 'a politician with a human face' (Lorbit 2010). The informal atmosphere of some of the messages cannot help in contributing to the positive image of the president, stimulating social trust and support. It may be a result of the specific character of the BM audience or participants outlined above. It may also be a reflection of a broader media discourse of an innovative and sociable leader, which is an important asset in the accumulation of symbolic power.

Ironically, according to the nationwide survey conducted by FOM (Public Opinion Foundation) in May 2009, Russians knew very well that President Medvedev had a blog (49% respondents), but only one third of the population knew what a blog was (Lebedev 2009). This is because the mass media had repeatedly emphasised that Medvedev had been using the internet every day and 'rather intensively' (Ashirova 2009). His blog was referred to on the television news as 'allowing the effective solution of problems' (Channel One 2009). This statement has probably reached a much bigger audience than the blog itself.

The problem-solving potential of the presidential blog seems to be the key frame for its interpretation in the mainstream media, which is a strategic symbolic asset in the network society as defined in this essay. I provide two examples of such media events drawn from the blog's comments. In the first case bloggers reported illegal casinos functioning all over the country. The complaints followed a BM video that focused on the elimination of casinos in Russia. Since its publication at the end of December 2009, the entry has received over 500 comments. In some of them bloggers mention addresses and other details of illegal casinos (Blog_d_medvedev 2009d). The topic was used in television news programmes to show how the communicative channel with the population helps in 'the fight against criminals' (Ntv.ru 2009). In the other case, there was a direct response from the president to a comment about the poor conditions at the children's hospital in Ryazan', a city about 200 kilometres south-east of Moscow. The blogger who posted the complaint admitted it was connected with Medvedev's upcoming visit to the city (Zubarev-a 2009). The story reached the president and was conveyed to the mainstream media by Medvedev's press office. Later on, Channel One, the main Russian television channel, reported on how the poor conditions at the hospital had been improved, framing the news as an example of BM effectiveness (Channel One 2009). As the state largely controls television in Russia, the story obviously illustrates mediatisation of politics, where the media event, if not staged entirely, seems to be deliberately emphasised.

As a rule Medvedev does not react explicitly to concrete comments in his blog. At best comments provoke discussions among the readers of the blog. Although commenting facilitates participation within the community, the interaction often seems disconnected from the team that produced the main content. Rather exceptional in this respect is the situation when one of the comments got a quick response from the moderators. This happened on 10 May 2009 when one blogger posted a complaint about the slow reconstruction of a war memorial in Krasnodar. Five days later, an answer came from the moderators which showed a scanned letter addressing the local governor and signed by Medvedev. The letter was an instruction 'to sort out the problem, punish those responsible and report back within three days' (Blog_d_medvedev 2009e). After Medvedev's interference, the reconstruction works were finished within a month and ahead of schedule. In another example of ideological use of a media event, the state owned *Rossiiskaya Gazeta* framed the story as 'assistance from the blog' (Pavlovskaya 2009).

Occasional moments of interaction and the effects of visual communication take place under specific conditions of time and space. Although timeless time and space of flows that characterise the media content of BM dissociate the blog further from broadcast strategies, the complete association with post-broadcast communication is not possible either. BM presents a case of rupture of conventional norms and their replacement by a new—hybrid—logic. This is because BM combines the elements of broadcasting and post-broadcasting communication. As a result of this merger, the presidential blog has managed to stimulate social consent and support in a generally hostile environment.

Furthermore, despite the valid criticisms, BM should be considered in a large context where its merits are distinct. I claim that one of the positive effects of the presidential blog is that it has stimulated the advent of other politicians in the blogosphere.[16] The blog also enhances the image of Medvedev as an innovative, progressive and at the same time sociable leader. Mass media circulate the idea that Medvedev is the 'main blogger in the country' (Ntv.ru 2009) who has conquered the *Runet* and become closer to the people (Beluza 2009).

Another positive implication of the blog is its usage as a publicity tool in the Russian blogosphere. Even negative opinions about BM contribute to its popularity as references and links to the blog may have a reverse effect when followed by an online audience. Finally, the blog by the president alters the overall perception of the internet and the blogosphere's political dimension. It is no longer just a space for furious and angry reflections on government policies or ideological propaganda. BM sets the pattern of political communication which is completely new to the Russian establishment.

## *Conclusions*

In her analysis of intensive state propaganda hidden in blogs, Fossato (2009) provides proof of the politicised character of the Russian blogosphere that attracts Kremlin ideologists. Strukov (2009) demonstrates mechanisms of the legislative internet control in Russia. These two studies refer to the period prior to the emergence of BM, and my

---

[16]For a discussion of blogs of Russian regional governors, please see Toefpl, in this collection.

analysis of it indicates that BM introduces a new strategy of attracting public attention. This alternative strategy implies, for example, that the government wishes to ensure 'that the amount of the "right information" outweighs the amount of the "wrong information"' (Strukov 2009, p. 218). BM somewhat follows this pattern with the help of symbolic violence and effective media representations as outlined in the essay.

At the same time BM occupies its position within the space of flows, having thus modified the habitual flows of information. In a sense, the government has acquired a new communication channel to guide the focus of public attention. Even the most antagonistic reactions eventually build up the publicity of the community and spread the traces of the state agenda. The blog introduces its own logic in the political discourse by balancing the limitations of broadcast tradition and advantages of post-broadcast form of communication. BM also increases the outreach of its own representation through connections with media networks both traditional and online.

Discussions within the commenting threads (subject to the symbolic violence of the guidelines) might turn out to be less antagonistic than elsewhere in the blogosphere. Although almost impossible in other blogs, censorship is part of BM's functioning. Restrictions on sensitive issues constitute the blog's capacity to control the overall process of agenda setting. At the same time, as has been evidenced on numerous occasions, the Russian blogosphere is free enough to pose critical questions beyond the outreach of BM. Therefore, pre-moderation of comments in BM may be even a positive phenomenon because it disables the use of offences and obscenities and thus constructs a different space for public discussion.

The distinctive trait of BM as a media tool is a combination of professional production (a team of scriptwriters, video editors and feedback watchdogs) with the grassroots technology of delivery and participation. The approach allows for carefully constructed representations and increases the level of appeal and trust among new audiences. To some extent, BM proves the new media logic to be efficient, as it endows the state with symbolic power stretching beyond its traditional media forms. The strategy contrasts with the media policy of Putin in his first term as president, when he seemed to underrate post-broadcast forms of communication. As noted above, about 20% of the entries in BM are attempts to establish an imaginary dialogue with the online community. Given the additional efforts and time consumption, the number of such entries is significant and shows a clear interest in addressing the internet audience directly. However, the tendency to use this opportunity has been in decline. Nevertheless such cases are crucial for our understanding of the president's approach to communication. I believe the approach decisively contrasts with the media strategies of Medvedev's predecessor, Vladimir Putin, who almost exclusively relied on television to communicate with the nation.

The powerful position of BM stems from a discourse in which the president appears to be close to the people, and an increased media exposure helps to maintain this discourse. BM attracts people's support with the help of post-broadcast forms of communication (a directed address to specific online audiences and the resulting feedback). This is achieved, however, at the expense of the relative passivity of participation. The members of the presidential blog community seem to be less active than is common for the Russian blogosphere in general. Bloggers tend to deal with the medium in terms of resembling the practices of traditional media, though the specific spatio-temporal dimensions of web applications—a timeless time and space of flows—allow for the newer logic. Indeed, the case of the presidential blog, as an example of the information channel between the state and its citizenry, marks a kind of hybrid model of blogging that is positioned between the official press service and the informality of participatory media. At the same time the presidential blog is equally limited in its scope. It has an advantage of being part of a wider media network because of the increased media attention. For the same reasons, the bloggers who decided to join the community and follow its rules are able to enjoy greater publicity as well. BM and bloggers mutually benefit from one another: the former gains political rating by occasionally responding to the complaints of the latter.

*Tampere University*

## References

Albrecht, S., Lübcke, M. & Hartig-Perschke, R. (2007) 'Weblog Campaigning in the German Bundestag Election 2005', *Social Science Computer Review*, 25, 4, pp. 504–20.

Alexanyan, K. & Koltsova, O. (2009) 'Blogging in Russia is not Russian Blogging', in Russell, A. & Echchaibi, N. (eds) (2009) *International Blogging: Identity, Politics, and Networked Publics* (New York, Peter Lang).

Arendt, H. (1958) *The Human Condition* (Chicago, IL, University of Chicago Press).

Arendt, H. (1972) *Crises of the Republic* (New York & London, A Harvest/HBJ Book).

Ashirova, E. (2009) 'S novym blogom', *Rossiiskaya Gazeta*, 22 April.

Balashova, Y. (2009) 'Tsepnye sobaki zony Ru', *Novaya Gazeta*, 23 October.

Be_joy (2009) Blog Comment, *LiveJournal*, 3 June, available at: http://blog-medvedev.livejournal.com/27490.html?thread=7312994#t7312994, accessed 25 September 2010.

Beluza, A. (2009) 'Dmitrii Medvedev pobedil Runet', *Izvestia*, 1 June.

Bentivegna, S. (2002) 'Politics and New Media', in Lievrouw, L. A. & Livingstone, S. (eds) (2002) *Handbook of New Media: Social Shaping and Consequences of ICTs* (London, Sage Publications).

Blog_d_medvedev (2009a) 'O razvitii interneta v Rossii', Blog Entry, *LiveJournal*, 22 April, available at: http://blog-medvedev.livejournal.com/2009/04/22/, accessed 15 July 2010.

Blog_d_medvedev (2009b) 'O Velikoi Otechestvennoi Voine, istoricheskoi istine i o nashei pamyati', Blog Entry, *LiveJournal*, 7 May, available at: http://blog-medvedev.livejournal.com/25564.html, accessed 15 July 2010.

Blog_d_medvedev (2009c) 'Beregite shkolu', Blog Entry, *LiveJournal*, 3 June, available at: http://blog-medvedev.livejournal.com/2009/06/03/, accessed 5 May 2010.

Blog_d_medvedev (2009d) 'Ostavshiesya igornye zavedeniya nado prosto prikhlopnut'', Blog Entry, *LiveJournal*, 29 December, available at: http://blog-medvedev.livejournal.com/43409.html, accessed 17 September 2010.

Blog_d_medvedev (2009e) 'Memorial "Vechnii Ogon"' v Krasnodare', Blog Comment, *LiveJournal*, 15 May, available at: http://blog-medvedev.livejournal.com/25564.html?thread=4633308#t4633308, accessed 17 September 2010.

Blog Dmitriya Medvedeva (2009) Videoblog, *LiveJournal*, available at: http://community.livejournal.com/blog_medvedev, accessed 12 December 2009.

Bourdieu, P. (2005) 'The Political Field, the Social Science Field, and the Journalistic Field', in Benson, R. & Neveu, E. (eds) (2005) *Bourdieu and the Journalistic Field* (Cambridge, Polity Press).

Burrett, T. (2011) *Television and Presidential Power in Putin's Russia* (New York, Routlege).

Castells, M. (2009) *Communication Power* (New York, Oxford University Press).

Cehutq91 (2009) 'O svobode slova v bloge grazhdanina Medvedeva', Blog Entry, *LiveJournal*, 21 September, available at: http://cehutq91.livejournal.com/7779.html, accessed 20 September 2010.

Channel One (2009) 'Prezident Rossii podvodit itogi', 8 September, available at: http://www.1tv.ru/news/polit/18126, accessed 18 August 2010.

Couldry, N. (2008) 'Mediatization or Mediation? Alternative Understandings of the Emergent Space of Digital Storytelling', *New Media and Society*, 10, 3, pp. 373–91.

Danowski, J. A. & Park, D. W. (2009) 'Networks of the Dead or Alive in Cyberspace: Public Intellectuals in the Mass and Internet Media', *New Media and Society*, 11, 3, pp. 337–56.

Davis, R. (2005) *Politics Online: Blogs, Chatrooms, and Discussion Groups in American Democracy* (London & New York, Routlege).

Davis, R. (2009) *Typing Politics: The Role of Blogs in American Politics* (Oxford & New York, Oxford University Press).

Deibert, R. & Rohozinski, R. (2010) 'Control and Subversion in Russian Cyberspace', in Deibert, R., Palfrey, J., Rohozinski, R., Zittrain, J. & OpenNet Initiative (eds) (2010) *Access Controlled: The Shaping of Power, Rights, and Rule in Cyberspace* (Cambridge, MA, MIT Press), pp. 15–34.

Dillman, D. A. (2007) *Mail and Internet Surveys: The Tailored Design Method* (Hoboken, NJ, John Wiley and Sons).

Emperators (2009) 'Davno pora', Blog Comment, *LiveJournal*, 19 June, available at: http://blog-medvedev.livejournal.com/28151.html?thread=8488183#t8488183, accessed 25 September 2010.

Europe Internet Statistics (2009) *InternetWorldStats*, available at: http://www.internetworldstats.com/europa2.htm#ru, accessed 10 August 2010.

Farrell, H. & Drezner, W. D. (2008) 'The Power and Politics of Blogs', *Public Choice*, 134, 1–2, January, pp. 15–30.

Fossato, F. (2009) 'Web Captives', *Index on Censorship*, 38, 3, pp. 132–38.

Fossato, F., Lloyd, J. & Verkhovsky, A. (2008) *The Web That Failed. How Opposition Politics and Independent Initiatives are Failing on the Internet in Russia* (Oxford, Reuters Institute for the Study of Journalism of the University of Oxford), available at: http://reutersinstitute.politics.ox.ac.uk/fileadmin/documents/Publications/The_Web_that_Failed.pdf, accessed 10 January 2009.

Gazeta.ru (2009) 'Medvedev rekomenduet chinovnikam vesti blogi', *Gazeta.ru*, 9 December, available at: http://www.gazeta.ru/techzone/2009/12/09_n_3296399.shtml, accessed 4 August 2010.

Google (2011) Google Transparency Report, available at: http://www.google.com/transparencyreport/, accessed 25 September 2011.

Gorny, E. (2006) 'Russian *LiveJournal*. The Impact of Cultural Identity on the Development of Virtual Community', in Schmidt, E., Teubener, K. & Konradova, N. (eds) (2006) *Control + Shift: Public and Private Usages of the Russian Internet* (Norderstedt, Books on Demand GmbH).

Gramsci, A. (1971) *Selections from the Prison Notebooks* (New York, International Publishers).

Gramsci, A. (2001) *Further Selections from the Prison Notebooks* (eBook).

Hjarvard, S. (2008) 'The Mediatization of Society. A Theory of the Media as Agents of Social and Cultural Change', *Nordicom Review*, 29, 2, pp. 105–34.

Holmes, D. (2005) *Communication Theory: Media, Technology and Society* (London, Sage Publications).

Igrick (2011) 'Kak zakalyalsya ZhZh', Blog Entry, *LiveJournal*, 5 April, available at: http://igrick.livejournal.com/501798.html, accessed 26 September 2011.

Kozhevin, I. (2009) 'Medvedev prizval ne iskazhat' istoriyu Velikoi Otechestvennoi voiny', TV Report, *Vesti.ru*, 8 May, available at: http://www.vesti.ru/doc.html?id=283517, accessed 10 December 2010.

Kurdakov (2009) 'Rasshifrovka efira "Dnevnika gubernatora"', Blog Comment, *LiveJournal*, 9 November, available at: http://belyh.livejournal.com/343699.html?thread=8992915#t8992915, accessed 20 September 2010.

Kuz'min, V. (2009) 'www.Kreml', *Rossiiskaya Gazeta*, 13 February.

Larina, Y. (2010) 'ZhZh—eto shans byt' uslyshannym', *Ogonek*, 12 April.

Lebedev, P. (2009) 'O bloge D. Medvedeva znayut bol'she lyudei, chem ob interv'yu presidenta tsentral'nym kanalam', *FOM: Public Opinion Foundation*, 28 May, available at: http://bd.fom.ru/report/map/blog280509, accessed 23 August 2010.

Lorbit (2010) Blog Comment, *LiveJournal*, 10 February, available at: http://blog-medvedev.livejournal.com/45329.html?thread=23120913#t23120913, accessed 20 June 2010.

Lundby, K. (ed.) (2009) *Mediatization: Concepts, Changes, Consequences* (New York, Peter Lang).

Navalny (2011) 'Sekretnoe oruzhie Edinoi Rossii', Blog Entry, *LiveJournal*, 19 February, available at: http://navalny.livejournal.com/554817.html, accessed 1 November 2011.

Ntv.ru (2009) 'Medvedev poblagodaril internet-pol'zovatelei', *Ntv.ru*, 8 October, available at: http://www.ntv.ru/novosti/177608/, accessed 5 December 2010.

Pavlovskaya, T. (2009) 'Assistance From the Blog', *Rossiiskaya Gazeta*, 25 June, available at: http://www.rg.ru/2009/06/25/reg-kuban/blog.html, accessed 6 October 2010.

Podshibyakin, A. (2010) *Po zhivomu. 1999–2009: LiveJournal v Rossii* (Moscow, KoLibri).

Samigullina, A. & Bocharova, S. (2010) 'Otpostilis', *gazeta.ru*, 23 April, available at: http://gazeta.ru/politics/2010/03/22_a_3341418.shtml, accessed 12 August 2010.

Street, J. (2004) 'Celebrity Politicians: Popular Culture and Political Representation', *The British Journal of Politics & International Relations*, 6, 4.

Strukov, V. (2009) 'Russia's Internet Media Policies: Open Space and Ideological Closure', in Beumers, B., Hutchings, C. S. & Rulyova, N. (eds) (2009) *The Post-Soviet Russian Media: Power, Change and Conflicting Messages* (New York, Routlege), pp. 208–22.

Strukov, V. (2010) 'Russian Internet Stars: Gizmos, Geeks, and Glory', in Goscilo, H. & Strukov, V. (eds) (2010) *Celebrity and Glamour in Contemporary Russia: Shocking Chic* (New York, Routlege), pp. 144–71.

Teisejas (2010) 'Zakonoproekty', Blog Comment, *LiveJournal*, 17 October, available at: http://blog-medvedev.livejournal.com/56446.html?thread=35434878#t35434878, accessed 20 October 2010.

Thompson, B. J. (1995) *The Media and Modernity: A Social Theory of the Media* (Stanford, CA, Stanford University Press).

Trifonov, A. (2004) '"Elektronnaya Rossiya" stanovitsya real'noi', *Rossiiskaya Gazeta*, 26 February, available at: http://www.rg.ru/2004/02/26/minsvyaz.html, accessed 22 August 2010.

van Dijk, J. (2006) *The Network Society. Social Aspects of New Media* (London, Sage Publications).

Webplanet.ru (2009) 'V bloge Medvedeva zabanili dvesti chelovek', *Webplanet.ru*, 15 January, available at: http://webplanet.ru/news/life/2009/01/15/medvedev_comments.html, accessed 20 August 2010.

Yandex (2010) Yandex Blog Ratings, available at: http://blogs.yandex.ru/top/com/, accessed 14 August 2010.

Zubarev-a (2009) 'Gosudarstvo i grazhdanskoe obshchestvo', Blog Comment, *LiveJournal*, 21 April available at: http://blog-medvedev.livejournal.com/22187.html?thread=167595, accessed 20 June 2010.

# Blogging for the Sake of the President: The Online Diaries of Russian Governors

FLORIAN TOEPFL

*Abstract*

Many Western researchers have hailed blogs of politicians as new, interactive and 'inherently democratic' tools of political communication. Yet, as this essay illustrates, blogs can be of comparatively even greater appeal to politicians in semi-authoritarian political contexts. In Russia, 29 out of 83 regional leaders (roughly 35%) were keeping a weblog in May 2010. This essay provides a comprehensive content analysis of all governors' blogs and, subsequently, fleshes out a typology of three characteristic types. It is argued that politicians' blogs are playing a far greater role in generating legitimacy for the Russian political system than they do in democracies, because the semi-authoritarian Russian system lacks other mechanisms which generate (input) legitimacy in developed democracies, such as highly competitive elections.

I came to like this phrase, it is beautiful. I'd like to repeat it: Losing the initiative online will result in losing the initiative offline. (President Dmitry Medvedev, Blog Entry, 31 May 2010)

BLOGGING IS CURRENTLY HIGHLY *EN VOGUE* AMONGST THE political leaders of Russia's regions. In May 2010, 29 out of 83 regional leaders kept a so-called 'weblog' or 'blog'. So roughly 35% of all Russian governors[1] made use of this new tool of political communication which is a surprisingly high quota in a country that, firstly, according to some Western observers (Freedom House 2010) has a semi-authoritarian rule, and where, secondly, internet penetration is still relatively low in comparison with most developed countries. According to data presented by the Russian Public Opinion Research Centre (VTSIOM), in April 2010 only 34% of all Russians accessed the internet at least once a week (VTSIOM 2010; Alexanyan 2009, pp. 1–4).

This work was supported by research fellowships of the Fritz Thyssen Foundation (Cologne, Germany), the Harriman Institute (Columbia University, New York, USA), and the Aleksanteri Institute (University of Helsinki, Finland). I owe thanks to Vlad Strukov for constructive critique of two previous versions of this essay.

Supplementary material for this essay can be found in the online version available at: http://www.tandfonline.com/loi/ceas20 (Figures S1–S5).

[1]In the regional constitutions, the leaders of the 83 Russian federal subjects are referred to as 'governors', 'presidents' or 'heads' of their jurisdiction. In this essay I refer to regional leaders uniformly as 'governors'.

Nonetheless, in terms of blogging activity, Russian governors by far outperformed politicians in most Western democracies. In Germany, for instance, not a single leader of a '*Bundesland*'[2] kept an online diary at the time of the research. In Canada, the UK and the USA, the proportion of blogging members of parliament seemed to fluctuate around the 10% mark (Pole 2010, pp. 77–78; Williams 2009; Francoli & Ward 2008; Ott 2006). These findings are puzzling, given the fact that many Western researchers have hailed blogs as new, interactive, inherently democratic tools of political communication. Blogs of politicians, in particular, have been most often discussed with regard to their 'democratic potential' (Coleman 2005b, pp. 279–80) and their possibility to reconnect an increasingly passive electorate with its representatives (Coleman 2005a, pp. 12–14). Yet, as the figures quoted above indicate, this new tool of political communication has turned out to be of comparatively greater appeal to politicians in the semi-authoritarian Russian context. In May 2010, some Russian governors dedicated considerable amounts of their time weekly to keeping online diaries; for others, having a blog published by their press team seemed at least to be a mandatory part of their communication mix. Although most Russian governors started to blog as early as 2008, and the phenomenon of blogging Russian politicians is extremely important for an understanding of Russian political communication, there has been very little research produced on the topic.

The aim of this essay is to contribute to a deeper understanding of the exceptional phenomenon of the blogging Russian governors, its role within the vibrant Russian blogosphere, and its significance and implications for the semi-authoritarian regime as a whole. Why, how and with which effects for the political system did Russian governors blog in May 2010? To approach the three central questions of research, the essay starts out by reviewing the existing research on Russian blogging politicians and outlines some key characteristics of the Russian blogosphere. The second section addresses the question of why so many Russian governors have started a blog; my claim here is that the vast majority of governors set up an online diary primarily to showcase their loyalty to then-President Dmitry Medvedev who—as it transpires from the epigraph to this essay—regarded political communication via the internet as central to his political profile. The third section of the essay enquires to what degree governors emulated the president's style of blogging—and in which aspects they did not and why. The fourth section develops a typology of blogs of Russian governors, singling out three different types of blogs: public relations (PR) blogs, 'effective statesmen' blogs and '*internetchik*' blogs. I contend that these three types of blogs differ not only in the degree of interactivity and the style of discourse between politician and voters, but also in the extent to which they strengthen the perceived responsiveness of regional governments and the perceived legitimacy of the Russian political system as a whole. In terms of methodology, this essay combines a series of qualitative and quantitative approaches. The latter includes a comprehensive analysis of all governors' blogs in existence in May 2010 and compares such features as the blogging platforms chosen by the governors, the number of posts per month, the maximum number of comments to a post and the type of posted materials (text, photo or videos). The qualitative analysis is based on an

---

[2]The Federal Republic of Germany consists of 16 partly sovereign states, so-called '*Bundesländer*' (singular: *Bundesland*).

interpretative approach (Yanow 2006) and draws on newspaper articles, government documents, and posts and comments published in the blogs.

## Overview of blogs of Russian politicians

The word 'blog' is a contraction of 'weblog'. Both terms are usually understood as a form of 'online diary' of a single person or a group. In a rather challenging definition, Drezner and Farrell conceive of a blog as 'a web page with minimal to no external editing, providing on-line commentary, periodically updated and presented in reverse chronological order, with hyperlinks to other online sources' (2008, p. 2).

Some of the blogs of Russian politicians do not feature hyperlinks or lack a comment function; others are edited by press teams. Therefore, for the purpose of a more holistic approach and inclusion of all varieties of Russian politicians' blogs, I will resort to a more general definition provided by Coleman and Wright, who think of a blog simply as 'a regularly updated webpage ... with information (textual, photographic or video) presented in reverse chronological order' (2008, p. 1).

For some scholars of political communication, the 'remarkable rise' of weblogs has been 'one of the most unanticipated developments of the first years of the twenty-first century' (Sunstein 2008, p. 87). From an estimated 50 blogs in 1999 (Drezner & Farrell 2008, p. 3), the number of blogs grew to over 150 million worldwide in November 2010 (Nielsen 2010). The phenomenon of blogging politicians has been best researched in the American and British political environment (Kerbel & Bloom 2005; Bichard 2006; Trammell 2006; Wright 2009; Coleman 2005b; Coleman & Moss 2008; Coleman & Wright 2008; Wright 2009). However, most studies centre on campaign blogs[3] rather than on permanent blogs (Howard 2006; Trammell 2006; Stanyer 2006) or on the role of the internet in election campaigns in a more general perspective (Bimber & Davis 2003; Ward & Davis 2008; Williams & Tedesco 2006; Kluver 2007).

The only study that has investigated the blogs of Russian politicians to date is an analysis by Goroshko and Zhigalina (2009) which is based on data gathered in autumn 2008 during the South Ossetia War. At that time, blogging politicians were a rather marginal phenomenon in Russia, for example, the blog of the president did not yet exist. In this context, the article compares 16 online diaries, including nearly all politicians' blogs active at that time (Goroshko & Zhigalina 2009, p. 93), using such criteria as Google-page rank, frequency of blog entries, number of comments, number of friends and user friendliness. Goroshko and Zhigalina find four politicians to be particularly active in blogging in August 2008: Vladimir Zhirinovsky, the leader of the Liberal Democratic Party of Russia (*Liberal'no-demokraticheskaya partiya Rossii*, LDPR); the leaders of the democratic opposition, Nikita Belykh and Boris Nemtsov; and the speaker of the Federation Council, Sergei Mironov—all of whom updated their blog daily. In terms of interactivity, Zhirinovsky's blog was found to lead the field, with an average of 230 comments per blog entry (Goroshko & Zhigalina 2009, p. 94). In their conclusion, Goroshko and Zhigalina suggest that Russian politicians' blogs could 'become a rather powerful PR-tool, directed primarily towards the target

---

[3]Campaign blogs are blogs that are set up by politicians exclusively for an election campaign and not continued thereafter.

group of opinion leaders', with one of their key features being that they 'blur the borderline between the public and the private, thus helping to create the illusion of an intimate and open discussion with the audience' (2009, p. 97).

After Medvedev took office in 2008, permanent blogs of politicians became a widespread and heavily discussed phenomenon in Russian politics; they received broad coverage in the Russian mass media.[4] Yet, little academic research has been produced on the topic so far. This essay seeks to contribute to the study of these blogs by scrutinising the role of the permanent blogs of a specific group of Russian politicians, the governors. While the analysis of this study deals exclusively with the Russian case, the implications of the essay also aim to enrich the broader, currently ongoing, academic debate on the question of whether internet-mediated communication in a non-democratic state should be seen rather as a 'technology of liberation' or as one of 'control' (Deibert & Rohozinski 2010; Diamond 2010).

Blogging in Russia started on the platform LiveJournal in the early 2000s (Alexanyan 2009, pp. 4–5; Gorny 2006a, pp. 228–75; 2006b). In spring 2009, a report by the leading Russian search engine Yandex counted 7.4 million Russian language blogs (Yandex 2009, p. 2). Of these blogs, only 12% or 890,000 were 'active', i.e. they had been updated at least once in the last three months. While blogs are often regarded as the least trusted and least reliable political media in Western countries (Drezner & Farrell 2008, p. 5), the blogosphere plays a rather different role for many Russian citizens (Toepfl 2011, forthcoming). With the central television stations acting as the mouthpiece of the government and with independent print and online media being, at least supposedly, under continuous pressure by vested interests, the blogosphere emerges as a promising and prominent space for politically highly motivated individuals to search for unmediated, first hand, credible information.

In spring 2009, more than 76% of all active Russian language blogs were hosted on one of the four leading blogging platforms: http://www.livejournal.com, http://ya.ru, http://mail.ru and http://www.liveinternet.ru (Yandex 2009, p. 3). A specific characteristic of the Russian blogosphere is that all of these four leading blogging platforms are 'social network system hybrids' (SNS-hybrids) (Etling *et al.* 2010, p. 12). SNS-hybrids combine features typical of open blogging platforms like Blogspot or Wordpress with features of closed social network services like Facebook or Myspace. With regard to politicians' blogs, the most momentous difference is that SNS-hybrids allow bloggers to maintain a network of 'friends' or 'followers' (that is, permanent readers). Consequently, Russian blogs are often received through a 'friends' page' similar to the 'News Feed' section of a Facebook account. By contrast, blogs in the USA and in most Western countries are usually read by directly accessing the blogs' URL or subscribing to an RSS feed. As a result, the 'macro structure of the Russian blogosphere features a network divided into largely separate camps, each based upon a large SNS hybrid, with strong internal and weak external links' (Etling *et al.* 2010, p. 13).

Of the four major blogging platforms, Yandex is the leading platform for political, intellectual and public-affairs related discourse (Etling *et al.* 2010, p. 13). As an analysis of outgoing links showed (Etling *et al.* 2010, p. 13), Yandex bloggers were far more active than those on other platforms in linking to news and other online content.

---

[4]See for example Bilevskaya (2010), Sazonov and Stolbun (2010) and Vrazhina (2010).

Moreover, Yandex hosted the highest number of active blogs. In 2009, for example, roughly 100,000 Yandex blogs were updated at least once a week; 250,000 blogs had been updated at least once in the three months before the data collection (Yandex 2009, p. 3). In July 2010, 49 Russian LiveJournal bloggers had more than 10,000 friends, with two bloggers being followed by more than 50,000 friends (Yandex 2010a). The most popular LiveJournal blogger was Artemii Lebedev (nickname 'tema'), a Moscow-based designer who posted mostly cynical reflections on social realities in Russia in a rather vulgar language. The second LiveJournal blogger with more than 50,000 friends was Rustem Agadamov (nickname 'drugoi'), at that time already an employee of the Russian internet company SUP which owns LiveJournal. In his blog, Agadamov was mostly reposting topical photos from the leading news agencies Reuters and AFP, adding short personal comments. The majority of the remaining LiveJournal bloggers in the 'reiting blogov Runeta' 'top 50' were artists, writers, poets, journalists, intellectuals, comedians or photographers. There was no politician listed in this top 50 ranking.

Although in November 2010, roughly 15,000 LiveJournal users 'watched' the videoblog of President Medvedev, his online diary did not appear in the top 50 ranking because it was registered on LiveJournal not as an individual blog but as a 'moderated community' (Medvedev 2010a). In addition, Medvedev's blog was also being published and read on the website Kremlin.ru. The president's Twitter account, opened only in June 2010, had already gathered 115,000 followers. By contrast, Prime Minister Vladimir Putin had neither set up a Twitter account nor a blog by November 2010.

At that time, by far the most successful blog of a politician by number of followers was that of Vladimir Zhirinovsky, the radical-populist leader of LDPR. According to the Yandex ranking of leaders, Zhirinovsky's blog was followed by 540,000 users in November 2010 (Yandex 2010a). However, Zhirinovsky did not blog within the network of the SNS-hybrid LiveJournal but on mail.ru (Goroshko & Zhigalina 2009, pp. 92–96). Compared with Zhirinovsky's online diary, the blogs of other political figures were rather unpopular with Russian internet users. For instance, in November 2010 the LiveJournal blogs of the leaders of the democratic movement *Solidarnost'*, Boris Nemtsov and Il'ya Yashin, were followed by only 6,800 and 5,700 friends, respectively. The blog of Sergei Mironov, the chair of the party Just Russia (*Spravedlivaya Rossiya*), had fewer than 2,700 friends while Gennady Zyuganov, the leader of the Communist Party (*Kommunisticheskaya partiya Rossiskoi Federatsii*, KPRF) did not keep a blog.

How did Russian governors blend into this picture? Table 1 presents a list of all blogging governors in May 2010, providing the information on their regions, the URL of their blogs and the dates when they started to blog. It is sorted by the date of the first entry in each blog of a governor, in ascending order. I compiled this list in three stages on the basis of three sources. First of all, I searched for newspaper articles on the phenomenon of blogging politicians,[5] compiling a preliminary list of the blogs of governors mentioned in these articles. During the second phase, I supplemented this list with governors' blogs quoted on the Russian internet portal goslyudi.ru. The portal is operated by the online platform polit.ru and aims at making politicians' blogs from all over the country accessible. Finally, I carried out searches on yandex.ru and

[5]These include Bilevskaya (2010), Sazonov and Stolbun (2010) and Vrazhina (2010).

## TABLE 1
## BLOGGING GOVERNORS (AS OF MAY 2010)

| Name | Federal subject | Address | First entry |
|---|---|---|---|
| Belykh, Nikita | Kirovskaya *Oblast'* | http://belyh.livejournal.com | 27 January 2006 |
| Chirkunov, Oleg | Permskii *Krai* | http://chirkunov.livejournal.com | 26 June 2008 |
| Komarova, Natalya | Chanty-Mansiiskii *Avtonomnyi Okrug* (Tyumenskaya *Oblast'*) | http://n-komarova.livejournal.com | 15 October 2008 |
| Bogomolov, Oleg | Kurganskaya *Oblast'* | http://www.kurganobl.ru/6477.html | 25 December 2008 |
| Boos, Grigorii | Kaliningradskaya *Oblast'* | http://www.gov39.ru/index.php?option=com_content &view=category&layout=blog&id=45&Itemid=71 | 26 December 2008 |
| Denin, Nikolaii | Bryanskaya *Oblast'* | http://www.bryanskobl.ru/news/videoblog | 19 January 2009 |
| Gaevskii, Valerii | Stavropol'skii *Krai* | http://www.gubernator.stavkray.ru/?go = video | 25 February 2009 |
| Pozgalev, Vyacheslav | Vologodskaya *Oblast'* | http://www.vologda-oblast.ru/photos.asp?LNG=RUS&V=52 | 25 March 2009 |
| Dudka, Vyacheslav | Tul'skaya *Oblast'* | http://blog-dudka.livejournal.com | 22 April 2009 |
| Slyunyaev, Igor' | Kostromskaya *Oblast'* | http://slunyaev.livejournal.com | 29 April 2009 |
| Morozov, Sergei | Ul'yanovskaya *Oblast'* | http://community.livejournal.com/blog_morozov | 14 May 2009 |
| Kozhemyaka, Oleg | Amurskaya *Oblast'* | http://www.portamur.ru/blogs/kozhemyako | 17 May 2009 |
| Volkov, Aleksandr | Udmurtskaya *Respublika* | http://aifudm.net/blog/volkov | 1 July 2009 |
| Artamanov, Anatolii | Kaluzhskaya *Oblast'* | http://www.artamonovguber.ru/blog.php | 7 July 2009 |
| Zhylkin, Aleksandr | Astrakhanskaya *Oblast'* | http://alexandr-jilkin.livejournal.com | 10 July 2009 |
| Tkachev, Aleksandr | Krasnodarskii *Krai* | http://a-n-tkachev.livejournal.com | 14 July 2009 |
| Dudov, Nikolaii | Magadanskaya *Oblast'* | http://www.gubernator.magadan.ru | 17 July 2009 |
| Zelenin, Dmitrii | Tverskaya *Oblast'* | http://dzelenin.livejournal.com | 31 July 2009 |
| Savchenko, Evgenii | Belgorodskaya *Oblast'* | http://www.savchenko.ru | 6 August 2009 |
| Katanandov, Sergei | *Respublika* Kareliya | http://www.gov.karelia.ru/gov/Blog/index.html | 14 August 2009 |
| Gordeev, Aleksei | Voronezhskaya *Oblast'* | http://blog.govvrn.ru/blog | 8 November 2009 |
| Nagovitsyn, Vyacheslav | *Respublika* Buryatiya | http://baikal-daily.ru/blog/61 | 11 November 2009 |
| Kress, Viktor | Tomskaya *Oblast'* | http://kress.tomsk.ru/ru/videoblog | 19 November 2009 |
| Tolokonskii, Viktor | Novosibirskaya *Oblast'* | http://sibkray.ru/blog | 9 February 2010 |
| Serdyukov, Valerii | Leningradskaya *Oblast'* | http://serdyukov-vp.ru/blog | 14 February 2010 |
| Brovko, Anatolii | Volgogradskaya *Oblast'* | http://anatoliy-brovko.livejournal.com | 1 April 2010 |
| Bochkarev, Vasilii | Penzenskaya *Oblast'* | http://blog-bochkarev.livejournal.com | 1 April 2010 |
| Mitin, Sergei | Novgorodskaya *Oblast'* | http://mitinsg.livejournal.com | 7 April 2010 |
| Turchak, Andreii | Pskovskaya *Oblast'* | http://turchak.ru/blog | 19 April 2010 |
| Shantsev, Valerii | Nizhegorodskaya *Oblast'* | http://shantsevvp.livejournal.com | 28 April 2010 |
| Jurevich, Michail | Chelyabinskaya *Oblast'* | http://yurevich-m.livejournal.com | 19 May 2010 |
| Kanokov, Arsen | *Respublika* Karbadino-Balkariya | http://blog.president-kbr.ru | 24 May 2010 |

*Source*: compiled by the author.

google.ru using the search words 'a governor's blog' (*blog gubernatora*) and 'a president's blog' (*blog prezidenta*) and screened the first 150 hits for blogs of governors that were not on my original data sheet. In the course of these three steps, not only did I find blogs of governors that contained only two or three entries, but also came across announcements by governors concerning the launch of blogs that did not yet exist. Thus I have good reason to believe that the list of blogging governors in Table 1 is comprehensive. As the data indicate, by May 2010, 32 out of 83 (or 38%) regional leaders had opened a blog. Three of these blogs (those of the Governors Dudka, Tkachev and Komarova) must be considered aborted by the time of this research, because these blogs had not been updated for more than three months. A total of 29 governors were actively blogging, which equalled a proportion of 35% of Russia's regional leaders.

How big was the audience that the governors reached out to with their blogs? Of those governors who blogged on a SNS-hybrid and thus could accumulate followers, the field was led by Governor Chirkunov with 4,420 friends. He was followed by the Governors Belykh (4,306 friends), Zhilkin (1,719), Yurevich (666), Brovko (333) and Dudka (99). Another measure for the audience of a blog is the Yandex index on blog 'authority' (*avtoritetnost'*), a figure calculated on the basis of the number of links to a blog, the number of comments, the number of readers and other data (Yandex 2010b). In the Yandex ranking of all Russian language blogs according to their 'authority', in December 2010, Governor Chirkunov ranked highest at position 304, followed by Belykh at 366 (Yandex 2010c). Zhirinovsky's blog was listed at position 455, probably because of a lack of interlinkage with other popular bloggers.

Therefore, the data indicate that the blogs of Russian governors are a phenomenon of rather recent origin that has gained rapidly in importance only since the end of 2008. By the time of this research in November 2010, however, the leading governors' blogs had managed to enter the discussion core of the vibrant Russian blogosphere, being amongst the top 500 most influential members of this networked public sphere according to the Yandex ranking of 'blog authority' (Yandex 2010c). The next section aims to explore why so many governors had decided to resort to this new tool of political communication.

### *Demonstrating allegiance: why did so many Russian governors set up blogs?*

Why did 38% of Russia's regional leaders make an effort to set up an online diary? This section claims that most governors started to blog primarily to demonstrate their allegiance and loyalty to then-President Medvedev who was known for his internet enthusiasm. So I aim to provide evidence for this hypothesis in three stages. First, I will discuss how central the internet was to the political profile of Medvedev and how he repeatedly called upon officials to follow his line. Secondly, I will demonstrate that governors had strong incentives to respond to the presidential demands. Thirdly, I will provide empirical evidence documenting that 93% of the blogging governors started to do so only after the president had started his blog. Finally, I will address possible counter-arguments that might undermine my central thesis.

After taking office Medvedev heavily propagated the notion of 'modernisation of Russia' making it central to his political profile and turning the notion into a real buzzword (Kamyshev 2010). As many observers have pointed out, Medvedev regarded the internet not only as a means but also as a symbol for his endeavours to 'modernise' the country and make it less dependent on natural resources. In line with that thinking, Medvedev opened a personal videoblog on LiveJournal as early as October 2008 and a Twitter account in June 2010. By contrast, Prime Minister Vladimir Putin did not make an effort to keep a blog or a Twitter account. The buzzword central to Putin's political profile seemed to be 'stability', a concept that did not necessitate a continuous emphasis on the pivotal importance of new communication technologies to the future of the country. These differences in the ideologies of the two leading Russian political figures are also visible in their communication strategies. Whereas Medvedev tried 'to give his image as a tech-savvy modernizer a broader appeal' (Bratersky 2010) by organising semi-annual online conferences with citizens, Putin continued to stick to the more traditional format of call-in TV shows.

In line with his political profile as a 'moderniser', Medvedev has called repeatedly upon his officials to familiarise themselves with the internet and the new tools of political communication. Speaking directly to governors at a State Council[6] meeting dedicated to the so-called programme of 'Electronic Government' in December 2009, Medvedev announced that in the future 'internet-activity' would be one of the evaluation criteria for the performance of regional leaders (Bilevskaya 2010). During a meeting in March 2010, he stated publicly that he considered a 'basic computer literacy' (*elemntarnaya komp'yuternaya gramotnost'*) to be very important for political leaders (vesti.ru 2010). He explicitly called upon governors to follow his example and to 'dig on the internet' (kovryat *internet*) (vesti.ru 2010). He also contended that:

> Only those who are able to do so are contemporary managers and those who cannot—I am sorry for having to say this—are just not fully prepared. If I watch [the internet], all others should watch [it] as well. I hope this will be noted not only by the government head offices but by the leaders of the regions. (vesti.ru 2010)

In an article published in the newspaper *Nezavisimaya Gazeta* on 21 January 2010, an anonymous source from within the Kremlin details the rationale behind these public calls of the President (Bilevskaya 2010). According to this source, the Kremlin was determined to 'inoculate the political elites of the country with internet-culture' (*privit' politicheskoi elite strany internet-kul'turu*) because the 'growth of internet users in Russia had brought about new challenges for Russian politicians' (Bilevskaya 2010). In particular, the Kremlin was worried about the decreasing trust in the broadcast media (Bilevskaya 2010). As the source stated, the assumption was that in the future the presence of politicians in blogs and social networks would have 'real impact' on the outcome of elections. Consequently, the governors were expected to 'permeate the blogosphere and become ringleaders of political discussions' (Bilevskaya 2010).

How strong were the incentives for governors to respond to these expectations of the Kremlin by setting up a blog given that Russia's regional leaders were not elected

---

[6]The State Council (*Gosudarstvennyi Sovet*) is an advisory body to the President of the Russian Federation, established in 2000. The council is made up mostly of leaders of Russia's federal subjects and meets four times a year.

directly by citizens but appointed (and dismissed) by the president? Only during the second and politically less important step were they confirmed by their regional assemblies (Zhuravskaya 2010; Turovskii 2010; Chirikova 2010). Thus, first and foremost, the political fate of a governor depended on how the president evaluated his or her performance. According to the analysis by Turovskii (2010, pp. 68–72), the federal centre based its decisions to appoint or dismiss a governor mainly on two criteria: the governor's control of the regional situation, and above all his or her capacity to generate votes for the 'party of power', United Russia (*Edinaya Rossiya*), and to prevent intra-elite conflicts; and his or her 'controllability', or in other words, the governor's willingness to act as the junior partner of the central power and to follow the rules of the administrative hierarchy.

My central claim here is that, for Russian governors, the need to adhere to the central power rather than the control of the situation in the region is the dominant incentive to set up a blog. In other words, Russia's governors established their blogs primarily to showcase their allegiance to the president and their willingness to act as the 'junior partners of the center' (Turovskii 2010, p. 69), whereas the intention of generating votes seemed of secondary importance. The reason for such an assumption is that the audience of most governors' blogs could be expected to be rather limited, with the local TV channels remaining a much more important medium to convey political messages to the mass audience. Secondly, the political fate of the governors was only indirectly dependent on citizens' vote, but directly on a decision of the president which, in turn, was based on a wider array of evaluation criteria and political agenda (Turovskii 2010, pp. 68–72). Thirdly, the proportion of bloggers amongst other types of Russian politicians that are not directly dependent on the president is still comparably low. For instance, only 5.5% (25 out of 450) of the members of the Lower Chamber of the Russian parliament, the State *Duma*, were keeping a blog in November 2010 (goslyudi.ru 2010). Finally, the proportion of bloggers amongst politicians in most Western, competitive democracies is far smaller, even though internet penetration in these countries is higher and incentives to generate electoral support can be assumed to be even stronger in these more competitive political environments. In the United Kingdom, for instance, according to a study by Williams (2009; see also Francoli & Ward 2008, p. 27), no more than 11% of all MPs kept a blog in 2008. In Germany, in November 2010, not a single regional leader of a *Bundesland* made an effort to keep an online diary. Thus, it seems very unlikely that gaining the support of voters could have been the predominant incentive that motivated an astonishingly large proportion of 38% of Russian governors to set up a blog.

Figure 1 shows how the number of blogging governors started to grow rapidly after October 2008, the month when the new president, who had taken office only half a year earlier, started an online diary. Only two governors had blogs before Medvedev started his own: Nikita Belykh of Kirovskaya *Oblast'* and Oleg Chirkunov of Permskii *Krai*. Nikita Belykh, the most experienced blogger, was the leader of the opposition party Union of Right Forces (*Soyuz pravykh sil*) when he started blogging in January 2006. It was only in December 2008 that Belykh was nominated as a governor by Medvedev— to the surprise of both pro-Kremlin and opposition politicians. Chirkunov was already governor of Permskii *Krai* when he started experimenting with his blog in June 2008.

Even though there has been no evidence of a governor having been dismissed for a lack of 'internet activity' or 'computer literacy' the president's statements must have made a strong impression on Russia's regional leaders. Political consultant Marat Gel'man links the sudden enthusiasm for blogging amongst Russian governors to a long-standing tradition of Russian officials emulating the leisure activities of their political leader: 'When Yel'tsin played tennis, everyone played tennis. Putin took up judo, and everyone took up judo. Medvedev started a blog and officials started developing the Russian internet space' (Fedina 2010).

In other words, Russian officials were mimicking Medvedev's blogging activities to express their loyalty to and their respect for their political leader.

Table 2 provides further empirical evidence, comparing the geographical distribution of blogging governors across Russia's eight federal districts as of May 2010. Whereas 55% (six out of 11) governors kept a blog in the North Western Federal

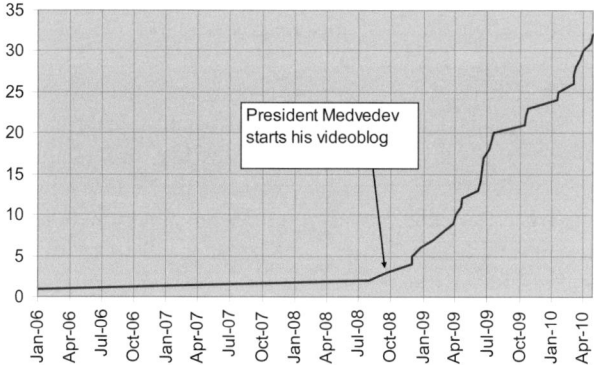

FIGURE 1. NUMBER OF RUSSIAN GOVERNORS KEEPING A BLOG.
*Source*: compiled by the author.

TABLE 2
THE DISTRIBUTION OF BLOGGING GOVERNORS ACROSS FEDERAL DISTRICTS (AS OF MAY 2010)

|  | Number of blogging governors in May 2010 | % of blogging governors in May 2010 | % of citizens accessing internet at home in 2009[a] |
|---|---|---|---|
| North Western Federal District | 6 of 11 | 55 | 37 |
| Southern Federal District | 3 of 6 | 50 | NA[b] |
| Urals Federal District | 3 of 6 | 50 | 23 |
| Volga Federal District | 6 of 14 | 43 | 27 |
| Central Federal District | 7 of 18 | 39 | 24[c] |
| Northern Caucasus Federal District | 2 of 7 | 29 | NA |
| Siberian Federal District | 3 of 12 | 25 | 21 |
| Far Eastern Federal District | 2 of 9 | 22 | 21 |
| Total | 32 of 83 | 39 | 30 |

*Notes*: [a]data for 2009 according to GfK (2009, p. 2). [b]The North Caucasian Federal District was split from the Southern Federal district on 19 January 2010. Separate data on internet penetration were not yet available at the time of research. For both districts combined, the figure was 20%. [c]Without the capital Moscow (52%).
*Source*: compiled by the author.

District, the proportion of blogging governors was the smallest in the Siberian and the Far Eastern Districts (21%). The data indicate a strong correlation between the level of internet penetration and the number of blogging governors. In other words, the higher the internet penetration, the higher is the proportion of governors who keep online diaries. However, the correlation between internet penetration and the proportion of blogging governors does not contradict the central point made in this section. Rather I conclude that the soft pressure on governors to establish a blog is higher in those regions where internet penetration is high.

While attempting to emulate Medvedev's political agenda, only very few Russian governors referred to the president directly in explaining why they set up their online diaries. For example, Aleksandr Tkachev of the Krasnodarskaya *Oblast'* writes in his blog: 'I had a long talk with Dmitrii Anatol'evich [Medvedev] and he recommended that I should pay attention to the internet. I have good relations with our president and I am sure that he is leading the country to a new level in many spheres of life' (Tkachev 2009). A more typical statement—without reference to Medvedev—comes from Anatolii Brovko, governor of the Volgogradskaya *Oblast'*, who writes in his permanent welcome post: 'I hope that my blog will be an effective communication platform, where I will learn about your proposals to improve the quality of governing, social and economic projects and ideas, and constructive criticism' (Brovko 2010).

Similarly, Governor Slyunyaev (2009) promises his readers that 'on my blog in LiveJournal, we can discuss, as we call it "without ties"'. Governor Yurevich (2009) appeals to his audience: 'Friends, not always does the government know about the real problems. You can help!'; and Governor Kanokov (2010) raves: 'Here and today, thanks to the internet I have the opportunity to discuss with you interactively events in the republic and government decisions and to put forward new proposals and express opinions and so help Karbadino-Balkariya quickly move towards our common aims' (Kanokov 2010).

In their blogs most politicians emphasised motivations aimed at improving communication with the electorate, learning more about the problems of citizens, getting closer in touch with citizens and discussing issues frankly and informally and passing on unmediated, unfiltered information. Thus, while the aim to demonstrate loyalty to the president appears to have been a central incentive for many Russian governors to set up a blog, they stressed different reasons in front of their readers.

### *Emulating the president: how did Russian governors blog?*

If many governors set up their blog to demonstrate their allegiance to the president, to what degree did they emulate the president's style of blogging—and in which aspects did they deviate? To answer these questions, I will first outline the specific style of Medvedev's blog. Then, I will evaluate to what extent governors followed the president's blogging style.

Since its launch in October 2008, Medvedev's blog has adhered consistently to the specific style of a 'videoblog'.[7] In November 2010, the blog consisted of 199 entries.

---

[7]Also, see Yagodin's essay in this collection. See also supplementary material for this essay which can be found in the online version at http://www.tandfonline.com/loi/ceas20 (Figure S1).

Without exception, all posts were videos, each lasting between two and five minutes. Most entries were recordings of Medvedev's meetings with government officials or speeches held in front of invited audiences, with only roughly every tenth video being arranged specifically for the blog. The only texts provided in the posts are the transcripts of the videos. In 2010, between two and nine entries were published per month. In terms of content, the messages usually dealt with political issues of the day, for instance corruption, reform of the police forces and forest fires. In nearly all the posts, Medvedev appeared wearing a suit and tie. Only very rarely did the president speak about non-political issues, with one of the few exceptions being a post on his hobby—photography (Medvedev 2010b). His family life has never been shown explicitly in the blog. The posts are simultaneously published on two platforms: on the government-administered platform kremlin.ru, a site that was previously used by Putin when he was president, and on the SNS-hybrid LiveJournal. On both platforms, readers were allowed to comment. However, all comments were checked by moderators before being published. In the typology delineated in the next section, this blog would fall into the category of a 'PR blog'. Before discussing the impact of this type of blog on politics, I would like to address the question of to what extent the governors were emulating the president's blog.

With regards to the style, only 12 of the 32 governors closely followed the role model of the president and kept their blog in the style of an exclusive 'videoblog'. Just like Medvedev's blog, these online diaries unexceptionally consisted of video clips of between approximately two and five minutes in length. The only texts provided were the transcripts of the clips. Most videos were recordings of meetings with other officials, public speeches or footage taken from local TV channels. Very few clips were recorded exclusively for the blog; and even these videos addressed viewers in a rather formal language.[8]

In terms of the choice of the blogging platform, the governors' blogs could be divided into two groups: the 14 politicians who opened their blogs on the SNS-hybrid LiveJournal, the leading platform for public-affairs related discourse in the Russian blogosphere, and the remaining 18 governors who chose to set up their blogs on privately administered websites, either the websites of their regional administrations, their personal websites or the websites of regional information portals. In most cases, these blogs were accessible through the main menu of these websites, by clicking on a button captioned as 'blog' or 'videoblog'. The webpages employed customised software solutions which allowed for publishing posts or videos in a reverse chronological order and, in some cases, also featured a comment function. To conclude, with regard to the choice of a platform, not a single governor fully copied the approach of the president by publishing his blog simultaneously on two platforms: on a private website and on LiveJournal, yet none of the governors completely deviated from the president's model by opening a blog on an alternative blogging platform, be it a Western platform like WordPress or one of the other three big

---

[8]See also supplementary material for this essay which can be found in the online version at http://www.tandfonline.com/loi/ceas20. Figures S1 and S2 illustrate the striking similarities between the layout of the blog of the president and that of a typical videoblog of a governor, in this case of Governor Denin of Bryanskaya *Oblast'*.

Russian SNS-hybrids (ya.ru, mail.ru or liveinternet.ru). Instead, governors decided to publish their blogs either only on LiveJournal or only on a privately administered website.

The decision by many governors to open their blogs on privately hosted websites and not on LiveJournal seems to have been, first of all, one in favour of control, i.e. for the possibility to disallow comments. Disallowing and moderating comments is technically possible on LiveJournal but is not common in the culture of the network. Blocking comments completely would have been regarded as a grave *faux pas* by the vast majority of the LiveJournal community, discrediting and ridiculing such a blog right from the start. In total, eight of the 32 governors' blogs did not allow the public to comment. All of these eight blogs were hosted on private websites and not on LiveJournal. Secondly, the decision to establish a blog on a private website seems to have been one in favour of the form of a 'videoblog'—which is not at all common amongst bloggers on LiveJournal. Without a single exception, all 12 of the 32 governors who established a 'videoblog', modelled closely after the blog of the president, did not choose to do so on LiveJournal but on a private platform. Thus, whereas the president obviously had the authority to enter the LiveJournal-community with a 'videoblog' that largely deviated from the blogging culture of the platform, no governor dared to do the same.

Conversely, those governors who chose to blog on LiveJournal aligned their blogging style with the blogging culture of the platform. On average, LiveJournal bloggers amongst the governors posted more frequently than non-LiveJournal bloggers. The seven bloggers who posted more than 10 entries in May 2010 hosted their blog on LiveJournal (see Figure 2). Governor Chirkunov led the field with 33 posts. He is followed by Governors Belykh (23), Brovko (21), Morozov (17), Zhilkin (15), Zelenin (13) and Shantsev (11). By the same token, while many blogs on private platforms were obviously administered by press teams, all LiveJournal bloggers seemed to keep their blogs personally (judging from how they blended in recent personal experiences and expressed rather personal thoughts). As a result, LiveJournal bloggers seemed to be more successful in actually getting in touch

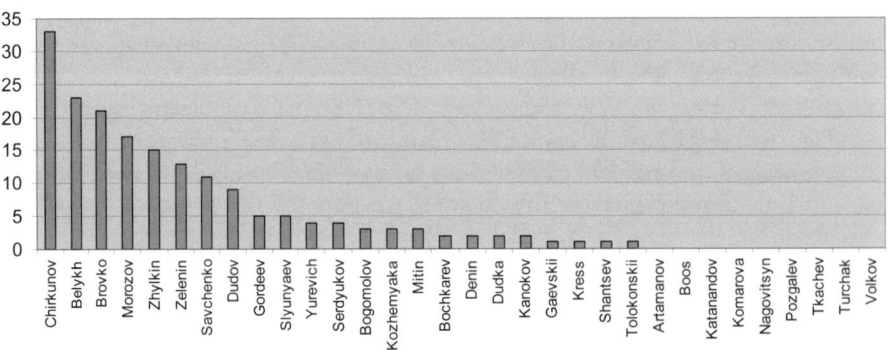

FIGURE 2. Blogging Activity: Number of Entries of Governors in May 2010.
*Source*: compiled by the author.

with citizens and generating feedback, i.e. their posts attracted more comments than those of non-LiveJournal bloggers (see Figure 3). The six governors that attracted the largest numbers of comments to a post in May 2010 were, without exception, blogging on LiveJournal. In the next section, the characteristics of the 29 governor blogs active in May 2010 and their impact on politics shall be discussed in greater detail.

### Political impact: a typology of governors' blogs

What effects can politicians' blogs have on politics? For many researchers of Western democracies, the primary hope associated with politicians' blogs is that these blogs would contribute to reconnecting political elites with the demos. For example, Coleman attests that, 'the problem faced by contemporary democracy is horribly simple. ... Governments have come to believe that the public don't know how to speak; the public has come to believe that governments don't know how to listen' (2005a, p. 1). In this dilemma, blogs are seen as a new tool of political communication to alleviate mutual misconception. Bucy and Gregson (2001, p. 375) introduce the concept of 'media participation' as a new 'form of participation that provides symbolic empowerment' to resolve the 'dilemma of the civic decline'. With this conception, suddenly, despite decreasing turnouts at votes in Western democracies,

> there is a form of participation in which a growing segment of the public regularly engages . . . .
> Even if only symbolically empowering for the individual, the experience of media participation
> is pivotal to maintaining the perception of system responsiveness and thereby serves as an
> important legitimizing mechanism for mass democracy. (Bucy & Gregson 2001, p. 375)

Can the blogs of Russian governors fulfil similar functions? Can they help increase the 'perception of system responsiveness' and thus serve as an 'important legitimizing mechanism' (Bucy & Gregson 2001, p. 375) for Russia's semi-democratic regime?

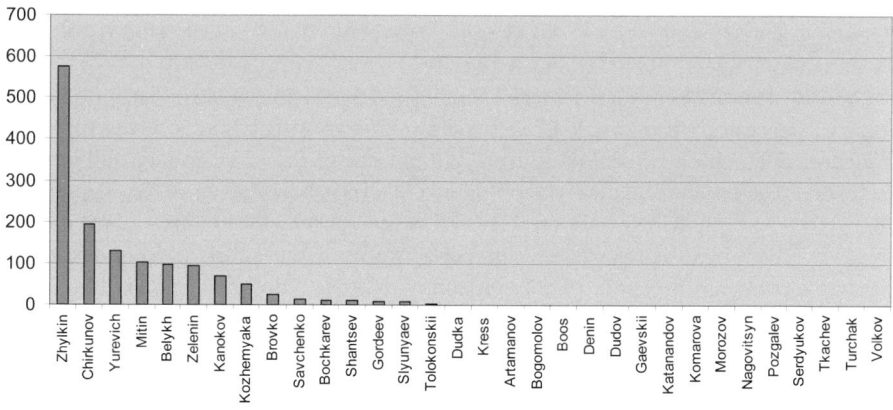

FIGURE 3. AUDIENCE RESPONSE. MAXIMUM AMOUNT OF COMMENTS TO AN ENTRY IN MAY 2010.
*Source*: compiled by the author.

To increase the perceived legitimacy of the Russian political system, the blogs of governors would have to fulfil one or more of the following criteria amongst their audience: criterion 1: that the politician is seriously listening to citizens (Coleman & Moss 2008, pp. 16–18)—the politician should author his or her blog in person, there should be a possibility to comment and the politician should at least pick up and refer to some of the comments in his or her posts; criterion 2: that the issues addressed in the comments are tackled by the politician in real life; criterion 3: that the politician is 'just like you' (Coleman & Moss 2008, pp. 10–14)—able to 'represent' and decide for his or her audience because he or she is perceived as 'one of them'. This aim can be pursued by politicians talking about non-political, private issues and creating emotional closeness.

By strengthening one or more of these three impressions amongst their audience, the blogs of Russian politicians increase the perception of responsiveness of the regional authorities and the 'perceived legitimacy' of the Russian political system as a whole. Against this backdrop, I suggest distinguishing the following three types of blogs: first, 'PR blogs' that do not fulfil any of the three criteria and are kept by a press team; second, 'efficient statesmen' blogs that fulfil criteria 1 and 2 above; and third, '*internetchik* blogs' that fulfil all three criteria above.

## PR blogs

PR blogs are obviously not kept by the governors themselves but by their press teams. On many blogs, only press releases or recordings of public speeches by the governor are published. The blogs draw very few comments, because the audience clearly feels that the governor is not seriously listening and only very rarely takes a look at the blog. On some of these blogs comments are not even allowed. On others they are to be filled in on a special form and thus do not become visible to the public. Many of these blogs are kept as explicit videoblogs, being close imitations of the president's blog. Others publish textual posts but follow the same communicative pattern. However, while Medvedev's blog, because of the enormous power he wields, frequently draws several hundred comments, the PR blogs of governors remain largely uncommented on. Examples of such blogs are those of the following 14 governors: Artamanov, Boos, Denin, Gaevskii, Bogomolov, Savchenko, Morozov, Tolokonskii, Dudov, Volkov, Katanandov, Kress, Pozgalev and Kanokov.

All PR blogs are kept on privately administered platforms, with the exception of the blog of Governor Morozov of Ul'yanovskaya *Oblast'* which is kept in the form of a moderated 'community' on LiveJournal. All but one of the 12 videoblogs fall into this category. The exception is the videoblog of Governor Nagovitsyn of the Republic of Buryatiya who published only messages recorded specifically for the audience of his blog, in which he also referred to comments. Eight of the 12 PR blogs did not even feature a comment function. In comparison with other types of blogs, PR blogs drew the smallest number of comments. The only governor who generated a substantial amount of comments (83) to a post in May 2010 was Kanokov (see Figure 3); however, this post was exceptional as it was the first entry to Kanokov's blog. All other PR blogs did not attract more than a maximum of two comments to a post in May 2010. The frequency of posts on most PR blogs was found to be quite low.

Governor Gaevskii of Stavropol'skii *Krai*, for instance, seems to post only one video per month. However, there are also highly active blogs such as that of Governor Savchenko of Belgorodskaya *Oblast'* that featured four entries in May and 16 entries in October 2010.[9]

A typical PR-blog is that of Governor Denin of Brianskaia Oblast'. Denin's videoblog is clearly modelled on the online diary of the President. On average one entry per month is posted. Common titles are 'The All-Russian Census of the Population 2010', 'Visit to a Company in Bryansk', or 'September 17—The Day of the Liberation of Bryansk!' The only entry posted in May 2010 is captioned 'Congratulations on the 65-anniversary of the Victory' in World War II. This video clip is about one minute long. It shows Denin wearing a tie against the background of the flag of his *Oblast'*. In his formal address, he congratulates veterans on their victory. There are no reactions to this post, as the blog does not offer a facility to comment.

PR blogs do not fulfil any of the three criteria presented above. They do not succeed in creating an impression of the politician listening seriously to citizens. The level of interactivity seems to be rather low on PR blogs, and most posts do not draw a single comment; and the audience of these blogs is marginal. As a consequence, the potential of PR blogs to increase the perceived responsiveness and the perceived legitimacy of politics in Russia can be evaluated as rather low.

### Efficient statesmen blogs

In contrast to PR blogs, on efficient statesmen blogs no press releases or TV footage from local channels are published. These blogs are kept by the governors themselves—or at least they seem to be judging from the way personal experiences and thoughts are blended into the posts. In spite of this, these bloggers do not talk about non-political issues such as hobbies, family life, sports activities or music. Yet, the governors generate the impression of seriously listening to citizens. For example, all blogs feature a comment function, and all politicians at least refer to comments in selected posts. To varying degrees, issues addressed in the comments are actually tackled by the politicians in real life. This is why I call this type of blog that of the 'efficient statesmen'. The blogs of the following eight governors belong to this category: Turchak, Nagovitsyn, Bochkarev, Kozhemyaka, Yurevich, Slyunyaev, Serdyukov and Gordeev.

Of these blogs, three are kept on the platform LiveJournal and five are kept on privately hosted platforms. On average, they attract far more comments than PR blogs, usually several dozen per post. A typical example is the blog of Governor Serdyukov of Leningradskaya *Oblast'*. Serdyukov keeps his blog on a privately owned website.[10] He does not post photos, the entries are reminiscent of long, formal letters to his readers. The few pictures on the website show the governor

---

[9]See the supplementary material for this essay which can be found in the online version at: http://www.tandfonline.com/loi/ceas20. Figure S2 shows a typical PR blog, that of Governor Denin of Bryanskaya *Oblast'*. Denin's videoblog is clearly modelled on the online diary of the president (see Figure S1).

[10]See http://www.serdyukov-vp.ru, accessed 15 January 2011.

sitting in his office behind a desk, dressed in a suit and wearing a tie. Typical posts are captioned 'You Ask—I Answer', 'On Gas and Water Tariffs', or 'I have Enough Power to Put Things in Order'. In the entry 'You Ask—I Answer' Serdyukov addresses the consequences of irregularities related to invoices for public housing utilities that readers of his blog had uncovered in comments to previous posts. This post reads as follows:

> I thank all of you for bringing up the problem and telling me about the situation in the villages, and also some of you for taking up my calls, sending me copies of utility bills .... This information has been handed over to the office of the public prosecutor .... Gennadii [name of commenter], trust me, I have already started tough negotiations. (Serdyukov 2010a)

As this post exemplifies, the style of communication on these blogs is far more interactive and conversational than on PR blogs. The governor directly refers to one of the commenters by his nickname, and he reports on measures taken by him personally. The post quoted above drew 114 comments. The readers of Serdyukov's blog seemed to feel that the governor was seriously listening to them. Many of the comments raised specific problems related to public housing, roads, buses or kindergartens. A typical comment reads as follows:

> Dear Valerii Pavlovich,

> I would like to write to you about a certain problem. My great-grandfather, a disabled person and veteran of World War II, was provided with a car nine years ago. In the past, according to regulations, the cars of invalids were exchanged after seven years. After nine years, the car is now practically useless. The only request he [my great-grandfather] made on the occasion of the 65th anniversary of the end of World War II was to receive a new car. But this request was denied! They said that they have no instructions to exchange cars. He is unable to walk. I beg you to help my great-grandfather! He lives in the countryside: [address and telephone number provided]. (Gnatenko 2010)

From the Western perspective, it is remarkable that this commenter—as many other authors of comparable 'complaint comments' that seem to thrive especially on the blogs of efficient statesmen—does not bother to relate her claim to any existing formal provision of law. Instead, she argues by referring to the conventions of 'former times' and to the fact that officials obviously do not have 'instructions' to exchange cars. Moreover, the author of the comment also seems convinced that the Governor Serdyukov can put the perceived grievance right, if he so wishes, independently of legal provisions. To make it easier for the governor to show mercy, she even provides both telephone number and address of her great-grandfather. To sum up, the patterns of interactions between politicians and citizens which can be observed in these 'complaint comments' seem to be rather Russian specific and an intriguing topic for further research. Presumably, these interactions can only be fully understood when interpreted in the context of the very specific Russian administrative culture, the roots of which can be traced back to the nineteenth century (Heusala 2005). However, this is a task that lies far beyond the scope of this essay.

More central to this study is the question to what extent the blogs of efficient statesmen fulfilled the second criterion quoted above: To what extent were politicians actually tackling grievances brought up by the readers of their blogs in real life? In the first few weeks after opening his blog in February 2010, Governor Serdyukov actually tried to react to some of the complaints made public on his blog, as for instance the irregularities related to bills for public utilities mentioned above. However, after barely three months of blogging in the time period between February and May 2010, the governor apparently abandoned his blog. By November 2010, he had not posted a single blog entry for more than five months. In one of his posts (Serdyukov 2010b) he states that he does not want to 'decline any responsibilities' but that many of the complaints published on his blog should better be addressed by municipal leaders. He proposes the simple solution to involve municipal leaders in the practice of blogging. Serdyukov provides links to six blogs of municipal leaders that were at the time establishing a blog. However, a visit to these websites in August 2010 showed that these blogs seemed to fail as well. Thus, the rise and fall of Serdyukov's blog illustrates how opening a blog in the style of an 'efficient statesman' can put a governor under an unbearable pressure to address an ever-growing list of complaints published in the comments section.

An example of how to cope more successfully with a large amount of complaints is the online platform Turchak.ru. On this platform, Governor Turchak of the Pskovskaya *Oblast'* encourages his regional officials and the heads of cities to respond to questions and grievances brought up by internet users. The personal blog of the governor is embedded in this website. The 'internet reception room' (*internet priemnaya*) is divided in the sections 'Blog of the Governor', 'Problems', 'Questions' and 'Proposals'. By November 2010, the section 'Problems' contained approximately 600 complaints. Turchak's officials had reacted to nearly all of these complaints. At least according to the conversation threads between officials and citizens published on the website, a series of these complaints were redressed (Turchak 2010b). Amongst these were: Svetlana Ural'skaya's plea to repair a fountain on a public square (Ural'skaya 2010); veteran Pavel Semenov's request for a new apartment (Semenov 2010); and Anna Smirnova's complaint about her grandfather being denied free medication (Smirnova 2010).

On Serdyukov's blog as well as on Turchak's platform, the style of language remains rather formal. The governor does not share any details of his private life—a typical feature for efficient statesman blogs. In a blog entry, Turchak voices this strategy explicitly as follows: 'The purpose of my site is not to engage officials and citizens in a football game. This is a platform that citizens can use to get timely information on problems of their everyday life and to help resolve the issues' (Turchak 2010a).

As these examples illustrate, efficient statesmen bloggers fulfil criteria 1 and 2 quoted above as they create the impression of listening seriously to citizens, and they do address selected issues brought up by their commenters in real life. In contrast to the first type of PR bloggers, they—most probably—author their blog posts personally. Their blogs attract a far bigger audience and far more comments than the first type of PR blog. Thus their potential to increase the perceived responsiveness and the perceived legitimacy of politics in Russia seems far greater than those of PR blogs.

## Internetchik *blogs*

In sharp contrast to efficient statesmen bloggers *internetchik* bloggers discuss private topics and tend to use informal language. These bloggers are usually even more active than the efficient statesmen type. On some blogs, new entries appear roughly every other day (see Figure 2). All *internetchik*-governors use LiveJournal as a platform. As governors are unveiling their private lives and talking about issues of human interest, these blogs on average attract a greater audience. Entries draw a high number of comments, usually several dozen and in some cases even hundreds (see Figure 3). The seven governors that can be assigned to this category boast the largest numbers of friends amongst the bloggers under investigation: Belykh, Chirkunov, Brovko, Zhilkin, Zelenin, Mitin and Shantsev.

The degree to which private issues are included and informal language is used varies within the group. For instance, Governor Shantsev talks in very few posts about non-political issues. However, after watching an ice-hockey game he titles a post 'OUR GUYS DEFEATED THE CANADIANS!!!' (Shantsev 2010), using only upper-case characters. In the post, he tells his readers about how he was able to predict the outcome of the game. To express his joy, he uses the smiley symbol ';)))'. In the comment section of the post, he replies three times to the readers' comments. Information on the private life of these politicians is common on these blogs: Governor Mitin posts a video of him rafting on a Russian river (Mitin 2010a) and pictures of him playing badminton (Mitin 2010b). Governor Zelenin shares with his readers his successful strategy to stop smoking (Zelenin 2010), and Governor Zhilkin publishes a post under the title 'Extreme fishing' that documents in a series of pictures how he caught several enormous fish (Zhilkin 2010a).[11]

An exceptional example of this type of blogger is Nikita Belykh, governor of the Kirov Region. Belykh tells his readers about his private life, including his experience of losing a tooth (Belykh 2010a) or his successes in teaching his son how to play poker (Belykh 2010b). In another post (Belykh 2010c), the governor links to a webpage that shows an extremely unfavourable photo of him: sunburnt, seemingly drunk, with a baseball cap, smiling dizzily. In the post, Belykh complains ironically about the fact that the media always pick up the most unfavourable pictures of him. This example illustrates the degree to which Belykh is making fun of himself and of the media in public, and how he reduces the emotional distance between himself and his voters by chatting with them as if they were his closest friends. The comments by his readers mirror the extreme emotional closeness between audience and political leader. For instance, one commenter asks Belykh if he has ever thought of losing weight, because then he 'would look better in photos' (Olga_Strelkova 2010).

The blog post that drew the most comments in May 2010 was published on the blog of another *internetchik*, that of Governor Zhilkin (2010b). In this blog entry, the governor announces a competition asking his readers to come forward with ideas for a social advertising campaign. As prizes for the best ideas, Zhilkin promises three iPads, a couple of T-shirts and an illustrated book with pictures of the region of Astrakhan'. Subsequently, the competition is also promoted on the news programme of a regional

---

[11]Supplementary material for this essay can be found in the online version at http://www.tandfonline.com/loi/ceas20 (see Figure S4).

television channel. Similar forms of 'crowdsourcing' can be occasionally observed on the blogs of other *internetchiks* and efficient statesmen. Other governors call for ideas for what to say at certain meetings or conferences, how to spend a part of the budget or how to rename streets in their cities. Moreover, just as efficient statesmen, *internetchiks* respond to selected complaints from readers. On the blog of Governor Slyunyaev, for instance, a user complains that a medal commemorating the 65th Anniversary of the Great Victory has not been delivered to his grandmother. The medal is then presented within less than 24 hours (Jusup0v 2010). In this respect, the behaviour of both *internetchik* and efficient statesmen bloggers mirrors that of President Medvedev who also reacted publicly to selected complaints. In reaction to comments on the president's blog, for instance, a casino in Balashikha was closed, a person from Saratov who was sacked was hired again and a businessman from St Petersburg got a loan (Sidorenko 2010).

Thus, the blogs of *internetchik* governors fulfil all of the three criteria stipulated above: with these blogs, governors create the impression of seriously listening to citizens, they tackle issues brought up by their readers in real life, and they succeed in reducing the emotional distance to the electorate. Consequently, *internetchik* blogs are, most probably, superior to PR blogs and efficient statesmen blogs in terms of raising the perceived responsiveness and legitimacy of politics in contemporary Russia.

### *Conclusion: blogs as a means to strengthen the legitimacy of the Russian regime*

This essay set out to explore the exceptional phenomenon of the blogging Russian governors who, in May 2010 in terms of blogging activity, by far outmatched politicians in many developed democracies. However, I maintain that most Russian governors did not set up their blog primarily with the intention of gaining electoral support. Rather, for these officials, starting a blog seemed to be, first and foremost, a symbolic action that showcased their allegiance and loyalty to the president, who was widely known for his internet enthusiasm. Medvedev's passion for the internet can, in turn, be interpreted as a pivotal pillar of his overarching political strategy of 'modernisation'.

Yet, even though most Russian governors may have started their online diaries primarily to appeal to the president, this does not suggest that their blogs, once in existence, had no impact on politics. In this essay, I have argued in particular that these blogs were, to various degrees, capable of strengthening the perceived legitimacy of the Russian semi-authoritarian regime. Drawing on Scharpf's (1997) distinction between 'input' and 'output' legitimacy and adopting a radical-constructivist perspective (Glasersfeld 2001), I would like to suggest a further differentiation between: 'perceived input legitimacy', understood as the portion of legitimacy belief that citizens draw from the perception that their voice is being heard in the decision-making process; and 'perceived output legitimacy', conceived of as the portion of legitimacy belief that citizens derive from their perceived satisfaction with the outputs (decisions and policies) of the political system.

It is important to note that according to these definitions it is not the actual design or efficacy of the input mechanisms at disposal that are decisive, nor are the actual outputs of the political system. Instead, my radical-constructivist notion of legitimacy exclusively refers to the degree to which the electorate perceives that it makes significant inputs to the decision process, respectively the way it perceives the outputs of the political system.

Viewed from this theoretical perspective, I argue that Russian governors strengthened the input legitimacy of their government, for instance: by referring to comments in their blog posts; by asking the blog audience for advice on specific political decisions; by generating emotional closeness; by publicly referring to complaints by citizens when justifying political decisions; or by generating the impression of listening seriously to the public and considering their demands in the decision-making process. Moreover, many governors have contributed to the output legitimacy of the Russian regime, mainly by presenting 'success stories' on their blogs about how selected grievances were brought up by citizens and subsequently redressed by the politician.

It is crucial to remark here that these endeavours by the governors have reached audiences far beyond the regular readership of their blog. First of all, the archive of blog posts, conversations and 'success stories' have remained a persistent 'online testimony' of the efforts of the governor. Thus even internet users who visited the governors' blog only once and skimmed through the website were—to the degree that the governors' efforts appeared convincing to them—left with the impression that the authorities were taking the needs of citizens seriously.

Far more momentous, however, must be considered the fact that the blogging efforts of Russian governors were extensively amplified and propagated by the Russian mass media, which can be regarded as largely loyal to the ruling elites. A TV report about a governor making an effort to blog can already increase the belief in the input legitimacy amongst Russian citizens—even amongst those citizens who have never accessed the internet. Moreover, the same effect can be assumed for reports about how governors dealt successfully with various grievances presented in blogs.

In this theoretical perspective, Medvedev's political strategy of inducing officials to blog can be interpreted as being aimed at increasing the perceived input and output legitimacy of the Russian political system. Given the fact that the proportion of blogging politicians in Russia is higher than in most Western democracies, I would even argue that blogs are playing a far greater role in generating legitimacy for the semi-authoritarian Russian political system than they do in Western democracies. In contrast to Western democracies, the Russian political system lacks a series of mechanisms commonly perceived as major sources of (particularly input) legitimacy in democratic states, such as highly competitive elections, deeply rooted party systems or well-organised interest groups. As a consequence, it may not come as a surprise that other channels of creating input legitimacy (such as politicians' blogs) have gained major importance in the Russian context. Thus, paradoxically, politicians' blogs seem to actually live up to the hopes of many Western researchers by reconnecting the political elites with the demos. However, their potential to establish closer ties between the representatives and the represented appears to be of particular appeal to political elites in the semi-authoritarian Russian context.

*London School of Economics and Political Science and LMU University Munich*

## *References*

Alexanyan, K. (2009) 'Social Networking on Runet', *Digital Icons: Studies in Russian, Eurasian and Central European New Media*, 1, 2, pp. 1–12, available at: http://www.digitalicons.org/issue02/karina-alexanyan/, accessed 4 January 2012.

Belykh, N. (2010a) 'Vyrvali zub. Bol'no…', Blog Entry, Dnevnik Gubernatora, 15 May, available at: http://belyh.livejournal.com/441281.html, accessed 15 November 2010.

Belykh, N. (2010b) 'PokerÄ', Blog Entry, Dnevnik Gubernatora, 13 May, available at: http://belyh.livejournal.com/440014.html, accessed 15 November 2010.

Belykh, N. (2010c) 'Nespravedlivost'', Blog Entry, Dnevnik Gubernatora, 7 May, available at: http://belyh.livejournal.com/436329.html, accessed 15 November 2010.

Bichard, S. (2006) 'Building Blogs: A Multi-Dimensional Analysis of the Distribution of Frames on the 2004 Presidential Candidate Websites', *Journalism and Mass Communication Quarterly*, 83, 2, pp. 329–45.

Bilevskaya, E. (2010). 'On-line politika. Regional'nym nachal'nikam porekomendovali vnedrit'sya v blogi i sotsial'nye seti', *Nezavisimaya Gazeta*, 21 January, available at: http://www.ng.ru/politics/2010-01-21/1_online.html?mthree=1#, accessed 6 July 2010.

Bimber, B. A. & Davis, R. (2003) *Campaigning Online: The Internet in US Elections* (Oxford, Oxford University Press).

Bratersky, A. (2010) 'Medvedev Plays "Good Tsar" Online', *The Moscow Times*, 23 November, available at: http://www.themoscowtimes.com/news/article/medvedev-plays-good-tsar-online/424152.html, accessed 5 December 2010.

Brovko, A. (2010) 'Privetstvie Posetitelyam Bloga', *Permanent Opening Post, Blog Anatoliya Brovka*, available at: http://anatoliy-brovko.livejournal.com/, accessed 15 November 2010.

Bucy, E. & Gregson, K. (2001) 'Media Participation: A Legitimizing Mechanism of Mass Democracy', *New Media & Society*, 3, 3, pp. 357–380. doi:10.1177/1461444801003003006

Chirikova, A. E. (2010) 'Regional Elites in Contemporary Russia', *Russian Politics and Law*, 48, 1, pp. 21–39.

Coleman, S. (2005a) *Direct Representation. Towards a Conversational Democracy* (London, IPPR), available at: http://www.ippr.org.uk/members/download.asp?f=%2Fecomm%2Ffiles%2FStephen%5FColeman%5FPamphlet%2Epdf, accessed 6 July 2010.

Coleman, S. (2005b) 'Blogs and the New Politics of Listening', *Political Quarterly*, 76, 2, pp. 272–80.

Coleman, S. & Moss, G. (2008) 'Governing at a Distance—Politicians in the Blogosphere', *Information Policy*, 13, 1–2, pp. 7–20.

Coleman, S. & Wright, S. (2008) 'Political Blogs and Representative Democracy', *Information Polity*, 13, 1, pp. 1–6.

Davies, T. & Gangadharan, S. (eds) (2009) *Online Deliberation: Design, Research, and Practice* (Chicago, IL, University of Chicago Press), available at: http://odbook.stanford.edu/viewing/htmldocument/75, accessed 15 November 2010.

Deibert, R. & Rohozinski, R. (2010) 'Liberation vs. Control: The Future of Cyberspace', *Journal of Democracy*, 21, 4, pp. 43–57.

Denin, N. (2010) 'S 65-letiem pobedy', Blog Posting, Videoblog Gubernatora, 5 May, available at: http://www.bryanskobl.ru/news/videoblog/texts.php?id=12, accessed 15 November 2010.

Diamond, L. (2010) 'Liberation Technology', *Journal of Democracy*, 21, 3, pp. 69–83.

Drezner, D. & Farrell, H. (2008) 'Introduction: Blogs, Politics and Power: A Special Issue of *Public Choice*', *Public Choice*, 134, 1, pp. 1–13.

Easton, D. (1957) 'An Approach to the Analysis of Political Systems', *World Politics*, 9, 3, pp. 383–400.

Etling, B., Alexanyan, K., Kelly, J., Faris, R., Palfrey, J. & Grasser, U. (2010) *Public Discourse in the Russian Blogosphere. Mapping RuNet Politics and Mobilization* (Cambridge, MA, Berkman Center), 19 October, available at: http://cyber.law.harvard.edu/sites/cyber.law.harvard.edu/files/Public_Discourse_in_the_Russian%20Blogosphere_2010.pdf, accessed 3 November 2010.

Fedina, O. (2010) 'Rossiiskie chinovniki osvaivayut Twitter', *RiaNovosti*, 15 July, available at: http://rian.ru/technology/20100715/255048045.html, accessed 5 December 2010.

Francoli, M. & Ward, S. (2008) '21st Century Soapboxes? MPs and their Blogs', *Information Polity*, 13, 1–2, pp. 21–39.

Freedom House (2010) *The Map of Freedom. Russia 2010*, available at: http://www.freedomhouse.org/template.cfm?page=363&year=2010, accessed 6 July 2010.

GfK (2010) *Internet uslugi v Rossii. Mart*, available at: http://www.gfk.ru/filestore/0071/0079/694/2.pdf, accessed 11 November 2010.

Gnatenko, E. (2010) 'Veteran i invalid VOV', Comment, Blog Valeriya Serdyukova, 30 May, available at: http://serdyukov-vp.ru/blog?post_id = 6, accessed 15 November 2010.

Gorny, E. (2006a) *A Creative History of the Russian Internet* (London, University College), available at: http://www.ruhr-uni-bochum.de/russ-cyb/library/texts/en/gorny_creative_history_runet.pdf, accessed 14 July 2010.

Gorny, E. (2006b) 'Russian LiveJournal. The Impact of Cultural Identity on the Development of a Virtual Community', in Schmidt, H., Teubener, K. & Konradova, N. (eds) (2006) *Control and Shift: Public and Private Usages of the Russian Internet* (Norderstedt, Books on Demand GmbH), pp. 73–90.

Goroshko, O. & Zhigalina, E. (2009) 'Quo Vadis? Politicheskie kommunikatsii v blogosphere Runeta', *Digital Icons: Studies in Russian, Eurasian and Central European New Media*, 1, 1, pp. 81–100, available at: http://www.digitalicons.org/issue01/issue1/goroshko_and_zhigalina.php?lng=English, accessed 25 June 2012.

Goslyudi.ru (2010) 'Blogi po Kategoriyam. Deputat Gosudarstvennoi Dumy', available at: http://www.goslyudi.ru/role/feddeputy, accessed 5 November 2010.

Heusala, A. (2005) *The Transition of Local Administration Culture in Russia* (Saarijärvi, Gummerus Printing), available at: http://ethesis.helsinki.fi/julkaisut/val/yleis/vk/heusala/thetrans.pdf, accessed 4 January 2012.

Howard, P. N. (2006) *New Media Campaigns and the Managed Citizen* (Cambridge, Cambridge University Press).

Jackson, N. (2008) '"Scattergun" or "Rifle" Approach to Communication: MPs in the Blogosphere', *Information Polity*, 13, 1, pp. 57–69.

Jusup0v (2010) 'Pro prazdnik pobedy', Comment, Blog Gubernatora Slyunyaeva, available at: http://slunyaev.livejournal.com/4813.html?thread=26829#t26829, accessed 15 November 2010.

Kamyshev, D. (2010) 'Kremlevskii polusrochnik', *Kommersant'-Vlast*, 17–18, available at: http://www.kommersant.ru/doc.aspx?DocsID=1363987, accessed 15 July 2010.

Kanokov, A. (2010) 'Obrashchenie Presidenta KBR k posetitelyam bloga', Blog Post, Videoblog Arsena Kanokova, 24 May, available at: http://blog.president-kbr.ru/?p=137, accessed 15 November 2010.

Kerbel, M. R. & Bloom, J. D. (2005) 'Blog for America and Civic Involvement', *The Harvard International Journal of Press/Politics*, 10, 3, pp. 3–27.

Kluver, R. (ed.) (2007) *The Internet and National Elections: A Comparative Study of Web Campaigning* (London, Routledge).

Lavrushin, E. (2008) 'Novaya Politika—Setevye Vlastiteli Dum', *Novaya Politika*, available at: http://www.novopol.ru/-setevyie-vlastiteli-dum-text36242.html, accessed 13 July 2010.

Lipset, S. (1959) 'Some Social Requisites of Democracy', *American Political Science Review*, 53, 1, pp. 69–105.

Medvedev, D. (2010a) 'Blog Dimitriya Medvedeva', Blog, available at: http://community.livejournal.com/blog_medvedev, accessed 15 November 2010.

Medvedev, D. (2010b) 'Gryadet epokha vozvrashcheniya neposredstvennoi demokratii', Blog Posting, Videoblog Dimitriya Medvedeva, 31 May, available at: http://blog.kremlin.ru/post/81, accessed 15 November 2010.

Medvedev, D. (2010c) 'Smysl Fotografii', Blog Posting, Videoblog Dimitriya Medvedeva, 10 February, available at: http://blog.kremlin.ru/post/66, accessed 15 November 2010.

Mitin, S. (2010a) 'Rafting', Blog Entry, Blog Sergeya Mitina, 3 June, available at: http://mitinsg.livejournal.com/5728.html, accessed 15 November 2010.

Mitin, S. (2010b) 'Badminton', Blog Entry, Blog Sergeya Mitina, 14 April, available at: http://mitinsg.livejournal.com/3715.html, accessed 15 November 2010.

Nielsen (2010) Blogpulse [Tool for Online Search], available at: www.blogpulse.com, accessed 5 November 2010.

Olga_Strelkova (2010) 'A Vy ne dumali', Comment, Blog of Governor Nikita Belykh, Dnevnik Gubernatora, 7 May, available at: http://belyh.livejournal.com/436329.html?thread=12607081#t12607081, accessed 6 December 2010.

Ott, R. (2006) 'Weblogs als Medium politischer Kommunikation im Bundestagswahlkampf 2005', in Holtz-Bacha, C. (ed.) (2006) *Die Massenmedien im Wahlkampf* (Wiesbaden, VS Verlag), pp. 213–33.

Pole, A. (2010) *Blogging the Political: Politics and Participation in a Networked Society* (London, Routledge).

Sazonov, A. & Stolbun, Y. (2010) '20 samykh populyarnykh chinovnikov-bloggerov', *Forbes Russia*, available at: http://www.forbesrussia.ru/ekonomika/lyudi/41195-reiting-samyh-populyarnyh-chinovnikov-bloggerov, accessed 3 July 2010.

Scharpf, F. W. (1997) 'Economic Integration, Democracy and the Welfare State', *Journal of European Public Policy*, 4, 1, pp. 18–36.

Semenov, P. (2010) 'Zhil'e Veteranam', Comment, Blog Gubernatora, 25 April, available at: http://turchak.ru/problemu/156, accessed 15 November 2010.

Serdyukov, V. (2010a) 'Sprashivali? Otvechayu: ZKKH I ZHKU', Blog Posting, Blog Valeriya Serdyukova, 2 April, available at: http://serdyukov-vp.ru/blog?post_id=11, accessed 15 November 2010.

Serdyukov, V. (2010b) 'O blogakh glav administratsii raionov', Blog Posting, Blog Valeriya Serdyukova, 29 April, available at: http://serdyukov-vp.ru/?post_id=14, accessed 15 November 2010.

Shantsev, V. (2010) 'Nashi Obygrali Kanadtsev!!!', Blog Entry, Blog Valeriya Shantseva, 21 May, available at: http://shantsevvp.livejournal.com/4780.html, accessed 15 November 2010.

Sidorenko, A. (2010) 'Russland: Vom Präsidenten zum Bloggen verdammt', *politik-digital.de*, available at: http://www.politik-digital.de/russland-weblogs-web20, accessed 14 July 2010.

Slyunyaev, I. (2009) 'Ob otkrytkom obshchenii', Blog Post, Blog Gubernatora Kostromskoi Oblasti Igorya Nikolaevicha Slyunyaeva, 29 April, available at: http://slunyaev.livejournal.com/2009/04/29/, accessed 15 November 2010.

Smirnova, A. (2010) 'Poluchenie l'gotnykh lekarstv', Comment, Blog Gubernatora, 20 April, available at: http://turchak.ru/problemu/97, accessed 15 November 2010.

Stanyer, J. (2006) 'Online Campaign Communication and the Phenomenon of Blogging: An Analysis of Web Logs during the 2005 British General Election Campaign', *Aslib Proceedings*, 58, 5, pp. 404–15.

Sunstein, C. (2008) 'Neither Hayek nor Habermas', *Public Choice*, 134, 1, pp. 87–95.

Tkachev, A. (2009) 'informatsionnyi President', Blog Entry, Tkachev's Journal, 26 July, available at: http://a-n-tkachev.livejournal.com/993.html, accessed 15 November 2010.

Toepfl, F. (2011) 'Managing Public Outrage: Power, Scandal, and New Media in Contemporary Russia', *New Media & Society*, 13, 8, pp. 1301–19.

Toepfl, F. (forthcoming) 'Making Sense of the News in a Hybrid Regime. How Young Russians Decode State TV and an Oppositional Blog', *Journal of Communication*.

Trammell, K. (2006) 'The Blogging of the President', in Williams, A. P. & Tedesco, J. C. (eds) (2006) *The Internet Election: Perspectives on the Web in Campaign 2004* (Oxford, Rowman & Littlefield), pp. 133–46.

Turchak, A. (2010a) 'Prinuzhdenie k Blogu', Blog Entry, Blog Gubernatora, 3 June, available at: http://turchak.ru/blog/1718, accessed 15 November 2010.

Turchak, A. (2010b) 'Problemy', *Turchak.ru*, available at: http://turchak.ru/problemu/reshennie/#filter, accessed 15 November 2010.

Turovskii, R. F. (2010) 'How Russian Governors Are Appointed', *Russian Politics and Law*, 48, 1, pp. 58–79.

Ural'skaya, S. (2010) 'Fontana na pl. Kalinina', Comment, Blog Gubernatora, 20 April, available at: http://turchak.ru/problemu/86, accessed 15 November 2010.

Vesti.ru (2010) 'Neradivykh chinovnikov otpravyat "na ulitsu"', available at: www.vesti.ru/doc.html?id=347563, accessed 15 May 2010.

von Glasersfeld, E. (2001) 'The Radical Constructivist View of Science', *Foundations of Science*, 6, 1–3, pp. 31–43.

Vrazhina, A. (2010) 'Dezavuiruite eto', *lenta.ru*, available at: http://www.lenta.ru/columns/2010/02/05/blogs/, accessed 13 July 2010.

VTSIOM (2010) 'Internet v Rossii', available at: http://wciom.ru/arkhiv/tematicheskii-arkhiv/item/single/13386.html?no_cache=1&cHash=70ad9d4c0a, accessed 6 July 2010.

Ward, S. & Davis, R. (2008) *Making a Difference: A Comparative View of the Role of the Internet in Election Politics* (Lanham, MD, Lexington Books).

Williams, A. (2009) *MPs Online: Connecting with Constituents*, available at: http://www.hansard society.org.uk/blogs/publications/archive/2009/02/24/mps-online-connecting-with-constituents.aspx, accessed 5 December 2010.

Williams, A. P. & Tedesco, J. C. (eds) (2006) *The Internet Election* (Lanham, MD, Rowman & Littlefield).

Woodly, D. (2008) 'New Competencies in Democratic Communication? Blogs, Agenda Setting and Political Participation', *Public Choice*, 134, 1, pp. 109–23.

Wright, S. (2009) 'Political Blogs, Representation and the Public Sphere', *Aslib Proceedings*, 61, 2, pp. 155–69.

Yandex (2009) 'Blogosfera Runeta', available at: http://download.yandex.ru/company/yandex_on_blogosphere_spring_2009.pdf, accessed 6 July 2010.

Yandex (2010a) 'Reiting Blogov Runeta po chitatelyam', available at: http://blogs.yandex.ru/top/?sort=readers, accessed 6 July 2010.

Yandex (2010b) 'Indikatory blogosfery', available at: http://help.yandex.ru/blogs/?id=1112101#blogs, accessed 5 December 2010.

Yandex (2010c) 'Reiting Blogov po avtoritetnosti', available at: http://blogs.yandex.ru/top/?username= &sort=rank, accessed 5 December 2010.

Yanow, D. (2006) 'Thinking Interpretively: Philosophical Presuppositions and the Human Sciences', in Yanow, D. & Schwartz-Shea, P. (eds) (2006) *Interpretation and Method: Empirical Research Methods and the Interpretive Turn* (New York, ME Sharpe), pp. 5–26.

Yurevich, M. (2009) 'Itogi Bloga', Blog Post, Blog Gubernatora Chelyabinskoi Oblasti, 3 June, available at: http://yurevich-m.livejournal.com/2010/06/03/, accessed 15 November 2010.

Zelenin, D. (2010) 'No Smoking', Blog Entry, Blog Zelenina, 31 May, available at: http://dzelenin.livejournal.com/14899.html, accessed 15 November 2010.

Zhilkin, A. (2010a) 'Ekstremal'naya Rybalka', Blog Entry, Blog Alexandra Zhilkina, 14 June, available at: http://alexandr-jilkin.livejournal.com/21987.html, accessed 15 November 2010.

Zhilkin, A. (2010b) 'Sotsial'naya Reklama', Blog Entry, Blog Alexandra Zhilkina, 31 May, available at: http://alexandr-jilkin.livejournal.com/19265.html, accessed 15 November 2010.

Zhuravskaya, E. (2010) Federalism in Russia. Working Paper No. w0141, Center for Economic and Financial Research (CEFIR), available at: http://www.cefir.ru/papers/WP141.pdf, accessed 15 August 2012.

# Political Challengers or Political Outcasts?: Comparing Online Communication for the Communist Party of the Russian Federation and the British Liberal Democrats

SARAH OATES

*Abstract*

This article compares the web presence of the Communist Party of the Russian Federation and the British Liberal Democrats in order both to analyse the ability of the internet to strengthen parties as political institutions as well as to reflect upon the relative democratic value of parties online in different types of regimes. The article compares the party websites in early 2010 through an analysis of online audience, web links, content posted by parties and user-generated material linked to the two parties. The research found that the online potential of party communication, despite the universal availability of powerful tools of information distribution and social networking opportunities, was far more closely tied to national political culture than to cyber-culture in general. The Communist Party web activity tended to parallel the party activity offline, failing to craft the appearance of a more modern or inclusive party. At the same time, supporters of the British Liberal Democrats made greater use of external social networking and were apparently more connected with the broader political sphere, although perhaps at the expense of party branding and control. The findings demonstrate the need to understand how national political organisations and attitudes can play a much stronger role than technological potential in shaping the democratising forces of the online sphere.

WHAT IS THE FUNCTION OF A RUSSIAN POLITICAL party website in a system that offers little scope for parties to actually aggregate the interest of the voters? Through manipulation of the rules on elections, media bias and government backing for pro-Kremlin parties, there would seem to be little scope for party websites to redress the imbalance of power between elites and citizens in the Russian Federation. At the same time, it is questionable whether one can import Western notions of political party activities and internet studies into the Russian case because of a lack of effective party

The author gratefully acknowledges the support of grants from the British Academy (*International Potential, National Limits: Investigating the Role of the Russian Internet in Constraining the Social Agenda*), the Carnegie Trust for the Universities of Scotland (*Decoding the Online Sphere in the Post-Soviet Region*), and the UK Economic and Social Research Council (RES-000-22-4159, *The Internet and Everyday Rights in Russia*) in research for this article.

organisations. In particular, it is important to avoid assumptions that political institutions will follow a Western path, as that was very much a part of the examination of the post-Soviet political sphere in the 1990s. It was then assumed by many that the world was moving into a new political era and that a change in political communication could mean a convergence of both political institutions and behaviour towards Western models of democracy.

History has demonstrated that the idea of a path-dependent understanding of post-Soviet development was not only flawed, but in many ways it obscured and stymied the understanding of the significance and direction of events in post-Soviet society. The same problem is inherent in the study of the Russian internet (or *Runet* as it is known by its nickname). As the internet develops as a formidable political tool in the United States, scholars have been quick to assume that this is a global phenomenon. In fact, even in the wake of the Arab Spring, there is compelling evidence that the nature of online interaction tends to be more a mirror of both national and personal attitudes than an expression of a relatively homogeneous, globalised phenomenon. However, as appealing as it may be to place (for example) US, British and Russian web use on a spectrum from fully engaged citizens to citizens who are not engaged at all, this is falling into the same unhelpful pattern as thinking about Russian political institutions such as parties and the media by comparing them to Western models. If we can avoid the idea of using the American or British development of online politics as the dominant template, it is possible to have a richer and more useful understanding of the internet's potential in the political sphere in Russia. This is challenging, because the central paradigms and research tools have been developed primarily in a liberal Western setting. However, a study of the Russian case is the inspiration for some interesting methodologies—and also suggests that the focus on the Western model leaves relatively little scope for meaningful intellectual debate about the role of media in societies around the globe. A study that can compare a Russian and Western party web presence at one juncture in history should be able to shed light not only on the specifics of Russian party life online, but also help us to think more broadly about the different forms that online party engagement might take.

This article examines a relatively small, but intriguingly important corner of the online sphere in Russia and the United Kingdom—the web presence of the Communist Party of the Russian Federation (CPRF, *Kommunisticheskaya Partiya Rossiiskoi Federatsii*) and the Liberal Democrats in the United Kingdom in March 2010. This adds to existing studies of the role of parties online in the post-Soviet sphere (March 2006; Semetko & Krasnoboka 2003). These two cross-sectional cases were chosen on the criteria that while both have political relevance in their respective countries, neither party had a realistic chance of assuming a majority in national politics in early 2010. In the case of the CPRF, this marginalisation has come about in part through targeted media bias as well as via the dominance of state-funded, state-backed parties that support the Kremlin (Oates 2006). The British Liberal Democrats have also served as an oppositional voice, generally on the Left, in particular as the only major British party that did not support the country's participation in the Second Gulf War. At the time of the study, the Liberal Democrats held 62 out of the 646 seats (9.6%) in parliament, which was close to the Communist Party's share of the *Duma* seats in 2010 (57 out of 450 seats or 12.7%). Methodologically, there were concerns in that the Liberal Democrats were preparing for the May 2010 British parliamentary elections during data collection for this project (in March 2010) while the Russian *Duma* elections were not scheduled until late 2011; however, there was relatively little electoral activity online, as will be discussed below. The fortunes of the Liberal Democrats were soon to change radically, as the party became coalition partners to form a government with the Conservatives after the 2010 British elections, despite quite significant ideological and policy differences. While the internet no doubt reflected and intensified the reaction of the party, its supporters and the electorate in general, this is outwith the scope of this study.

How did these parties, with significant electoral support but with minority positions in their national legislatures, use the internet as a party tool? The communication features of the internet can enhance the democratic effects of party politics by allowing parties to provide much more information about their organisation, history, structure, values, ideology and policies (Gibson 2010; Gibson, Nixon & Ward 2003). However, there is evidence that parties will adapt the web to existing party strategy rather than use the communicative potential of the web to transform their political strategy (Römmele 2003). If parties are elected into office or form part of the government, they can also provide more policy information with far greater detail to a broader range of citizens (Chadwick 2006). Technically, the interactive capabilities of the online sphere even make it possible to discuss policies in depth and use crowd sourcing as ways to foster ideas for political engagement, although in practice cyber-communication often devolves to the party loyal or the party elite (Linaa Jensen 2006). All of this enhanced communication is also less expensive than previous methods of communicating through the mass media or direct campaigning, helping parties to more efficiently seek support, recruit members or volunteers as well as avoid media bias or framing that could be detrimental to their fortunes. The advent of social networking and other many-to-many networking tools, including, micro-blogging sites such as Twitter and the ability to distribute videos with ease, should allow political ideas and actions to flourish among citizens, particularly as parties are able both to take advantage of pre-existing online social networks and reach the traditionally disengaged youth on their central medium (Chadwick 2006; Owen 2006; Anstead & Chadwick 2008).

Studies of how parties use the online sphere, however, suggest that any sort of cyber-utopian era for political parties is very far from realisation (Norris 2003). Attention has been drawn to the use of the internet in campaigns in the United States, particularly during presidential elections (Anstead & Chadwick 2008). However, the United States is rather exceptional in this case, as elections are centred more on candidates than on political parties. Thus, while lessons can be drawn in particular from the synergy between Barack Obama and the grassroots use of Web 2.0, the rise of the internet as a central source of political information as well as the youth appeal of his candidacy mean that this is not necessarily a lesson that can be translated into different political environments. As Kavanagh (1995) has pointed out, the American system of campaign communication is distinctive from European campaign communication. Americans have a greater focus on electoral campaigns over party policy, prioritise the coverage of candidates over the news of party organisations, as well as emphasise the 'horse race' aspects of the campaign at the expense of discussions of ideology or policy. The fundamental work for studying online communication by political parties in comparative perspective has its roots in Gibson and Ward (2000). Rachel Gibson and

Stephen Ward established a cross-national tool for analysing central features of political party websites (Gibson, Margolis *et al.* 2003; Gibson 2010). Studies using the Gibson and Ward scheme have found that while parties are becoming increasingly sophisticated in their online use, this use varies by country and between parties.

Beyond the measurement of how parties may use the internet is concern with how the audience might choose to engage with online parties. A study by Lusoli and Ward found that while British parties were creating more elaborate websites with impressive amounts of information resources, the audience was not particularly interested.[1] This parallels findings by others who have studied the online political sphere, with Hindman (2009) suggesting that most of the online audience is not interested in politics and Davis (2009) finding that those who discuss politics in blogs tend to stay in relatively self-contained groups. In addition, there does not appear to be compelling evidence that the online sphere fundamentally changes the level of openness or transparency in party organisations. A study of Russian political parties online in 2003 found that the parties were taking little advantage of the opportunities offered by online communication (March 2006). This was not surprising, not only due to the relatively weak development of Russian parties (discussed further below), but also because of the low penetration of the internet in Russia a decade ago (less than 10% of the population, according to March, which parallels other Russian internet audience figures). However, there are two quite significant factors relating to the internet and political parties in Russia that would suggest a fundamental shift in the role of online communication for parties since the study by March: the development of social networking online as well as the explosion of internet use in Russia.

While the demographics of support for the two parties differ significantly—with older, poorer, less educated and rural Russian citizens tending to support the Communists (White & McAllister 2008) and well-educated British citizens from a broad age range providing key Liberal Democrat support (Ford & Goodwin 2010, p. 11)—the Communists have compelling reasons to turn to the online sphere despite the relatively low internet use among their core supporters. In particular, the internet offers the Communists the only forum in which they have the ability to publicise their message without the hostile framing of most of the Russian mass media (especially television, see Oates 2006). In addition to broadcasting their points of view without traditional mass media bias, agendas or framing, these minority parties can also use the specific communication tools of the online world to solidify and attract new supporters. Unfiltered information aside, these features include the enhanced ability to present material (particularly through video) as well as opportunities of social networking through many-to-many communication tools ranging from forums to blogs to twittering (Gibson 2010).

As discussed below, this is where significant differences in the approach to the online sphere emerge between the Russian Communists and the British Liberal Democrats. The online sphere of the Communists was an insular place in March 2010. On the other hand, the Liberal Democrat discourse was tangential to the official party website and sprawled

---

[1]See 'Nice Website, Shame No-one Visits It: Politics Still a Turn-off, Even in Cyberspace', press release from the Economic and Social Research Council, 23 February 2005, available at: http://www.eurekalert.org/pub_releases/2005-02/esr-nws022305.php, accessed 17 January 2012.

across social networking sites. Yet, despite the Web 2.0 façade, were the supporters of the Liberal Democrats engaging in much meaningful political exchange? Or had the Communists retained a sense of identity and purpose that would be (possibly) diminished by the glittering array of information technology tools linking in and out of the Liberal Democrat central website? Also, was the lack of centrism in the Liberal Democrat party website, which stretched across hundreds of constituencies, creating useful ties to particular communities or further diluting the party's central strength? The current electoral design of both countries would encourage specific web strategies—the Liberal Democrats must win seats through constituencies while the Communists can now only gain seats through a national party list system as constituency seats have been eliminated in the Russian *Duma* races. Whether this reflects party strategic preferences would need more study of whether the parties perceive their strengths to lie in their local or national identity. The investigation of web content production as well as reception studies from the audience is beyond the remit of this article. Rather, this study focussed on content, both generated by the party and by forum participants, to provide evidence to compare the role of the web in party politics in the two countries.

This article will examine these issues in the following ways: first, online usage, both for the countries in question and with regard to the CPRF and Liberal Democrat websites. The notion of a 'digital divide' is a compelling and reasonable idea, i.e. that there is limited political promise from the internet if relatively few people are connected (Norris 2001). This article will consider both the amount and nature of general internet use, as well as the specific traffic relating to the central web locations for both parties. This will be measured through web link analysis. This research employs the IssueCrawler freeware (htpp://www.issuecrawler.net) to map incoming and outgoing links that relate to a central website. In addition, the content offered by the party organisations is examined. This section will use the Gibson and Ward (2000) quantitative website analysis scheme to provide a scale to measure the emphasis placed by parties on a range of functions, such as information provision, voter participation, campaigning for electoral support, resource generation (i.e. money and members) and networking with other like-minded organisations. Third, a qualitative review of a sample of user-generated content related to each party is discussed. In particular, this study looks at internal party forums. What is being said and who is speaking? Who is speaking with authority? What is the manner of the discourse and the dialogue?

*Online usage*

By the spring of 2010, Russia had overtaken the United Kingdom in terms of the sheer number of people online, although overall internet penetration remained much lower in the post-Soviet state (43% compared with 83% in the United Kingdom by June 2010, according to World Internet Stats). By June 2010, Russia had 59.7 million people online, compared with an estimated 51.4 million users in the United Kingdom (World Internet Stats). From a relatively low rate of internet penetration given its economic development (Cooper 2008), Russian internet use has exploded in recent years. With the exception of nations that started from very low penetration a decade ago (for example Belarus, Macedonia, Albania and Bosnia & Hercegovina), Russia experienced the highest growth in internet usage in Europe from 2000 to 2010 (World Internet Stats). By 2010, the only larger online population in Europe at the time was Germany, with 65.1 million people online (World Internet Stats).[2] The Russian government has supported the growth of the internet, noting that the most rapid expansion recently has been outside of the main cities and that penetration is set to reach almost 100% for those under 40 (Russian Federal Agency on the Press and Mass Communication 2011).

The growth in internet usage had not paralleled any overt signs of democratisation in Russia by 2010. This is not unusual, in that Freedom House reported in its 2011 study of the internet in 37 nations (including Russia) that rising internet use does not appear to parallel greater freedom of expression or other signs of democratic growth. Rather, Freedom House has identified trends of greater internet control and co-optation on the part of many states. In particular, its 2011 *Freedom on the Net* report highlights increasing blockages of online political content; a lack of transparency; more cyber-attacks against regime critics; more exploitation of centralised internet 'chokepoints'; more manipulation of online information; as well as more incidents of offline coercion linked to internet activities.[3] There have been a number of high profile cases in Russia in which individuals have been arrested and imprisoned for online activity, with some of the arrests clearly politically motivated (Agora 2011). This parallels concerns raised by the OpenNet Initiative that states are becoming much more effective at controlling the internet, evolving from blocking access and censorship to a 'third-generation' of control in which states proactively use the online sphere to monitor dissent, spread propaganda and attack critics (Deibert *et al.* 2009, 2011 ). At the same time, the online sphere allows regimes to track, detect, arrest and/ or subvert political dissidents (Morozov 2011).

Overall, Freedom House found that the level of online freedom in Russia decreased from 2009 to 2011 on measurements of internet obstacles to access, limits on content and violations of user rights. At the same time, the organisation noted that the internet still remained more free than the Russian media sphere in general, which is deemed 'not free' by the organisation and was tied with Gambia for 175th, listed just below Congo and ahead of Vietnam in the 2010 world media freedom rankings of 197 countries.[4] Although Freedom House highlights some concerns with 'expansive' libel laws that can 'stifle criticism', overall the United Kingdom was deemed to have a particularly free media environment (tied for 24th place in its rankings) and its online sphere was not an object of concern or scrutiny for Freedom House in its 2009 and 2011 studies of online freedom (Freedom House 2011). There is quite a robust debate about the relative levels of media freedom in a range of countries—particularly when comparing the more commercial system in the United States with the public funding of television in the United Kingdom (Oates 2008; Hallin & Mancini 2004; Street 2010)—but the key point for this comparison is that the Russian internet would appear to continue to offer a chance for more open and meaningful exchange of information than the traditional media in Russia. This point is highlighted by Freedom House in noting that the online sphere still

---

[2]World Internet Stats, available at: http://www.internetworldstats.com, accessed 3 July 2012.

[3]*Freedom on the Net*, available at: http://www.freedomhouse.org/report-types/freedom-net, accessed 3 July 2012.

[4]See http://www.freedomhouse.org/images/File/fop/2010/FOTP2010Global&RegionalTables.pdf, accessed 3 July 2012.

remains partly free, while it has deemed the overall media sphere not free at all in Russia. Thus, the expectation of the contribution from political party websites is even higher for Russian citizens than for British citizens as *Runet* can fill critical information voids in a non-free state. Indeed, there is evidence from a range of election coverage to show that Russian citizens receive particularly biased and unfair coverage of their political parties and candidates (European Institute for the Media 1994, 1996a, 1996b, 2000a, 2000b; Organization for Security and Co-operation in Europe/Office for Democratic Institutions and Human Rights 2004a, 2004b).

Perhaps unsurprisingly given the vast array of entertainment and light news available online, people do not flock to political websites in either the United Kingdom or Russia. According to the web-tracking company Alexa, neither the Communist Party nor the British Liberal Democrat websites were at all popular in their respective countries in March 2010.[5] This is unsurprising, as according to calculations by Hindman (2009), only a tiny percentage of web traffic is directed to political sites. While the Russian telecommunications report (2011) did find that reading the news was one of the most popular activities online, there was no mention of political interest in the report. Indeed, finding out about politics online did not figure in the top 10 interests for the online sphere, which included interest in cars, holidays, real estate, beauty, health, medicine and financial affairs (Hindman 2009, p. 57). Thus, neither political party website was likely to score highly in terms of overall usage. Alexa does not report the number of visitors to websites; rather it estimates the percentage of global internet users who visit a particular site. This is very low for both sites, but the Communist Party site was notably more popular than the Liberal Democrat website in March 2010 despite the higher percentage of citizens online and the greater degree of democratic transparency in Britain (see Table 1). The Liberal Democrats attracted an average of 0.00081% of the global internet traffic in the three months ending 16 March 2010 (when the statistics were checked).[6] The Communist Party attracted 0.00262% for the same time period.[7] This shows that the Communist Party website was attracting approximately three times as many visitors as the Liberal Democrats—despite the fact that the Liberal Democrats were approaching an election in a matter of weeks and Russian 2011 *Duma* elections were 18 months away at the time of this study. As noted above, the number of internet users in each country is roughly equal, although the percentage of internet use was far lower in Russia at the time. This counters the arguments that party websites are more popular and possibly more relevant in Western democracies than in semi-authoritarian states such as Russia. Alternatively, it could support the argument that the Liberal Democrat website was particularly irrelevant in the spring of 2010.

In order to address those issues, a comparison was made between competing party website traffic rankings on Alexa in both countries (see Table 1). Interestingly, the Alexa statistics suggest that the CPRF website is stronger than many other party websites in Russia. This is not surprising in comparison with the relatively weak political presence of parties such as Union of Right Forces (*Soyuz Pravykh Sil*, SPS), *Yabloko* or even the relatively resilient Liberal-Democratic Party of Russia (*Liberal'no Demokraticheskaya*

---

[5]See http://www.alexa.com, accessed 3 July 2012.
[6]See http://www.libdems.org.uk/home.aspx, accessed 3 July 2012.
[7]See http://kprf.ru, accessed 3 July 2012.

## TABLE 1
### Traffic, Rankings and Other Information on Party Websites in Russia and Great Britain

| Party | Share of global internet audience (change in past three months) | Global traffic rank | Traffic rank in country | Sites linking in | Online since | Most common website visited just before | Most common website visited just after | Nationality of audience |
|---|---|---|---|---|---|---|---|---|
| *RUSSIAN FEDERATION* | | | | | | | | |
| Communist Party of the Russian Federation (www.kprf.ru) | 0.00262 (+0.4%) | 50,862 | 1,807 | 508 | 30 October 1997 | Yandex.ru | Yandex.ru | 87.6% Russian |
| United Russia (official party website, www.edinros.ru) | 0.00148 (+32%) | 109,435 | 3,852 | 464 | 14 March 2002 | Yandex.ru | Mail.ru | 84.8% Russian |
| Liberal-Democratic Party of Russia (www.ldpr.ru) | 0.00048 (−44%) | 364,487 | 24,806 | 282 | 3 September 1997 | Yandex.ru | n/a | 83.8% Russian |
| *GREAT BRITAIN* | | | | | | | | |
| Liberal Democrats (www.libdem.org.uk) | 0.00081 (+13%) | 179,333 | 5,536 | 1,077 | n/a | Google.co.uk | Google.co.uk | 83.7% British |
| Labour (www.labour.org.uk) | 0.00098 (+25%) | 165,536 | 5,541 | 1,276 | 8 November 1994 | Google.co.uk | Google.co.uk | 85.9% British |
| Conservatives (www.conservatives.com) | 0.00172 (+29%) | 94,054 | 3,404 | 1,578 | 11 July 1996 | Google.co.uk | Google.co.uk | 74.1% British; 6.7% US |
| British National Party (www.bnp.org.uk) | 0.00496 (−10%) | 34,304 | 1,428 | 931 | 29 January 2000 | Google.co.uk | Google.co.uk | 65.3% British; 12.6% US |

*Notes:* figures from www.alexa.com as reported on 16 March 2010. Figures for the share of the audience and traffic rank are averaged over three months previous to 16 March 2010. Note that a lower traffic rank means a more popular website.

*Partiya Rossii*, LDPR, which is a personality party centred around the nationalist Vladimir Zhirinovsky) in 2010. However, it is arguably significant that in early 2010 the CPRF website was markedly more popular than the highly professional website of United Russia (*Edinaya Rossiya*), the 'party of power' that dominates in the Russian parliament and which supported Dmitri Medvedev for president in 2008. According to the Alexa rankings, http://www.edin.ru, had about half the penetration of http://kprf.ru, reaching a global internet penetration of 0.00148% in the same three-month period ending on 16 March 2010.[8] However, it should be noted that a website with the URL of http://www.er.ru was more popular than the CPRF website and, indeed, http://www.edin.ru had become a travel website by January 2012. In early 2010, http://www.er.ru was the 'United Russia Portal' and while clearly supportive of the party, also carried a range of news and interviews. In look and feel, the United Russia Portal was more similar to Western party websites that incorporate stories designed to look like news (rather than statements, manifestoes or press releases) and it is unsurprising that it rapidly became the 'main' party website. This underlines the point that it can be very difficult to identify the 'central' website of a political party. It should be noted that http://www.er.ru was growing in popularity very rapidly in early 2010. Unsurprisingly, the bulk of the traffic on Russian party websites was from domestic users.

Was the British Liberal Democrat website dwarfed in the United Kingdom by the online presence of the dominant Labour and Conservative parties in early 2010? A surprising finding was that the most dominant party website studied for this article in March 2010 was the far-right British National Party (BNP), with a website that was far more popular than either Labour or the Conservatives (see Table 1). This lends support to the idea that party websites are much more a reflection of a range of factors in a political system than a manifestation of real political popularity, electoral viability or a popular mandate. The most striking finding is that these figures suggest that the British Liberal Democrat website was more popular than that of Labour as the ruling party in March 2010. However, the British Conservatives (the clear favourites who went on to win the largest share of the votes in the May 2010 British parliamentary elections) were attracting almost twice as much traffic to their website as Labour during this time. Thus, there is no correlation of website traffic with actual seats in the British parliament, although it seemed clear that the online audience was more interested in the non-ruling parties. While across traditional media platforms the focus was on Labour as the ruling party and Conservatives as their main opponents and likely successors in the upcoming elections, the online traffic pattern was different. There are a range of possible explanations—from Conservative leader David Cameron's initiative with one of the first British leader video blogs with a relatively personal touch, to possibly a better web design (not part of this research), to an undefined group of internet users who were particularly attuned to the Conservative cause. The popularity of the BNP website fits much more easily into the theory that the web can provide a source for non-mainstream views that resonate with a relatively large group of people. As the BNP is seen to represent racist views (something it is careful to avoid on the public website and is indeed illegal in the United Kingdom), it is not covered in a meaningful way in the mainstream media. It remains an interesting comment on both the internet and the British party

[8]See http://www.alexa.com, accessed 3 July 2012.

landscape that the BNP website was five times as popular as the Labour party website in March 2010. It is also a comment on the relative popularity of the web for 'outsider' causes within democracies.

## Web link analysis

Web link analysis allows for the visualisation of how a website fits into both national and international online networks. This research uses IssueCrawler, a freeware tool available at https://www.issuecrawler.net.[9] IssueCrawler takes a list of 'seed' URLs, culled from those published on a website, and can search up to three sublevels (or clicks) from the original URLs to generate a map of links leading in and out of the home website. The map also represents the relative strength of these ties and the centrality of particular websites, producing a representation of hubs, nodes and satellites surrounding the home website. This allows researchers to make judgments about the reach and network structure of a political party website. However, the exercise is limited by the web links that are entered into the original crawl. As noted by the IssueCrawler designers, care has to be taken not to input too many web links and to avoid really large websites (such as google.com, bbc.co.uk, etc.) because the figure will become too cluttered and skewed by the presence of a web Leviathan. On the other hand, enter too few web links from the original website and the map generated will be too small to be of much use. In addition, it was necessary to search for the relevant webs links by hand for this project.[10]

Entering links by hand was relatively easy in the case of the Communists, who list web links to 10 regional party organisations, communist parties in foreign countries and supportive media outlets. It was more difficult to refine the list of web links for the British Liberal Democrats, because there were separate web pages for hundreds of constituencies and they were not on a single list. An examination of a sample of the constituency web pages suggested that most of them adhered to a central template provided by the party, with some local information added in. The Liberal Democrat web crawl was set with a list of parties worldwide that were listed on the website as linked to the party as well as a few organisations that were listed as well. The crawl was set to three levels because of the omission of constituency web links in the original seed list. To parallel this, the CPRF web crawl was also set at three levels (see Figures 1 and 2).

The figures suggest a fundamental difference between the web linkages of the two parties. The CPRF was centred around the party website while much of the online activity related to the British Liberal Democrats had migrated away from the party website onto the social networking site Facebook. In the case of the CPRF, the party website was clearly the central hub for the online sphere.[11] However, the pattern for www.libdems.org.uk was much more asymmetric, as the large hubs competing for web traffic relating to the Liberal Democrats were generic, rather than party, websites. In

---

[9]An extensive list of articles using the IssueCrawler tool can be found at http://www.govcom.org/full_list.html, accessed 23 January 2012.

[10]It is possible to do this automatically through a software tool, but this was not available to the author at the time. Finding the URLs manually was useful in any case, in that it served as a way to qualitatively review the structure and content of the websites.

[11]See http://kprf.ru, accessed 3 July 2012.

**Russian Communist party links 23 March 10**

**Co-link Map Details:**

| | |
|---|---|
| Author: | Sarah Oates |
| Email: | s.oates@lbss.gla.ac.uk |
| Crawl start: | 24 Mar 2010 - 08:19 |
| Crawl end: | 24 Mar 2010 - 11:02 |
| Privilege starting points: | off |
| Co-link Analysis Mode: | page |
| Iterations: | 1 |
| Crawl Depth: | 3 |
| Node count: | 50 |

Map generated from issuecrawler.net by the Govcom.org Foundation, Amsterdam.

**Legend:**

(org.cy) (.ru) (.org.za) (.org) (.org.uk) (.ie) (.org.br) (.cu)

(.gr) (.net.ua) (.co.uk) (.hu) (.nl) (.es) (.pt)

(.net) (.se) (.be) (.it)

**Statistics:**

○ **kprf.ru**

| | |
|---|---|
| Destination URL: | http://kprf.ru/dep/76768.html |
| Page date stamp: | 24 Mar 2010 - 08:27 |
| Links received from crawled population: | 5440 |

Links from network (1 - 20)

1. sovross.ru
2. communistpartyofireland.ie
3. inter.kke.gr
4. kke.gr
5. kprf-udm.ru
6. kprfnsk.ru
7. altkprf.ru
8. newworker.org
9. sacp.org.za
10. skp.se
11. munk.aspat.hu
12. moskprf.ru
13. mkkprf.ru
14. kprforel.ru

Links to network: 34

FIGURE 1. WEB LINK ANALYSIS FOR THE COMMUNIST PARTY.
*Note:* analysis conducted *via* IssueCrawler (www.issuecrawler.net), 23 March 2010.

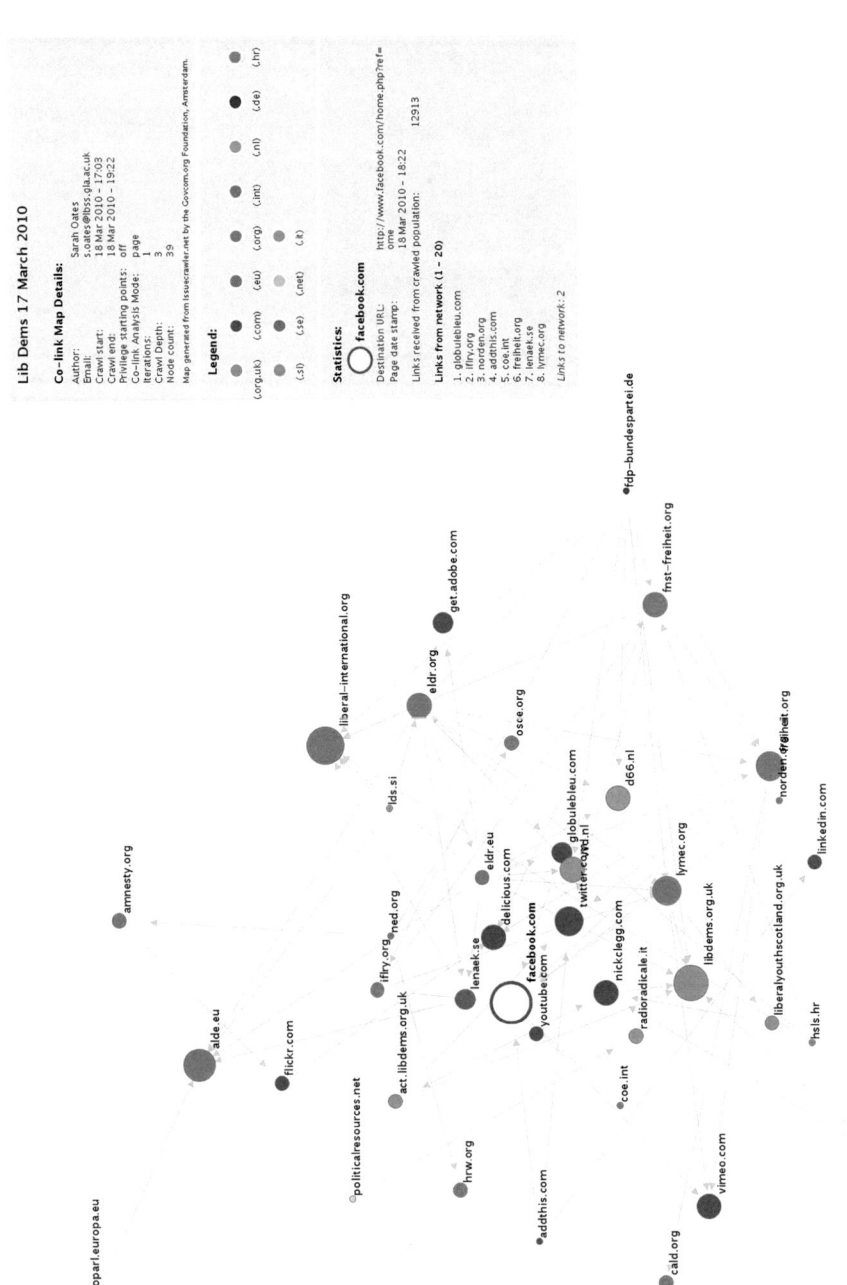

**Lib Dems 17 March 2010**

**Co-link Map Details:**

| | |
|---|---|
| Author: | Sarah Oates |
| Email: | s.oates@lbss.gla.ac.uk |
| Crawl start: | 18 Mar 2010 - 17:03 |
| Crawl end: | 18 Mar 2010 - 19:22 |
| Privilege starting points: | off |
| Co-link Analysis Mode: | page |
| Iterations: | 1 |
| Crawl Depth: | 3 |
| Node count: | 39 |

Map generated from Issuecrawler.net by the Govcom.org Foundation, Amsterdam.

**Legend:**

(org.uk)  (.com)  (.eu)  (.int)  (.org)  (.nl)  (.de)  (.hr)

(.si)  (.se)  (.net)  (.it)

**Statistics:** facebook.com

Destination URL:  http://www.facebook.com/home.php?ref=home
Page date stamp:  18 Mar 2010 - 18:22

Links received from crawled population:  12913

**Links from network (1 - 20)**
1. globulableu.com
2. iifry.org
3. norden.org
4. addthis.com
5. coe.int
6. freiheit.org
7. lenaek.se
8. lymec.org

Links to network: 2

FIGURE 2. Web Link Analysis for the Liberal Democrats.

*Note:* analysis conducted via IssueCrawler (www.issuecrawler.net), 17 March 2010.

particular, the presence of a relatively large concentration of links to Facebook was in contrast to the lack of the presence of a social networking site closely linked to the Russian Communists. As social networking via sites such as Vkontakte (In Contact) and Odnaklassniki (Classmates) are popular in the Russian domain (Goroshko & Zhigalina 2009; Russian Federal Agency on the Press and Mass Communication 2011), this was not because there was an absence of such spaces on the Russian internet in March 2010. Rather, it would appear that the web traffic on the Russian party site remained within the domains set up and controlled by the party itself; in the British web domain the discourse surrounding the party had shifted—at least to a degree—to external sites not under direct control of the political party. These are relatively tentative conclusions, not least because it would be enlightening to examine the web linkage position of Russian and British political parties in general. In other words, were the Russian Communists and the British Liberal Democrats outliers in their own national web spheres? However, to a certain degree this is irrelevant, as we are thinking here about the comparison across, rather than within, national party systems.

## Content

Gibson and Ward (2000) established a cross-national tool for analysing central features of political party websites.[12] The scheme assigns points for features of websites in the categories of information provision, resource generation, internal networking, external networking, participation, campaigning and delivery (encompassing 'glitz' factor, access, navigability, freshness and visibility). Once the points are assigned, websites can be scored for their performance in each category. Although the scheme remains one of the best ways to quantify web performance (of parties and other organisations), some elements have become less relevant with the rapid development of web technologies as video has become standard, pages of large political organisations are updated constantly and the number of web links has grown astronomically (Oates 2008). This study uses the Gibson and Ward measures of information provision, resource generation and participation to compare the Russian and British party websites. Measurements of networking are replaced by the web link analysis provided above. The web link analysis also embraces much of the measure of visibility, although it is fair to say that it cannot really comment on navigability and freshness. However, in the competitive political marketplace of both Britain and Russia, it is clear that both party organisations devoted significant resources to their party websites and information was posted routinely throughout a given day.[13] This renders the Gibson and Ward measurement of how often the page is edited somewhat obsolete. In addition, the Gibson and Ward scheme does not really draw a distinction between information generated by the party and information generated by the users. With the advent of forums and especially direct links to social networking sites maintained by supporters (as opposed to party members), both the pace and nature of

---

[12]See also Gibson, Margolis *et al.* (2003) and Gibson (2010).

[13]Each party organisation was adding news and updates at least daily and sometimes several times a day in early March 2010, according to observation by the author. This study did not attempt to make an in-depth study of different versions or updates as the focus was on a cross-sectional study.

content have shifted, although more so in the British case. That being said, the three categories of the Gibson and Ward scheme used for this analysis provide some important illumination to compare the party websites (see Table 2).

What emerges from this analysis is that political party websites in both countries are a repository of a lot of information, much of it not readily available from other authoritative sources. It is easy to forget that parties exist between elections when they are not playing a visible part in government or vying for office. By the same token, both the CPRF and the Liberal Democrats were functioning in March 2010 in their respective societies as influential political outsiders, forcing parties in power to react to their significant constituencies and relatively coherent ideologies as both sat markedly to the left of mainstream politics.

It is worth noting that the look and feel of the party homepages were quite different. The CPRF homepage was cluttered by web standards and was quite long. The font was

TABLE 2

QUANTITATIVE ANALYSIS OF THE RUSSIAN COMMUNIST AND BRITISH LIBERAL DEMOCRAT WEBSITES, MARCH 2010

**Information provision**

| Feature | *Score* | |
|---|---|---|
| | *CPRF* | *Liberal Democrats* |
| Organisational history | 1 | 0 |
| Structure | 1 | 0 |
| Values/ideology | 1 | 1 |
| Policies | 1 | 1 |
| Documents (manifesto, constitution) | 1 | 1 |
| Newsletters | 1 | 0 |
| Media releases (speeches, statements, interview transcripts, conferences) | 1 | 1 |
| People/Who's who | 1 | 1 |
| Leader focus | 1 | 1 |
| Candidate or MP profiles | 1 | 1 |
| Electoral information (statistics, information on past performance) | 0 | 1 |
| Event calendar (prospective or retrospective) | 1 | 0 |
| Conference information | 1 | 1 |
| Frequently asked questions | 0 | 1 |
| Privacy policy | 0 | 1 |
| Article archive or library | 1 | 1 |
| TOTAL | 13 | 12 |

**Resource generation**

| Feature | *Score* | |
|---|---|---|
| | *CPRF* | *Liberal Democrats* |
| (i) *Donation index* 0–4 | 1 | 4 |
| (ii) *Merchandise index* 0–4 | 0 | 0 |
| (iii) *Membership index* 0–4 | 2 | 4 |
| (iv) Associate membership/volunteer solicitation 1=present 0=absent | 0 | 1 |
| TOTAL | 3 | 9 |

*Note: Code:* for resource generation, (1) if present, (0) if absent. For resource generation, (1) reference made and postal address listed; (2) download form and post; (3) online enquiry (specific email or online form); (4) online transaction; (0) no references made.

relatively small and there was a lot of text to absorb. While there were links on the top banner and along the side of the page, they were relatively small and lacked visibility. On the other hand, the Liberal Democrat homepage was very simple, short and uncluttered. The links appeared as bright 'buttons' to press. A quick review of other party websites in the United Kingdom (and indeed British websites in general) showed that this followed a national trend. While the Russian online sphere had added features and incorporated graphics, photos, etc., it was still very text-heavy. Meanwhile, British party websites had adapted a simplified, user-friendly style that was heavy on slogans and easy to navigate. One could argue that the Russians were offering more intellectual fodder on their websites. On the other hand, though, one could point out that the British were attempting to make politics more user friendly for a broader range of citizens.

In terms of information provision as defined by Gibson and Ward, the Russian Communist Party received a fractionally better score—13 out of 16, compared with 12 out of 16 possible points for the Liberal Democrats (see Table 2). The Communists provided a great deal of detail about their organisational history and structure. Students of Russian and Soviet history would have noted that the Communists continued to take formal rules very seriously in their party organisation. While the Communist Party of the Soviet Union was criticised for adhering to democratic form over function, it would appear that the modern CPRF was trying to adhere to both form and function with an open forum, etc. It also is arguably easier to be democratic and open when the stakes are relatively low, as the CPRF has little political power. On the other hand, it was rather disappointing that the British Liberal Democrats offered nothing obvious in the way of organisational history or structure.[14] The Liberal Democrats did provide a way for individuals to find out about how to run for parliament on the party ticket, but one wonders how relevant this is to actually running for office. An individual would have to be in very good standing with the party and have a lot of party support to become an official party candidate.

Another way in which the Communists outscored the Liberal Democrats was in providing a newsletter on the site, along with archives. However, it could be argued that the Liberal Democrats had moved on from a static newsletter to offering regular email and/or Twitter updates. Both sites gave users opportunities to input email addresses to receive regular updates (which was a separate function and simpler than 'joining' the sites, which will be discussed below). It was interesting to note that subscription to a regular paper newsletter from the Liberal Democrats cost £30 a year. Neither site had a section for 'frequently asked questions' that was relatively obvious. Unsurprisingly, the Liberal Democrats had an emphasis on the upcoming British campaign in May 2010, while the Communists had little to say about national elections slated for December 2011 in Russia.

There were no electoral performance statistics on either site, although the Communists included a link to a public opinion firm that offered several analytical reports on the unfair coverage of the party in the traditional Russian media.

---

[14]It should be noted that the author took the pragmatic view that if the information was not relatively obvious—i.e. could be found in 15 minutes by someone already fairly familiar with the website—that it was not included as present on the website. It is possible sometimes to find information via internet searches, but that is not the same as passive provision in a relatively clear way on a website.

Unsurprisingly, the media links on the CPRF website were to 'friendly' outlets, including television and radio reports that supported the Communist cause. Here, web links reflect the reality of the Russian media sphere, in which media outlets are strongly biased (Oates 2006). Although virtually all British newspapers retain ideological allegiances, most are not as biased as the Russian media. In addition, Russia lacks public-service television. As a result, it is not surprising that the web link analysis above showed the CPRF website embedded within a narrow sphere of supportive media while the Liberal Democrat website linked in and out of a range of more mainstream media sources. This suggests a heightened degree of isolation for the Russian Communist website and a limited ability to engage with a significant audience beyond supporters who would be drawn to the site by their pre-existing political preferences.[15] As such, despite narrowly outscoring the British Liberal Democrats in terms of information provision, the web link analysis suggested that the Communist Party website was 'preaching to the converted' (Norris 2003, p. 21). The way in which the Liberal Democrat site was more linked to generic or neutral information spaces online suggests that it was a much more effective information resource for the party to reach a significant sector of the electorate than the Russian communist website.

The Russian Communists should perhaps be commended for keeping commerce off their website to a large degree. It was relatively difficult to find a way to donate to the party, although bank details were available under 'contact information'. On the other hand, the Liberal Democrats streamlined the process with an online donation form. The Communists also continued their historic tradition of making it relatively hard to join or work for the party—with a long page discussing the rights and responsibilities of members. People were invited to send an email inquiry about joining the party. The Liberal Democrats were less discriminating, exhorting people to join the party (for a small fee) or volunteer through several different paths (including a link on the homepage). This is reflective of societal norms in general in terms of party membership. Party membership has been on the decline in the United Kingdom, although parties continue to solicit and encourage membership. Across the party spectrum in Russia, regulations stated by parties on their websites reflect that one must apply for and meet a list of requirements, often including references from existing party members, before joining one party exclusively (Oates forthcoming 2012).

In terms of the Gibson and Ward participation measurements, the Communists were not very keen on email contact. A search (using 'email' as the search term on each party homepage) found just six mentions of email addresses across the entire Communist website. A similar search on the Liberal Democrat site returned 7,957 hits in March 2010. This is no doubt a reflection in some ways of the more ubiquitous use of email in general for political and government organisations in the United Kingdom, but it cannot explain the almost complete absence of email addresses on the Russian Communist Party page (including a dearth of email contacts for CPRF members of parliament). This would appear to be a reflection of the difference in the nature of authority and openness between the two societies. In Russia, access to authority is

---

[15]This is also an issue to be considered when assessing the role of party websites in non-free states—can they become authoritative news voices for a segment of the population? What is the line between party information and news?

complex and uncertain; in Britain there is at least the appearance of the ability of citizens to interact regularly with their representatives through regular meetings ('surgeries'), petitions, the post and (increasingly) email as well. This view is challenged somewhat by Dmitri Medvedev's blog, although further study is needed to gauge the truly interactive or open nature of the blog and its comment area.

It is important here to note the strong role of regional organisations, an element of parties and websites that is not explicitly covered by the Gibson and Ward scheme. Both websites featured links to regional party organisations. There were far fewer for the Russian Communists, yet it is striking to see the ties between the parent and regional organisations (both in terms of web link analysis and in observation on the websites themselves). The regional organisations for the Liberal Democrats were far more numerous and hence less aggregated than those of the CPRF. The Liberal Democrats had websites for hundreds of constituencies, making their web presence very local indeed (although the format and much of the content appears to come from the parent site and further study would be needed to ascertain how much content is truly locally generated or relevant). As the famous Speaker of the US House of Representatives Tip O'Neill once said 'All politics is local' (1993)—and this suggests that the web might be most effective at supporting local communities of political interest rather than effecting national shift in party affiliation and support. These two case studies would suggest that a measurement of how much a political party website engages, builds, supports and reflects a local political movement and/or party support group would be useful.

This study did not employ the Gibson and Ward coding framework interaction index, which gives points for features such as bulletin boards, guest books, chat rooms and online debates with leaders. It has become too difficult to separate these elements from the general function of party forums (present on both websites). However, it is key to note here that while the Liberal Democrat website had links to external social networking tools, the Russian Communist website did not in March 2010. Hence, while there was no obvious, lively forum actually embedded within the Liberal Democrat website, there was a prominent one on the Russian Communist website. On the other hand, while an internet search turned up many opportunities for people to 'chat' with Liberal Democrat leader Nick Clegg online (such as through the parental advocacy group Mumsnet or sponsored by British newspapers), these opportunities did not seem to be readily available with CPRF leader Gennady Zyuganov during March 2010. Again, one should reflect on whether this is a measure of web strategy or evidence of the more formal nature of Russian politics, in which online chat could be seen as trivial or undignified rather than relevant to the political life of the party. It may also be a function of the relative proximity of the British 2010 elections. Finally, it might also be linked to the tastes and habits of the user community in each country, although that was beyond the scope of this study.

*Users and user-generated content linked to party websites*

The web analysis for the CPRF showed that the discussion had not migrated away from the party site onto more generic social networking sites such as VKontakte, Odnaklassniki or YouTube (which functions as a social networking tool when comment is linked to videos). Preliminary evidence, particularly the web link analysis, suggests that more discussion took place on the Facebook Liberal Democrat sites than on the internal Liberal Democrat forum. This article now compares the content on internal forums for both parties, but presents evidence about traffic and use of Liberal Democrats on Facebook.

## The Communist Party web forum

The CPRF website forum was very transparent with both general statistics and details on the participants.[16] As of 26 March 2010, the forum listed total posts of 193,884, with total topics of 2,909 and total members of 9,583 (see Tables 3 and 4). The website listed the most users ever online as 120 on 1 June 2006. The site also showed the user names of who was currently online, as well as identified users in one of five categories by colour coding of the name: administrators, advice moderators, CPRF supporters, participants and members of the CPRF. The website displayed a list of all members, with profile information that included the date the member joined, the number of posts the person had made and on which discussion threads the person participated. The list included a link to contact the individual via the stored email address (although the website did not display the address) and included a web link if one was provided by the forum member. As of 24 March 2010, the website listed a total of 9,577 users in the forum. Unsurprisingly, the vast majority (6,858 or 72%) had never posted even a single message. Only 1,817 members of the forum (19%) had more than one post. There were 10 highly active site members (see Table 4, in which these individuals are referred to as 'super-users'), each of whom had posted more than 2,000 times and 40 members with more than 1,000 posts. In fact, the single most active member ('Yarov') had 6,757 posts and was responsible for 3.5% of all posts on the CPRF forum by 24 March 2010. Most of the forum members also remained anonymous in their role—

TABLE 3

RUSSIAN COMMUNIST FORUMS (ALL LISTED) (AS OF 25 MARCH 2010 12:01 AM EST)

| Group name | Description as listed on website | Forum themes | Total posts |
|---|---|---|---|
| CPRF—Practical | Party news, legislation, elections, referendums, Deputy 'vertical' | 467 | 28,665 |
| CPRF—Theory | Party programme, rules, development of MLT | 335 | 42,356 |
| Discussion on important political, economic and social news | Urgent policy | 868 | 40,435 |
| Cognitive discussions | History, culture, science, education | 519 | 38,121 |
| Mosaic of life | Simply about daily existence | 450 | 28,708 |
| Administration | Functional questions and feedback with management | 69 | 2,616 |
| Political 'hospital' (*politklinika*) | Disputes, abuse, squabbles and various types of delirium | 79 | 3,966 |

*Note*: all groups were active on 25 March although some threads were closed to further discussion.

[16]In order to view details about users and various statistics, the author joined the site and marked the reason for joining as 'research'. There was no participation in the website forum beyond joining; for example, the author did not make comments, email members or join groups.

## TABLE 4
Top Ten Posters on www.kprf.ru Forum as of 26 March 2010

| Rank | Name | When joined | Posts | Description | Website? | % of total posts | Posts per day | Posting pattern |
|---|---|---|---|---|---|---|---|---|
| 1 | Yarov | 23 April 2003 | 6,816 | Guest | No | 3.51 | 3.79 | No specific posting details |
| 2 | Andrei Geran | 22 March 2004 | 4,410 | Council of Moderators | No | 2.27 | 2.01 | 38% of posts on 'Discussion about important news in politics, economics and society' forum |
| 3 | Igor Sh | 27 March 2005 | 3,746 | Participant | No | 1.93 | 2.05 | 32% of posts on 'Mosaic of life' forum (11% on 'Socialism or capitalism?' thread) |
| 4 | OLSRKhM | 15 January 2005 | 3,481 | Participant | No | 1.79 | 1.84 | 74% of posts in the 'CPRF/Theory' forum (12% in 'Second coming of Marx' thread) |
| 5 | Leonid Ilich | 21 July 2007 | 3,389 | Participant | No | 1.74 | 3.46 | 29% of posts on 'Discussion about important news in politics, economics and society' forum |
| 6 | Yu.M. | 21 May 2009 | 3,330 | Participant | No | 1.71 | 10.78 | 60% of posts on 'CPRF/Theory' forum (35% on 'New in theory and the present' thread) |
| 7 | Mamushkin | 5 June 2003 | 3,299 | Council of Moderators | Yes | 1.7 | 1.33 | 23% of posts in 'Mosaic of life' forum. Has a picture of Stalin as one of his photo icons |
| 8 | Yeok | 21 January 2003 | 2,462 | Participant | No | 1.27 | 0.94 | 35% of posts in 'Mosaic of life' forum (9% in 'The US dollar as the shadiest transaction of the 20th century' thread) |
| 9 | LAS | 5 August 2007 | 2,192 | Participant | Yes | 1.13 | 2.27 | 93% of posts in 'Cognitive discussions' forum (93% in the 'New in theory and the present' thread) |
| 10 | A.S. | 16 March 2009 | 2,091 | Participant | Yes | 1.08 | 5.58 | 78% of posts in 'CPRF/Theory' forum (54% in 'Nature of the cost' thread) |
| TOTAL | | | | | | 18.13 | 3.41 | |

only 585 were classified as participants (*uchastniki*), 51 as supporters (*storonniki*), 30 as members (*chleny*) and six as administrators.

The forum titles reflected a relatively broad and ideological approach to discussion. In terms of content, anyone who has been involved in an ideological discussion late at night in a Moscow kitchen would recognise the style and tone. Several of the forums, including 'capitalism *versus* socialism' had been running for years. An overview of dozens of pages found little in the nature of exchange or dialogue—rather, a small number of posters tended to reiterate their convictions. When posters were challenged or contradicted, there was little in the nature of thoughtful response—or indeed, any response at all. It would appear that, as in traditional Soviet newspapers, people stated their opinion without interest in reflection or response. Of course, this is not merely a characteristic of Russian forums; bloggers in general broadcast their thoughts and opinions often without much interest in argument or dialogue (Davis 2009). What was intriguing is that the choice of topic tended to be quite ideological and even philosophical in nature. There was little that was playful (or arguably engaging) on the CPRF forum postings. However, if one sought discourse on communist and/or socialist ideology, the link of the Russian past to the present political situation, the role of the media (especially on the forum on kprf.tv), there was a wealth of discussion in this area.

The detailed information on users via the CPRF internal forum (not available in the same way for the Liberal Democrat forum participants) allowed for a reflection on the nature of posting activity by individuals. The 10 most active posters on the CPRF website forum fell into two categories (see Table 5). Although there was no distribution of posting activity data on the most prolific poster ('Yarov'), out of the remaining nine,

TABLE 5

LIBERAL DEMOCRAT GROUPS ON INTERNET FORUM SITE, 10 MOST POPULAR (AS OF 25 MARCH 2010 9:45 AM, FROM HTTP://ACT.LIBDEMS.ORG.UK/GROUPS?SORT = MOSTPOPULAR)

| Group name | Members | Purpose | Most recent posting | Total postings |
|---|---|---|---|---|
| Liberal Youth | 228 | Youth, student wing of party | 17 March | 6 pages |
| NO2 ID | 178 | Group campaigning against government Storage of biometric data | 20 March | 5 pages |
| Nick Clegg Meets | 176 | Schedule of public meetings, photos with Liberal Democrat leader | 16 March | 3 pages |
| Approved Candidates | 170 | Not clear—not real content | 24 March | None |
| ALDC | 140 | Association of Liberal Democrat Councillors | 19 March | 1 page |
| Lib Dems on Twitter | 124 | Group for those who tweet | 20 March | 6 pages |
| LibDem Voice | 121 | Not stated | 25 March | 1 page |
| Federal Conference | 119 | For those adding Liberal Democrat conference | 15 March | 3 pages |
| Lib Dem Media Watch | 115 | Group formed when Liberal Democrat spokesperson was dropped from prominent news programme; discusses media coverage of Liberal Democrats | 25 March | 25 pages |
| Lib Dem Real Ale Drinkers | 100 | Ale drinking | 20 March | 4 pages |

four focused their posts in a specific forum. In two cases ('LAS' and 'A.S.'), the posters were further concentrated mainly on one thread within a forum. Thus, almost half of the most active users of the CPRF forum were significantly engaged in one area of discussion. For the other super-users, the posting activity was more diffuse. Overall, these 10 individuals accounted for 18% of the total posts on the CPRF website forum and were each participating at an average rate of 3.41 posts daily. Some of the super-users were waxing or waning in their activity. For example, 'Yeok' was uploading just under a post a day (on average), although he had amassed a significant archive of 2,462 posts over seven years. On the other hand, 'Yu. M.' was posting at a rate of 10.78 per day by March 2010, reaching the top 10 list of forum users less than a year after joining the forum in May 2009. Thus, there was a significant difference in the intensity of use for the most active forum participants. The next logical step would be for a structured content analysis of the posts of these users, in particular with a view to whether they tend to broadcast opinions, engage with other users (either in a civil or uncivil fashion), lead opinion or genuinely seek dialogue/input. In addition, it would be useful to look at Communist bloggers on external sites, although the lack of links from the CPRF web page (unlike the Liberal Democrats) might make them relatively hard to find.

### The Liberal Democrat web forum

This focus on the more ideological nature of politics was far less apparent in the user-generated content relating to the British Liberal Democrats. This is perhaps unsurprising as it was clear from the web link analysis above that the locus of the discussion had migrated from the Liberal Democrat site itself to Facebook. Indeed, although the Liberal Democrat website offered a forum area called the 'Lib Dem Network' or 'Act' that was linked from the homepage, it was not very popular.[17] Unlike the cluttered and busy homepage of the CPRF, the Liberal Democrat homepage was quite basic and simple, with links along the right-hand side to policy statements, press releases rolling across the top as well as web links to volunteer, join, renew and donate. This made the sign-in (or register) button for the Liberal Democrat network on the left-hand side of the homepage quite visible. Nonetheless, as of 25 March 2010, the largest group on the Liberal Democrat network was 'Liberal Youth' with only 228 members, followed by NO2ID (a group campaigning against the British government's plan to introduce identification cards with biometric data) with 178 members (see Table 5).

A brief qualitative analysis of the three most popular groups (by number of members) on the Liberal Democrat Forum showed that they followed slightly different formats. The Liberal Youth site contained several references or links to other sites (particularly Facebook and Twitter). There was little discussion of issues, although information was distributed at the beginning of the site. The tone was light-hearted (one member was teased about a hairstyle) but by the standards of youth internet exchange, it was rather earnest and serious albeit with little ideological content. NO2ID was a far more serious group, in which there were exchanges about differences in belief about the value of identification cards with biometric information for British citizens. Here, instead of links to other groups on Facebook and Twitter, there were more links to news articles about

---

[17]Again, the author joined the site only to observe.

ID cards and a national DNA database. The third-most popular group, Nick Clegg Meets, was brief and very functional, essentially just comments on meeting with the party leader or a desire to meet with the leader (this could be interchangeable with a 'Vladimir Putin Meets' or 'Dmitri Medvedev Meets' web page).

What is perhaps more telling was who was not exchanging ideas and thoughts on the Liberal Democrat internal forum. For example, the forum for Agents and Organisers (volunteers and staff) had only 97 members and just two pages of posts. Despite the upcoming and important May 2010 general election in the United Kingdom, the latest activity as of late March was on 21 February 2010. If the Liberal Democrats were using the web for serious political organisation, it was not visible here. In terms of political debate, the most interesting of the top 10 internal forums was the Liberal Democrat Media Watch, formed when Liberal Democrat parliamentarian Jo Swinson was dropped from the BBC *Question Time* show in November 2009. The group stated as its aims 'to encourage ACTivists to complain to the media when Lib Dems are not fairly represented'.[18] Although the group was sparked by a particular catalyst—a Liberal Democrat Member of Parliament being excluded from a high-profile political debate programme with Labour and the Conservatives at the last minute—the discussion was also about the general media coverage of the Liberal Democrats. The forum members voiced dissatisfaction with the way in which the Liberal Democrats were framed as political outsiders by the media and shared how they had filed complaints on the matter to the BBC.

If the internal 'Act' forum for the Liberal Democrats was relatively sparsely populated, what sort of activity was found on social networking sites such as Facebook? The official link from the Liberal Democrat homepage to Facebook brought up a site with 8,711 fans (compared with 7,906 for Labour and 26,224 for Conservatives).[19] The 'wall' comment area was far more active than the internal forums, with dozens of posts and links daily (this article will not include a detailed analysis of the material as it does not parallel the Russian case). There were other Facebook groups with 'Lib Dem' or 'Liberal Democrat' in the title but they were much smaller than the main Facebook site.

## Conclusions

This article has attempted to identify and deploy coding methods to better understand the relative role of political party websites in comparative perspective. It is acknowledged that the comparison of the Russian Communist Party and the British Liberal Democrats might at first seem too much of a 'different case' scenario. However, the two cases force us to consider methodology and the meaning of context quite carefully in the study of the internet. Perhaps what these two cases highlight even more, however, is the difficulty in taming the information tide when studying the

[18]See http://act.libdems.org.uk/group/wewantjoswinsononquestiontime, as of 10:51 am 25 March 2010.

[19]Checked on 25 March 2010. The Labour homepage did not have an obvious link to Facebook and only one note asked the user to click to become a 'fan' instead of just accessing the page. This might have put people off—on the Conservative homepage one had the choice of just viewing the Conservative Facebook page or becoming a fan with a single click.

internet. There is now such a vast range of data available to examine websites of political parties and virtually every other institution that it is now necessary to think much more carefully about which information to sample, what methods to use and how to contextualise that information. In particular, this article highlights the difficulties of coming to grips with the vast array of content in forums, both sponsored by party websites and allied to party causes on social networking sites and beyond. The analysis and understanding of content lies at the heart of rigorous study in political communication, but is it worth analysing the words of a handful of people who are not linked to the rest of society in particularly meaningful ways? While Liberal Democrat supporters seem trivial or even somewhat random at times, rarely engaging in meaningful dialogue online via the party website forums, the CPRF forum participants often seem to be preaching to the converted and failing to reach onto the broader social networking platforms that are relatively popular in Russia. Is either method of discourse really contributing to political empowerment and change for citizens, albeit in very different states? How much—or perhaps more importantly which—content should be analysed and how often? Perhaps the answer lies in thinking more about the participants on the websites (whether they are visitors, users generating content or even party functionaries choosing or creating content) as part of a community. This highlights the idea that even in the online world there is agenda setting and gate keeping as there is in traditional media, although it is not yet so obvious. By thinking about the relative power structure and asymmetric control over a website and/or its forums, perhaps we can get a better understanding of the role of the internet in the political sphere. For future research, it is important to continue to analyse how parties 'broadcast' information about themselves online and attempt to use online tools, but it is also critical to move off party websites and look for political dialogue woven throughout the web discourse, particularly on social networking sites. There is a huge surge of interest in this, particularly in using software that can detect words either singly or in relationship to one another or to even build testable families of words that can measure public sentiment as expressed in user-generated content online (Hopkins & King 2010).

Finally, this article has not mentioned the idea of a catalyst, although catalysts are generally understood to be important in transforming the internet from information sources into politicised media. This was a point made by Fossato *et al.* (2008) in a study of three prominent Russian political movements with blogs, but certainly a point that has been illuminated with new force and power through the role of online communication in the Arab Spring. In both countries, these party websites are still saying things that are not part of the mainstream media. For the Liberal Democrats in March 2010, this was about what they perceived to be an unjust winner-take-all electoral system and media coverage that unfairly diminished public support into an asymmetrically small number of parliamentary seats. Indeed, the Liberal Democrat website is a snapshot of a party at the end of a historical era, for a lively election campaign, the introduction of Nick Clegg into national televised debates and a surprising power-sharing coalition with the Conservatives changed the very nature and goals of the party by June 2010. For the CPRF, it is about a media that distorts, lies and omits critical information. On 26 March 2010, the CPRF website featured a story regarding the attack on Shatura Mayor Valery Larionov, a CPRF candidate recently elected in the Moscow Region in the face of United Russia opposition. In its headline, the CPRF website asks whether this was 'extremism or vengeance for United Russia's lost elections?' (Communist Party of the Russian Federation 2010). A scan of Russian media on the day found little coverage of the crime—in which Larionov was beaten with iron rods until his son scared off the assailants—much less the suggestion that it could have been a political crime. While there is a legitimate argument to fear the fracturing of the news sphere, in which audiences seek narrow sources of information and ignore the broader public arena, the emphasis on alternative news on the CPRF website could prove to be the website's largest contribution to civil society in Russia.

The intention of this study was to consider the relative role of party websites in the context of two very different regimes. However, there are several intriguing elements and evidence to note since the end of the study that might suggest that *Runet* is poised to play a more significant political role than many had predicted. The first is the example of the Arab Spring, which has created evidence that the speed and flexibility of the internet may create democratic opportunities in ways unimagined by scholars and states alike. For example, the Egyptian government had tightened its controls on the internet in the two years prior to the 2011 revolution, but had failed to understand the way in which the internet was both rooted in local communities as well as capable of creating a nimble and diverse network of opponents to the regime. It is also interesting to note that Russia and Egypt had very similar scores on internet freedom from Freedom House in 2011, both tolerating much more freedom in the online sphere than in the realm of the traditional media. However, it is important to point out that there are fairly significant indicators that Putin and the current party of power (United Russia) enjoy considerably more support than Mubarek's regime in early 2011. The December 2011 protests, which were reliant on *Runet* for organisation and information, have highlighted the strength and rapidity of how the internet may force political change in states such as Russia.

The final point is about the existence of politics in places in which both states and political scientists do not really expect it. This is becoming a rising area of research in the United Kingdom, such as by examining how people talk about politics in online spaces that are ostensibly more commercial in nature (Wright 2012). At the same time, the Occupy movement, street demonstrations in the United Kingdom and the English riots in the summer of 2011 all suggest that much of the political energy is found outside of 'traditional' political institutions such as parties, parliaments, unions and NGOs. Is the real political debate taking place in online spaces that parallel these new political movements that have very little to do with political parties? If so, it would be quite useful to use research methods developed by looking at political party websites online. It is no longer about parties 'online' or parties 'offline', but about how the interaction and exchange between citizens and traditional political organisations is challenged, changed and possibly enhanced by the online sphere. This makes understanding online political catalysts within their cultural contexts more important than ever.

*University of Maryland*

### References

Agora Human Rights Association (2011) *Threats to Internet Freedom in Russia, 2008–2011: An Independent Survey* (Moscow, Agora Human Rights Association), available in English at: http://www.openinform.ru/fs/j_photos/openinform_314.pdf and in Russian at: http://www.openinform.ru/fs/j_photos/openinform_313.pdf, accessed 23 January 2012.

Anstead, N. & Chadwick, A. (2008) 'Lessons from the US Digital Campaign', *Renewal: A Journal of Social Democracy*, 16, 3–4, September, pp. 86–98.

Chadwick, A. (2006) *Internet Politics: States, Citizens, and New Communication Technologies* (New York, Oxford University Press).

Communist Party of the Russian Federation (2010) 'Moskovskaya oblast': Soversheno zverskoe napadenie na glavu goroda Shatura kommunista Valeriya Larionova. Ekstremizm ili mest' proigravshikh vybory edinorossov?', available at: http://kprf.ru/rus_law/77335.html, accessed 17 November 2010.

Cooper, J. (2008) 'The Internet in Russia—Development, Trends and Research Possibilities', *CEELBAS Post-Soviet Media Research Methodology Workshop*, 28 March, Birmingham, University of Birmingham.

Davis, R. (2009) *Typing Politics: The Role of Blogs in American Politics* (New York, Oxford University Press).

Deibert, R., Palfrey, J., Rohozinski, R. and Zittrain, J.L. (Open Net Initiative) (2010) Access Controlled: The Shaping of Power, Rights, and Rule in Cyberspace (Cambridge: The MIT Press).

Deibert, R., Palfrey, J.G., Rohozinski, R. and n Zittrain, J. (Open Net Initiative) (2009) Access Denied: The Practice and Policy of Global Internet Filtering (Cambridge: The MIT Press).

Dutton, W. H., Helsper, E. J. & Gerber, M. (2009) *The Internet in Britain 2009* (Oxford, Oxford Internet Surveys, Oxford Internet Institute), available at: http://www.oii.ox.ac.uk/microsites/oxis/events/?id=11, accessed 25 February 2010.

European Institute for the Media (1994) *The Russian Parliamentary Elections: Monitoring of the Election Coverage of the Russian Mass Media* (Düsseldorf, European Institute for the Media), available at: http://www.media-politics.com/eimreports.htm, accessed 23 January 2012.

European Institute for the Media (1996a) *Monitoring the Media Coverage of the 1995 Russian Parliamentary Elections*, February (Düsseldorf, European Institute for the Media), available at: http://www.media-politics.com/eimreports.htm, accessed 23 January 2012.

European Institute for the Media (1996b) *Monitoring the Media Coverage of the 1996 Russian Presidential Elections*, September (Düsseldorf, European Institute for the Media), available at: http://www.media-politics.com/eimreports.htm, accessed 23 January 2012.

European Institute for the Media (2000a) *Monitoring the Media Coverage of the December 1999 Parliamentary Elections in Russia: Final Report*, March (Düsseldorf, European Institute for the Media), available at: http://www.media-politics.com/eimreports.htm, accessed 23 January 2012.

European Institute for the Media (2000b) *Monitoring the Media Coverage of the March 2000 Presidential Elections in Russia*, August (Düsseldorf, European Institute for the Media), available at: http://www.media-politics.com/eimreports.htm, accessed 23 January 2012.

Ford, R. & Goodwin, M. J. (2010) 'Angry White Men: Individual and Contextual Predictors of Support for the British National Party', *Political Studies*, 58, 1, pp. 1–25.

Fossato, F., Lloyd, J. & Verkhovsky, A. (2008) *The Web That Failed: How Opposition Politics and Independent Initiatives are Failing on the Internet in Russia* (Oxford, Reuters Institute for the Study of Journalism), available at: http://reutersinstitute.politics.ox.ac.uk/fileadmin/documents/Publications/The_Web_that_Failed.pdf, accessed 1 March 2010.

Freedom House (2011) Freedom on the Net 2011: Global Scores. New York: Freedom House. Available online at http://www.freedomhouse.org/template.cfm?page=664 (last accessed September 21, 2011).

Gibson, R. K. (2010) 'Open Source Campaigning?: UK Party Organisations and the Use of the New Media in the 2010 General Elections', *Annual Meeting of the American Political Science Association*, 1 September, Washington, DC.

Gibson, R. K., Margolis, M., Resnick, D. & Ward, S. J. (2003) 'Election Campaigning on the WWW in the USA and UK: A Comparative Analysis', *Party Politics*, 9, 1, pp. 47–75.

Gibson, R. K., Nixon, P. & Ward, S. (2003) *Political Parties and the Internet: Net Gain?* (London, Routledge).

Gibson, R. K. & Ward, S. (2000) 'A Proposed Methodology for Studying the Function and Effectiveness of Party and Candidate Websites', *Social Science Computer Review*, 18, 3, pp. 301–19.

Goroshko, O. I. & Zhigalina, E. (2009) 'Quo Vadis? Politicheskie kommunikatsii v blogosfere Runeta', *Russian Cyberspace*, 1, 1, pp. 81–100, available at: http://www.digitalicons.org/issue01/pdf/issue1/Political-Interactions-in-the-Russian-Blogosphere_O-Goroshko-and-E-Zhigalina.pdf, accessed 9 August 2010.

Hindman, M. (2009) *The Myth of Digital Democracy* (Princeton, NJ, Princeton University Press).

Hopkins, D. J. & King, G. (2010) 'A Method of Automated Nonparametric Content Analysis for Social Science', *American Journal of Political Science*, 54, 1, pp. 229–47.

Kavanagh, D. (1995) *Electioneering* (Oxford, Blackwell).

Linaa Jensen, J. (2006) 'The Minnesota E-democracy Project: Mobilising the Mobilised?', in Oates, S. *et al.* (eds) (2006), pp. 39–58.

March, L. (2006) 'Virtual Parties in a Virtual World: The Use of the Internet by Russian Political Parties', in Oates, S. *et al.* (eds) (2006), pp. 136–62.

Morozov, E. (2011) *Net Delusion: The Dark Side of Internet Freedom* (New York, Public Affairs).

Norris, P. (2001) *Digital Divide: Civic Engagement, Information Poverty, and the Internet Worldwide* (Cambridge, Cambridge University Press).

Norris, P. (2003) 'Preaching to the Converted?: Pluralism, Participation and Party Websites', *Party Politics*, 9, 1, pp. 21–45.

Oates, S. (2006) *Television, Elections and Democracy in Russia* (London, Routledge).

Oates, S. (2008) 'Comrades Online?: How the Russian Case Challenges the Democratising Potential of the Internet', in *Politics: Web 2.0*, International Conference, New Political Communication Unit, Department of Politics and International Relations, 17–18 April, Royal Holloway, University of London.

Oates, S. (forthcoming 2012) *Revolution Stalled: The Political Limits of the Internet in the Post-Soviet Sphere* (New York, Oxford University Press).

Oates, S., Owen, D. & Gibson, R. K. (eds) (2006) *The Internet and Politics: Citizens, Voters and Activists* (London, Routledge).

O'Neill, T. with Hymel, G. (1993) *All Politics is Local: and Other Rules of the Game* (Holbrook, MA, Bob Adams by arrangement with Random House).

Organization for Security and Co-operation in Europe/Office for Democratic Institutions and Human Rights (OSCE/ODIHR) (2004a) *Russian Federation Elections to the State Duma 7 December 2003 OSCE/ODIHR Election Observation Mission Report*, 27 January (Warsaw, Office for Democratic Institutions and Human Rights), available at: http://www.osce.org/item/8051.html, accessed 23 January 2012.

Organization for Security and Co-operation in Europe/Office for Democratic Institutions and Human Rights (OSCE/ODIHR) (2004b) *Russian Federation Presidential Election 14 March 2004 OSCE/ODIHR Election Observation Mission Report*, 2 June (Warsaw, Office for Democratic Institutions and Human Rights), available at: http://www.osce.org/odihr-elections/14520.html, accessed 23 January 2012.

Owen, D. (2006) 'The Internet and Youth Civil Engagement in the United States', in Oates, S. *et al.* (eds) (2006), pp. 20–38.

Rainie, L. (2010) *Internet, Broadband, and Cell Phone Statistics: As of December 2009, 74% of American Adults (Ages 18 and Older) Use the Internet* (Washington, DC, Pew Internet & American Life Project), available at: http://www.pewinternet.org/Reports/2010/Internet-broad-band-and-cell-phone-statistics.aspx?r=1, accessed 25 February 2010.

Römmele, A. (2003) 'Political Parties, Party Communication and New Information and Communication Technologies', *Party Politics*, 9, 1, pp. 7–20.

Russian Federal Agency on the Press and Mass Communication (2011) *Internet v Rossii: Sostoyanie, tendentsii i perspektivy razvitiya* (Moscow, Regional Public Centre for Internet Technology), available at: http://www.fapmc.ru/files/download/Print.pdf, accessed 16 July 2011.

Saunders, R. A. (2009) 'Wiring the Second World: The Geopolitics of Information and Communications Technology in Post-Totalitarian Eurasia', *Russian Cyberspace*, 1, 1, pp. 1–24, available at: http://www.digitalicons.org/issue01/issue1/robert-saunders.php?lng=English, accessed 9 August 2010.

Semetko, H. & Krasnoboka, N. (2003) 'The Political Role of the Internet in Societies in Transition: Russia and Ukraine Compared', *Party Politics*, 9, 1, pp. 77–104.

Street, J. (2010) *Mass Media, Politics and Democracy*, 2nd edn (New York, Palgrave Macmillan).

White, S. & McAllister, I. (2008) '"It's the Economy, Comrade!" Parties and Voters in the 2007 Russian Duma Election', *Europe-Asia Studies*, 60, 6, pp. 931–57.

Wright, S. (2012) Politics as usual? Revolution, normalization and a new agenda for online deliberation. New Media & Society, 14, 2, March, pp. 244–261.

# Mediating New Europe-Asia: Branding the Post-Socialist World via the Internet

## ROBERT SAUNDERS

WITHOUT QUESTION, THE VARIOUS COUNTRIES of the former Eastern Bloc have been the sites of rapid social, cultural and political transformations since the late 1980s, a period which also saw the explosion of new forms of digital media, including but not limited to personal web pages, SMS/texting, podcasts, streaming videos, (micro)blogs and file sharing. This period also saw many of the region's governments initiate programs intended to actively manage how their nation-states were viewed abroad (Kaneva 2012; Aronczyk 2013). Central to this process was the use of various forms of mass media to produce and project positive national images, including new media. The rise of the Internet has both extended the reach of pre-existing media (newspapers, radio, television, etc.) and communication platforms (mail, phone, fax, etc.) *and* created entirely new media forms and pathways of social interaction. The end of one-party rule in Eastern Europe and post-Soviet Eurasia created a fecund environment where new media flourished (Herron 1999; Emory and Bates 2001; Nune 2003; Bowles 2005; Schmidt and Teubener 2006; Imre 2009; Saunders 2009, 2010; Uffelmann 2011), and in cases where Eurasian-style 'managed democracy' began to place (or reinstate) limitations on free speech and journalistic practices, new media have emerged as a vital tool for maintaining hard-won political freedoms (Semetko and Krasnoboka 2003; Warf 2009; Morozov 2009; Mungiu-Pippidi and Munteanu 2009; Johnson and Kolko 2010; Nikiporets-Takigawa 2013), while also paradoxically serving the interests of quasi-authoritarian regimes (Kalathil and Boas 2003; Morozov 2011; Gorham 2011). This nexus of developments has created a dynamic milieu where a variety of different actors have sought to 'brand' the nations of new Europe-Asia.

New media have played a particularly important role in transforming the respective national images of the countries that make up new Europe-Asia, or what might be called the post-Second World (Saunders 2012), as well as shaping how the region as a whole is conceived, represented and consumed via digital and popular culture. This chapter explores the mediation of the 'new' Europe-Asia via new media, interrogating digital technologies' variegated roles in reconciling and harmonizing the former Eastern Bloc's place within Europe (and to a lesser extent Asia) following the cessation of one-party rule and the end of the Cold War. Concurrently, this chapter investigates the 'international mediatization' (Hitchcock 2003) of the post-Second

World countries, individually and collectively, and how such transformations inform larger political and diplomatic structures. Due to the deterritorialized nature of new media and information and communication technologies (ICTs), the domestic-foreign frontier is more permeable than ever, thus resulting in powerful influences from outside the region, as well as forcing media producers within the region to consider and often accommodate the demands of audiences overseas. Following the model of Peter van Ham in *Social Power in International Politics* (2010), this analysis focuses on three groups of actors: the state, domestic non-state actors and external/foreign non-state actors. Rooted in a constructivist paradigm of international relations theory (see, for instance, Hopf 1998; Price and Reus-Smit 1998; Duvall and Varadarajan 2003; Zehfuss 2003; Guzzini and Leander 2006) and influenced by the burgeoning analytical tools of critical (Ó Tuathail 1996; Bachmann and Sidaway 2008; Power and Campbell 2010; Ciută and Klinke 2010) and popular geopolitics (Sharp 2000; Dittmer 2007; Dodds 2008; Dunnett 2009; Grayson, Davies and Philpott 2009; Purcell, Scott Brown and Gokmen 2010; Downing 2013), the following analysis seeks to shed light on how new media (existing alongside and often directly influencing mainstream media) produces reality, and more specifically, how they reify national image for external consumption.

Following a brief overview of the relevant concepts associated with the meta-processes discussed herein, this chapter—which is structured as a survey of major themes and trends rather a case study—will examine how new media 'brand' the post-socialist world (see Szondi 2007; Jansen 2008; Marat 2009; Kaneva 2012; Aronczyk 2013).[1] The first section of the body investigates the ways in which the state has attempted to yoke new media technologies to inform, educate and manipulate external audiences, from the creation of national web sites, Facebook pages and Twitter feeds to the distribution of promotional videos and other ICTs which advance the 'party line'. The subsequent section examines how non-state actors within the region have employed digital technology to contest, interrupt and negate state-affiliated, new media-centric representations in cyberspace, i.e. the digital geographies of information created by and sustained through electronic interactions of humans over global computer networks (see Kitchin 1998). In order to elucidate the growing importance of globalization on national identities in the current era, the third segment assesses the work of cultural producers from outside new Europe-Asia who engage in the process of (re-) defining the region through parody, satire and pastiche, and also examines how such popular culture influences and informs national cultures (and the Jamesonian 'commodification of culture') in the post-Second World. By triangulating these contemporary efforts at constructing and redefining national imaginaries (state, internal non-state, external non-state), my aim is to synthesize the effects of these contentious, though often mutually reinforcing, flows of politicized information

---

[1] Less critical treatment of the practice can be found in the work of scholars who also work as nation brand consultants, including Olins (2002), Anholt (2007), and Dinnie (2008).

and images to arrive at a holistic assessment of the role of new media in 'branding' new Europe-Asia.[2]

This chapter, like others in this volume, treats post-socialist Europe-Asia (formerly known as the Second World) as a separate area of cultural production vis-à-vis new media (as well as nation branding). In doing so, there is the danger of reinforcing a variety of stereotypes about the region and perpetuating notions of an undifferentiated 'whole'; however, I would argue that by elucidating important, even pivotal, differences between these states, there is value in doing so. Given the large number of states covered in this study, i.e. 29 internationally recognized countries and three de facto states (Kosovo, South Ossetia and Abkhazia), it is impossible to present a genuinely empirical analysis on new media branding across the post-Second World. Instead, this study attempts to provide a high-level overview of trends in the region based on the author's experience in studying the phenomenon of nation branding in post-socialist countries during the past decade, including interviews with government officials, tourism and brand consultants, new media entrepreneurs and other relevant actors in (or representing) Russia, Ukraine, Georgia, Lithuania, Slovenia, Kazakhstan, Tajikistan and Uzbekistan.

### *New media and political communication: mediation, mediatization and nation branding*

The advent of new media has remapped the ecology of political communication (Gurevitch, Coleman and Blumler 2009), particularly through the structural capacity of the Internet, which is a technological platform that allows consumers to become producers of content with minimal cost and few barriers to distribution. This apparently simple change has rewritten the rules of who speaks for whom, as the formerly reliable (and nigh impermeable) barriers to entry associated with broadcast media (i.e. print, radio and television) have been shattered by blogs, podcasts and YouTube (though distribution channels on these platforms often lack large audiences, thus resulting in the neologism of 'narrowcasting'). Regardless, this transformation requires a major rethinking of the paradigm of *mediation* as it relates to politics and especially political communication, both on the national and international/global levels.[3] A rather slippery term on its own, mediation can mean *either* the process of intervening between parties in conflict (reconciliation) *or* the function or activities of transmission (interlocution). It is the second of these two meanings that has shaped the concept of mediation in the realm of political communication. To use a rather straightforward definition, mediation, in the words of S. Brent Plate, is the process whereby meaning is 'encapsulated and put into a format that we are taught to recognize, name and engage' (2012, 156). The rise

---

[2] One aspect of this phenomenon that is not explored in this analysis is how these states brand themselves for other countries in the region, e.g. how does Russia 'brand' itself to Poland? In fact, there has been very little research done on this subject, creating a rather enticing lacuna for other scholars to fill (many thanks to Vlad Strukov for pointing this out).

[3] I distinguish between *international* and *global* as follows: 'international' refers to direct and indirect relations between states (countries), whereas 'global' refers to a variety of formal and informal interactions between states and non-state actors (NGOs, IGOs, citizen groups, media organizations, etc.).

of broadcast media during the late nineteenth century and its global ascendency in the twentieth century allowed for the mass mediation of political communication through sound (radio) and sight (TV), triggering a revolution in political messaging and how politics is 'delivered' (see McLuhan 1960; Innis 1972; Anderson 1991; Deibert 1997; Starr 2004).

This differs somewhat from the concept of *mediatization*, defined by Knut Lundby as the meta-processes by which 'everyday practices and social relations' are transformed by media technologies, producers and organizations (2009, x), though both share the principle that through the introduction of new forms of media/tion, 'the very nature of our being is altered' (Plate 2012, 157). In his essay 'Four Phases of Mediatization', Jesper Strömbäck states:

> The first aspect of the mediatization of politics is the degree to which the media constitute the most important or dominant source of information on politics and society. A second aspect is the degree to which the media are independent from political institutions in terms of how the media are governed. A third aspect is the degree to which the media content is governed by a political logic or by media logic. A fourth aspect, finally, is the degree to which political actors are governed by a political logic or by media logic. (2008, 234)

The complexity of the relationship between these two processes has been explored by a number of cultural studies scholars, most notably Jay David Bolter and Richard Grusin (2000) in *Remediation: Understanding New Media*. New media have played an important role in challenging and reconfiguring older media forms, resulting in the decline of the (print) newspaper and magazine industry (Eastland 2005), the remaking of journalism (McChesney 2012), the rise of infotainment *qua* news (Jebril, Albæk and de Vreese 2013) and the growth of the 'YouTube effect' in global politics (Naim 2007). Returning to the notion of political communication, it quickly becomes evident that as the medium through which political messaging changes, so does the reception of such content, or what might be called the 'experiencing of the real' (Erll and Rigney 2009, 4); this is particularly important in the production, representation and consumption of national image in the international marketplace of ideas.

Nation branding, a unique form of political communication, is a much-talked about concept, though one which suffers from a certain level of ambiguity.[4] For some scholars, nation branding is little more than a twenty-first century form of political propaganda (Marat 2009; Sussman 2012), while others view the shift toward the 'brand state' as a positive development which symbolizes the waning of belligerent geopolitics and '*Blut und Boden*' nationalism (van Ham 2001, 2005; Gaither 2007). Taking a middle road, certain scholars see nation branding as a peculiarly postmodern form of public diplomacy, a form of power projection that is reflective of the interconnectedness of the current global era (Fan 2008; Barr 2012). Perhaps the most concise definition of the practice, and one which touches on all of the above-mentioned perspectives, is as follows:

---

[4]In my own work on nation branding, I define the concept as 'the practice of brand management applied to a country's national image' (Saunders 2012, 51).

[T]he attempt by a government or its nominated subsidiaries to communicate simplistic and normally utopian premises regarding that nation to domestic and international audiences. These communications are based on the prevailing ideological dogma of the nation's elite and are tasked with consolidating or inculcating nationalism amongst the citizens of the nation, attracting capital from overseas and improving the strategic position of the source government. (Alexander 2013, 137)

As this definition shows, political economy and consumption are tightly bound to the national imaginary (Lewis 2011), the result of dramatic transformations of the past 25 years which have created an environment where nations are seen to be operating in a competitive marketplace. Whereas nations once sparred on the battlefield, now they 'brand' operating in a giant neoliberal supermarket where companies, tourists and potential allies all seek the best 'product'. Tools of nation branding are varied, but chief among these are the media. Nations, small and large, pay top dollar to place multi-page spreads in influential publications such as *Foreign Policy*, while other countries have created sleek promotional commercials which regularly air on international broadcast networks like Euronews. As I explore below, nations are adopting slogans, crafting logos and engaging in multivalent forms of promotion across multiple media platforms, including motion pictures, television programs and print campaigns.

For the countries of the post-Second World, mediation, mediatization and nation branding come together in a nexus. The overlap of the new media revolution and the retrenchment of totalitarian controls of freedom of the press and speech have produced vibrant and dynamic spaces where the mediation of politics and the mediatization of the political sphere have occurred rapidly and concurrently,[5] stimulating a cacophony of voices seeking to (re)make the nation via the Internet and other forms of digital media, though the 'voice of the state' remains strong, particularly via efforts at nation branding. Melissa Aronczyk, one the most prolific analysts of the concept of nation branding, argues that post-socialist Eastern Europe has functioned as a region where nation branding is seen as an essential armature for achieving political normalcy, economic integration and a 'return to Europe' (2013, 87). However, the 'Cold War' hangover has created an environment where jaundiced representations of the post-socialist 'East' have become fodder for external actors seeking fame and/or fortune in the digital expanse of cyberspace (see Saunders 2008a, 2008b, 2012). Reflecting the radically altered state of affairs made possible by the open and horizontal nature of the Internet (which does very little to privilege the state vis-à-vis other actors from within or outside of a given country or region) and fluidity of new media forms, this chapter is structured in such a way that gives roughly equal treatment to each of these three groups of actors.

---

[5]New media as a catalyst for (abortive) revolutions in post-Soviet space is a particularly important marker of this trend, with Internet communications (especially social networking sites such as Twitter) playing a key role in political upheavals and widespread protests in Ukraine (2004–2005, 2013), Belarus (2006, 2011), Moldova (2009), Armenia (2011), Russia (2011) and Azerbaijan (2013).

*The state, national image and new media in new Europe-Asia*

In new Europe-Asia, the state's relationship to new media technologies is both complex and conflicted. In Central Europe, the Internet is an important component in developing 'knowledge economies' and increasing marketization, though there are wide gaps between countries in terms of access, infrastructure and digital literacy across the region. Certain post-Soviet republics, most notably the Baltic States of Estonia, Latvia and Lithuania, rushed to embrace the opportunities provided by the digital revolution. The northernmost of these republics revels in its 'E-stonia' identity (Jansen 2012; Mansel 2013), proudly serving as a model of computerization and e-government, even using the devastating effects of the 2007 cyberwar with (pro-)Russian hackers as a platform for reinforcing its overall 'digital nation' status and openness to new media. However, the spectrum is wide, including states such as the resource-rich Central Asian republic of Turkmenistan, long ranked as the one of the worst places in the world in terms of connectivity and Internet freedom. Even today, the country's new media continue to languish in the shadow of its deceased dictator, Saparmurat Niyazov, though private Internet access is slowly becoming a reality (Lomov 2013). The largest of the newly independent states, the Russian Federation, embodies the myriad contradictions of new media in new Europe-Asia. With more than 53 million Internet users, Russia now outstrips that of any other European country in total users (Economist 2012); however, the Kremlin has instituted a variety of laws regulating Internet use and hampering the development of online civil society (Strukov 2009). Regardless of the level of Internet access and freedom, the vast majority of the post-Second World states have come to recognize the importance of projecting a positive image in cyberspace.

New media are an increasingly important plank in any nation branding strategy, buttressing print-, television- and sport-centric campaigns, as well as 'soft power', public diplomacy and social outreach programs conducted through other means (cultural institutes, educational initiatives, relief/development aid, etc.). Positive, online representation of a given state's key 'brand assets'—including good governance, suitability for foreign investment, personal safety, security of property, desirability of export products, uniqueness of cultural heritage/geography and popular attitudes to the 'outside world'—are steadily becoming part of the state's mandate in the twenty-first century. Strategies to burnish a state's brand can be as simple as developing an attractive, easy-to-navigate and content-rich web site or portal for the national government (president, parliament, ministry of tourism, etc.), though any digital branding program should also include substantial 'push' content alongside 'pull' content like web sites. Key examples of the latter include the US Embassy of the Republic of Kazakhstan's use of YouTube and Twitter feeds by Russia's President and now Prime Minister, Dmitry Medvedev (see also Yagodin's chapter in this volume). In both cases, the aim of such online branding was to present a 'fresh face' of the country to a wider, typically English-speaking public to counter negative stereotypes of the respective countries. In the case of Kazakhstan, the aim was to negate false perceptions of the nation stemming from Sacha Baron Cohen's long running 'Borat' parody project (see Saunders 2008a), whereas in the case of Russia, Medvedev—demonstrating at least some understanding of the mediatization of politics in the contemporary milieu—presented himself (and his brand) as a calculated paragon of stability and legality, as opposed to purported capricious-ness of Vladimir Putin's regime which made Russia a dangerous place to do business. Operating from a radically different position of power, another effort worth mentioning is the widely viewed 'The Young Europeans' YouTube video which sought to use professional production values and popular music alongside images of attractive, young Kosovars 'building' their new nation. Such efforts represent a digital salve for healing what nation branding expert Keith Dinnie (2008) labels the 'identity-image gap' wherein negative stereotypes trump facts on the ground.[6]

Indubitably, any state's primary web site represents the vanguard of any 'new media' brand, conveying the desired impressions of a given country to the outside world. An analysis of the primary 'state' web sites (based on Internet search engine returns) demonstrates that there are stark contrasts across the region when it comes to representation via the World Wide Web. Employing English-language search terms (i.e. 'Czech Republic' rather than 'Česká republika') and the search engine Google as the agent for locating the 'face' of the various post-Second World nations, a wide spectrum is evident.[7] At one end of the continuum, countries like Slovenia and Estonia strive to present themselves as safe, friendly and attractive destinations for travel and tourism (as well as investment), focusing on unique historical sites, a robust cultural heritage and an inviting geography. At the other end of the spectrum, countries like Russia and Tajikistan avoid representations associated with travel and tourism, instead focusing on the mediation of the quotidian activities of the government (presidents, parliaments and other official organs of the state). Using a set of factors including the Google search index, the quality of the web site (specifically the use of rich Internet applications (RIAs)) and standard branding tools like a logos and slogans, Table 1 presents a tentative nation brand presence on the Internet.

[6]In the United States, there is an increasing recognition among prestigious online media outlets that virtual nation branding is highly profitable. *The Washington Post*, for example, has 'dedicated staff' who assist countries in 'crafting and presenting' their 'nation brands' via the Internet.

[7]I recognize that any attempt to establish a single web site as a country's primary destination for Internet traffic is hugely problematic. That being stated, it is possible to use a rather limited set of protocols to deter-mine which site most web users will most easily locate in cyberspace when searching for a given county by name. In determining a given country's 'flagship' web site for this study, the following parameters were employed: 1) the prevalence of the web site in a search in the global lingua franca, i.e. English (Mauranen 2005), conducted via the Internet's most popular search engine, i.e. Google (eBiz 2013); 2) the web site is produced by 'a government or its nominated subsidiaries', in line with the previously referenced precepts of nation branding discussed by Alexander (2013); and 3) the web site provides a wide variety of informa-tion about the country in question, including information about its geography, people, heritage, culture and political system. It should be stated that all searches were conducted in the United States during the autumn months of 2013, which might possibly influence the search results due to Google geospatial and advertising protocols.

## TABLE 1
BRAND PRESENCE ON THE INTERNET

|  | *Countries* | *Characteristics* | *Primary sites* |
|---|---|---|---|
| Tier I | Czech, Latvia, Slovenia, Mongolia | Logo, slogan, travel-centric, use of RIAs, high 'googleability' (all metrics) | czech.cz, latvia.travel, slovenia.info, mongoliatourism.gov.mn |
| Tier II | Poland, Hungary, Slovakia, Estonia, Lithuania, Croatia, Bosnia, Montenegro, Romania, Bulgaria | Logo, slogan, travel-centric, use of RIAs, high 'googleability' (most metrics) | poland.pl, gotohungary.com, slovakia. travel, visitestonia.com, lietuva.lt, croatia.hr, bhtourism.ba, montenegro. travel, romaniatourism.com, bulgariatravel.org |
| Tier III | Serbia, Macedonia, Albania, Belarus, Ukraine, Georgia, Turkmenistan | Logo, slogan, travel-centric, use of RIAs, high 'googleability' (some metrics) | serbia.travel, exploringmacedonia. com, albaniantourism.com, belarus. by, traveltoukraine.org, georgia. travel, turkmenistan.gov.tm |
| Tier IV | Moldova, Armenia, Kazakhstan, Uzbekistan | Government-centric, but with a travel-centric site/page, no logo, no RIAs, low 'googleability' | moldova.md, gov.am, government. kz, gov.uz |
| Tier V | Russian Federation, Azerbaijan, Kyrgyzstan, Tajikistan | Government-centric/exclusive, no logo, no slogan, no RIAs, low 'googleability' | kremlin.ru, azerbaijan.az, president. kg, president.tj |

While the factors assessed in Table 1 are few in number, they do represent some of the most important inputs for shaping a country's image as it is consumed by outsiders.[8] For those seeking basic information about a country, high 'googleability' is important since any search where the country's flagship web site is indexed lower than its Wikipedia, CIA Factbook, or other general information site signifies a lack of country over the (virtual) country brand. Likewise, the importance of presenting the country as a destination is one of the most effective ways to achieve a perception of positive quiddity (see, for instnce, Anholt 2002; Gilmore 2002; Morgan, Pritchard and Morgan 2002; Papadopoulos and Heslop 2002; Risen 2005; Sinha Roy 2007), something which can be negatively attenuated when the main site focuses on governmental actions, presidential power or other state-centric factors (particularly when such representation reflects embedded Soviet-era norms).[9] The use of logos and mottos/slogans (see Table 2) also buttress this uniqueness, though often at the risk of undermining the gravitas a sovereign nation-state is expected to maintain in the realm of foreign affairs.

---

[8]According to Rusciano (2003), there are two important distinctions in national image as it is related to international affairs: the first being self-image (*Selbstbild*), or how a people view themselves in the global sphere, and the second being how foreign nations or cultures perceive a given country and its peoples (*Fremdbild*).

[9]In addition to factors such as GDP, (positive) press mentions and other 'real world' factors, firms like Bloom Consulting utilize Internet presence and ease of use in obtaining online information about a country in annual 'nation brand' ranking reports.

## TABLE 2
### COUNTRIES WITH SLOGANS

| | |
|---|---|
| Czech Republic | 'Hello Czech Republic' |
| Slovakia | 'Little Big Country' |
| Estonia | 'Positively Surprising' |
| Latvia | 'Best Enjoyed Slowly' |
| Slovenia | 'I Feel S*love*nia' |
| Croatia | 'The Mediterranean As It Once Was' |
| Montenegro | 'Wild Beauty' |
| Bosnia | 'The Heart Shaped Land' |
| Romania | 'Explore the Carpathian Garden' |
| Armenia | 'Favourite Destination since Noah's Time' |
| Turkmenistan | 'Turkmenistan: The Golden Age' |
| Mongolia | 'Go Nomadic. Experience Mongolia' |

Lastly, the use of high-quality and attractive graphics is important for promoting the 'likeability' factor in online nation branding, particularly in the current era of social networking and online sharing of tastes. In fact, a number of countries in the region have buttressed their web page strategies with regularly updated Facebook pages that explicitly seek 'likes' from strangers on the Internet (see Table 3), highlighting the emerging trend toward 'fandom' in international relations (see Dittmer and Dodds 2008). All the Baltic republics maintain vibrant Facebook pages such as 'Visit Estonia', 'If you like Latvia, Latvia likes you' and 'We love Lithuania'. Nation branding wunderkind Slovenia also has a well-liked page at 'Feel Slovenia'.

Generally speaking, the smaller and 'closer' a country is to the European Union (both physically and in terms of membership), the more likely it will pursue a strategy based on destination branding replete with market-based tactics for projecting a positive national image to the outside world. Larger countries, and specifically those with few or no links to the European Union, tend to put forth a highly statist brand, and one which values the quotidian exercise of power over the intangible allure of place and space; however, deeper interrogation produces a few interesting paradoxes. Mongolia, a country that existed in comparative isolation from the Soviet takeover in 1924 until 1991, maintains a presence in cyberspace that rivals that of the most-effectively branded countries of Central Europe, with excellent graphics all the other telltale signs of sound online strategy for brand development; however, unlike most of the other states in the

## TABLE 3
### OFFICIAL FACEBOOK PAGES OF POST-SECOND WORLD COUNTRIES (2014)

| Country | Title | Approximate number of 'likes' |
|---|---|---|
| Estonia | 'Visit Estonia' | 105,000 |
| Latvia | 'If you like Latvia, Latvia likes you' | 49,000 |
| Lithuania | 'We love Lithuania' | 71,000 |
| Slovenia | 'Feel S*love*nia' | 48,000 |

*Note*: Facebook pages that replicate Wikipedia content and which are not managed by the country in question are not listed.

region, Mongolia has no designs on 'Europeanizing' their brand via the World Wide Web, and instead seeks to locate Mongolia in 'Inner' or 'East' Asia.

Equally surprising is the case of Belarus. Due to its maintenance of many Soviet-era political and economic structures and the authoritarian regime of Alexander Lukoshenko (1994–present), Belarus might be expected to adhere to practices that would place it in a peer group with countries like Uzbekistan or Tajikistan. Yet, in terms of its web site presentation, the country compares favourably with countries that command well-developed nation brands. With its smart design and travel-centric format, the web site (belarus.by) presents the country as a sporting dynamo that tantalizes travellers and investors alike. Belarus is thus a cogent manifestation of Strömbäck's mediatization model, even if some of the country's image 'branding' smacks of (postmodern) political propaganda. Even more curious is the high-quality site run by Turkmenistan, which employs the slogan 'The Golden Age' and shows a sort of digitized map of the country and the surrounding region (a bit of high irony, given the country's historically abysmal Internet freedom ranking). It should be noted that while Russia performs poorly in the matrix above, this is more than compensated for through the growing presence and influence of RT (formerly known as Russia Today).[10] The international, English-language broadcaster's ability to distribute robust, multi-faceted and (most importantly) re-usable content about the Russian Federation (including shows on geography, culture, industry, investment and commerce) via its programming represents a paradigm that is simply beyond the reach of other countries in the post-Second World.

In addition to the use of web sites and Facebook pages, streaming video content via YouTube and other forums, and politicians' use of social networking tools like Twitter, there is a bevy of other established and emerging tools and techniques for building nation brands online. As nation branding campaigns from Western and developing countries have demonstrated, nation brands are often aided by positive feedback loops created by key places, products and celebrities (including sporting figures and national or quasi-national teams). By associating popular commodities (whatever their form) with the nation, actors involved in online public diplomacy campaigns have been able to augment their respective national profiles in cyberspace, and—in many cases—tie these brands closer to 'Europe'.

Online place branding has proved a powerful complement to overall nation branding campaigns, with representative examples that include the Bran Castle web site (Romania), the 'Magical Cracow' site (Poland) and the digital collection of the Hermitage Museum (Russian Federation).[11] Traditional product branding has also been a boon to certain countries, most notably the Czech Republic, which benefits from online advertising campaigns by its world-renowned lagers, including

---

[10]While less influential on the global stage, the Kremlin's attempts at extending its 'soft power' via what Gorham (2011) calls 'virtual rusophonia', including a Cyrillic-language domain realm, the potential creation of a 'sovereign Internet space', and other initiatives that can be viewed as a form of nation branding, though not in the typical sense.

[11]See, respectively, http://www.bran-castle.com/, http://www.krakow.pl/ and http://www.hermitage museum.org/.

the geographically bound beers *Pilsner Urquell/Plzeňský Prazdroj* (Plzeň) and *Budweiser/Budvar* (České Budějovice). In the realm of sports, a charismatic athlete—when profoundly attached to his or her country of origin—can strongly influence the *Fremdbild* of a given state, e.g. the Serbia tennis champion Novak Djokovic, whose positions on the Serbian Orthodox Church, Kosovo and other issues often emerge as firebrand topics in the blogosphere, and whose close identification with Serbian symbols effectively 'flags' the nation in cyberspace.[12] In such cases, celebrities become both mediators and mediations of the state they call home.

Increasingly, the Eurovision Song Contest is migrating from a traditional media event to the digital environment, particularly since 2000 when it was first broadcast via the Internet. According to some scholars, the spectacle has in fact become primarily a 'new media event' (see Highfield, Harrington and Bruns 2013) since one half of all votes for the winner are cast via cellular telephones ('televotes'). Thus, the competition serves as a way for Europeans to effectively use new media to 'vote' for a country in a way that is not replicated in any other field of human experience. A host of post-Soviet nations—including Estonia (2001), Latvia (2002), Ukraine (2004) and Azerbaijan (2011)—have used the platform to 'brand' their way out of Russia's shadow by winning the competition and subsequently hosting the following year. Importantly, Russia has also won since its independence (2008), as has Serbia (2007), with both countries using the competition to ameliorate their own national images. However, Eurovision is a double-edged sword that can harm as well as help with image campaigns, as is explored in the subsequent section of this chapter. Taken together, new media affects country brands in increasingly meaningful ways, touching on nearly every aspect of how other nations or cultures perceive specific countries and the post-Second World region as a whole.

*Alternative narrations and (national) culture jamming on the Internet:*
*the challenge from within*

New media have proved to be powerful tools in the arsenal of those who wish to challenge the authority of the state to define the nation, citizenship, society and community. As I have explored elsewhere (Saunders 2010), the advent of the Internet and the perpetual expansion of cyberspace have proved a boon for marginalized polities which seek to advance alternative narratives of the nation. Such developments have been proved particularly impactful in the post-Second World, as the deployment of new ICTs and the development of a complex new media ecosystem have neatly paralleled the ebbing of one-party systems and other constraints on free speech. Counterintuitively, the attempted and/or partial re-imposition of (quasi-)authoritarian limits on discourse in states like the Russian Federation, Ukraine and Armenia have only increased the use of new media, as traditional media platforms like television and newspapers have either come under the control of political elites or been made irrelevant by new market dynamics. With its speed, ease of use and low cost, the Internet has emerged as the broadcast

---

[12]For more on the concept of 'flagging the nation', see Michael Billig's *Banal Nationalism* (1995).

platform of choice for those wishing to 'speak truth to power', advance alternative world views or challenge the statist monopoly of what it means to be 'Russian', 'Hungarian', 'Tajik', etc.

Given the geographical peculiarities of cyberspace, the influence of the independent actors is not trivial. As any Google search will tease out, the state is not particularly well-positioned to guarantee its voice is the only (or, in some cases, even the dominant) one 'heard' in the digital realm. Non-governmental organizations, US government agencies (particularly the CIA), engaged netizens, crowd-sourced sites like Wikipedia and a panoply of other online media producers command comparatively strong positions on the Web vis-à-vis the provision of information and the shaping of national images of the world's nearly 200 sovereign states. In such a complex and protean environment, diasporans, national minorities, media-savvy activists and politically inclined, committed 'culture jammers' (see Carducci 2006; Warner 2007; Sandlin and Milam 2008)—recognizing the inherent commercialism and commodification-based dynamics of nation branding—can deeply impact, and even interrupt, the nation branding efforts of a particular state via global new media. Furthermore, in a world where popular culture is increasingly prone to transborder transmission and consumption, cultural producers—even those who lack explicit political or cultural agendas—also exert strong, often negative (or at least negating), influences on the (contrived) national images of their home countries.

While often seen as 'outside' the region, diasporas active on the Web would be more accurately framed as 'internal' actors when analyzing the mediation of nation brands via new media and the Internet, though certainly they occupy a liminal space (see Figure 1). Armed with linguistic, geographic and cultural competencies, digital diasporans (Brinkerhoff 2009) possess a level of legitimacy tantamount to online actors operating from the homeland. Given the low levels of access to certain ICTs in the region during the first decade after the end of the one-party system, the diaspora functioned as the default 'voice of the nation' for many countries.[13] In the case of Russia, it was only around the year 2000 that Russians in Russia surpassed Russians outside of Russia in terms of Internet presence (Schmidt and Teubener 2006). For countries like Turkmenistan (Annasoltan 2010) and Albania (Saunders 2005), this skewed distribution still holds, whereas in other cases, such as that of Romania, diasporan online politics has evolved far beyond that of the 'national' sphere (see Trandafoiu 2013), creating complex negotiations and contestations of *Selbst-* vs. *Fremdbild* online. Consequently, much of cyberspatial 'branding' of national cultures has come from sources outside the state (though this is changing over time as national governments attempt to 'manage' these disparate and often dissonant threads).

As Figure 1 is meant to suggest, there is a partly hierarchical structure to the 'new media nation brand'; however, governmental actions on the Internet are only part of the larger picture. While any given state enjoys a privileged position online, it must compete against a vast array of other information producers. In terms of bits and bytes, Russia's primary state web site is not significantly different from a well-designed page

---

[13]Notable exceptions would include low emigration, digitally developed states like the Czech Republic, Estonia and Slovenia (see Saunders 2009).

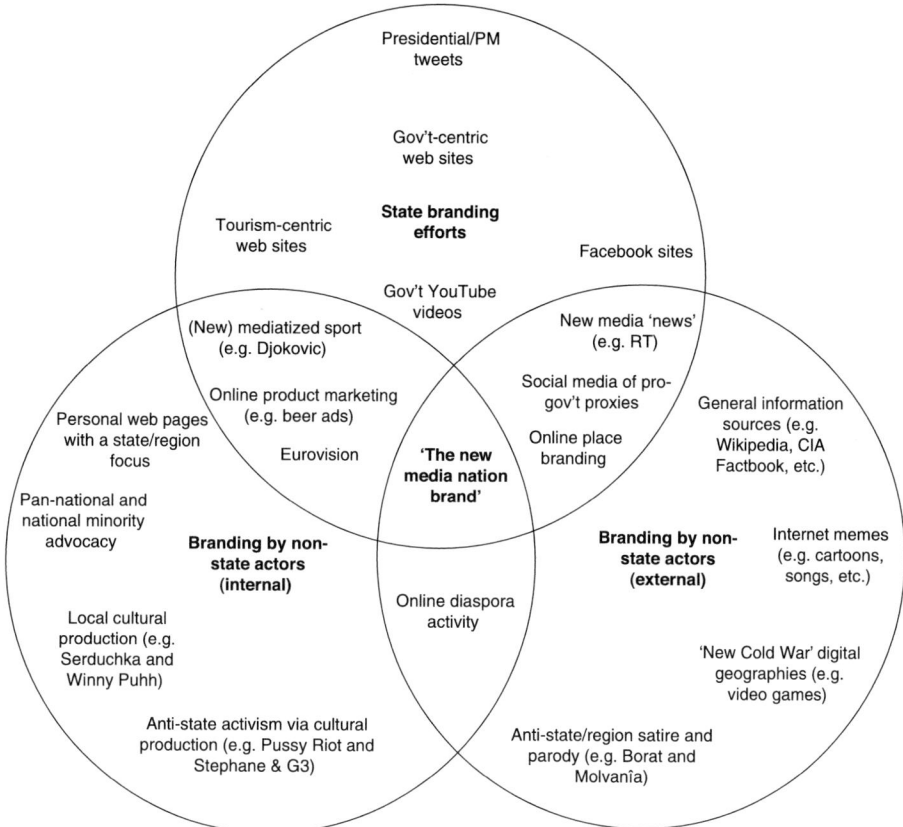

FIGURE 1 New Media Branding Meta-Categories with Examples.

developed by a non-state actor. In fact, a savvy web developer may be able to place his/her site above that by the Russian Federation by catering to the political economy of Google and other search engines (see Fuchs 2012).

Further complicating state-based efforts at presenting a unified and cogent 'brand' are sub-national actors, particularly national minorities and other groups wishing to put forth an alternative narration of the nation on the Internet. As a number of scholars have shown (Christensen 2003; Santianni 2003; Bernal 2006; Eriksen 2007; Candan and Hungar 2008; Ranganathan 2011), marginalized ethnic and religious minorities moved quickly to secure portions of digital space in the name of the 'nation', creating various 'narrational' problems (Bhabha 2001) which directly challenge the dominant, highly refined paradigm of national identity being articulated by their states of residence. As totalitarian controls on society abated in the late 1980s and early 1990s, the countries of the post-Second World saw a dramatic upsurge in national identity movements, resulting most notably in the breakup of the federal states of Czechoslovakia, Yugoslavia and the USSR, as well as ethnic conflict in Nagorno-Karabakh, Transnistria, the North Caucasus, Georgia and elsewhere. Such crises have also played out in the corners of cyberspace, with sustained campaigns to 're-write' history online.

Across the Russian Federation, Internet activists operating in the name of one or more of the country's one hundred ethnic minorities (Tatars, Chechens, Kalmyks, etc.) have used the Web, YouTube and other new media to market the uniqueness of their own culture apart from the pervasive image of Russia as exclusively Slavic, Eastern Orthodox and Russian-speaking, fracturing Russia's national image in the 'West' as it seeks to rebuild in the wake of the Soviet Union's dissolution in 1991. National minorities have done likewise in other countries in the region, including Estonia and Latvia (Russians), Bosnia (Serbs) and Bulgaria (Turks). Across East-Central Europe, Internet-savvy Roma have worked to establish a strong presence in cyberspace in an effort to both unite the widely scattered population and make use of the Web as a tool for achieving social justice in European Union member states and aspiring states (often resulting in extensive 'bad press' for countries like Slovakia where antiziganism is rampant).

In some cases, new media have even been weaponized in the name of the nation, as was the case in 2007 when the removal of the Bronze Soldier of Tallinn, a Soviet-era monument sanctified by the country's ethnic Russians, was relocated. A subsequent barrage of cyber-attacks by pro-Russian 'hactivists' caused serious damage to Estonia's Internet infrastructure and even triggered the development of a cyber-strategy for the North Atlantic Treaty Organization (NATO). Supranational actors operating online have similarly contributed to conflated, confusing and contentious 'brands' through efforts to maintain or restore elements of the past online, whether through attempts to sustain Yugoslavia or the USSR as cyber-imaginaries or online efforts to negate 'artificial' national boundaries between the Eastern Slavs or the Turkic peoples of Eurasia. Collectively, this new media cacophony works against the nation branding efforts of newly independent states like Ukraine, Kyrgyzstan and Montenegro and has made online nation branding a rather difficult undertaking indeed.

More recently, cultural producers in the region have migrated to new media platforms to articulate alternative narratives about their home countries, reinforcing the seemingly unbreakable link between art and politics. Perhaps the most emblematic example of the political-digital-popular cultural continuum is the viral nature of the dissident punk band Pussy Riot's music videos and the Internet-centric campaign to gain support for two of the band's jailed members following the infamous 'punk prayer' incident of February 2012.[14] During the summer of 2013, the band released a new video entitled 'Like a Red Prison' (Как в красной тюрьме), which condemns Russia as a corrupt petro-state like the United Arab Emirates and Putin as an 'ayatollah', (thus complicating Russia's efforts at 'Europeanization' of its brand). The video became a mainstream news story in the UK, the US and elsewhere, quickly moving beyond the realm of new media and the Russophone world. As Vlad Strukov (Strukov 2013) frames the issue, Pussy Riot functions as a 'global meme' via the Internet, a new

---

[14]In her essay on the 'punk prayer', Olga Voronina (2013) explores how the (re-)circulation of video of the protest 'represents a usurpation of virtual space' occupied by the Russian Orthodox Church, directly challenging its spiritual authority by linking its corrupt power politics of the Russian state.

'platform of performative appropriation' that directly informs the outside world about media, culture and politics within the Russian Federation.

Less overtly politicized, yet just as trenchant in terms of interrupting a nation in the throes of rebranding is the case of Verka Serduchka, Ukraine 2007 Eurovision entry (see Miazhevich's chapter in this volume). The creation of comedian and performer Andriy Danylko, Serduchka—a *Surzhyk*-speaking, railroad car steward—is a (post-)Soviet manifestation of a variety of tropes which counter Ukrainian efforts at distinguishing the newly independent state from its larger neighbour Russia (and the historical entity of the USSR). Though the character's embodiment of a 'common pool of cultural (mostly Soviet) symbols', Serduchka 'undermines' the Ukrainian nation brand (Miazhevich 2012, 1517). Danylko's transvestism and embrace of all things Sovietesque most sharply contrasted with the 'aggressively feminine' and 'quintessentially Ukrainian' former prime minister Yulia Tymoshenko, the 'Marianne' of 'Orange' (i.e. anti-Russian) Ukraine (Zhurzhenko 2013).

As alluded to earlier in this chapter, the Eurovision-YouTube nexus has served as a mechanism for intra-regional culture jamming, particularly on the part of the Georgian pop band Stephane & 3G—a pop act that was to have represented the South Caucasian nation at the 2009 competition in Moscow. However, the band's song 'We Don't Want a Put In'—clearly an anti-Russian anthem meant to rebuke the Kremlin for the 2008 invasion of the 'upstart' republic—was disqualified for its political content. The video, however, became an Internet sensation and served to further complicate Russia's efforts at branding itself as a force for stability in post-Soviet space. While less controversial than the anti-Putin Georgian entry, Estonia recently became the site of Eurovision kerfuffle when the outrageous rock band Winny Puhh generated a heated online debate in 2013 about presenting a 'poor image' of the otherwise-model-country to the outside world with its video for '*Meiecundimees üks Korsakov läks eile Lätti*' ('Our homeboy Korsakov just went to Latvia'), described by *Metal Insider* as the 'weirdest video of the year'.[15] Viewed from the perspective of critical geopolitics, which attempts to deconstruct the accepted 'truths' about the world around us, such developments demonstrate just how many inputs there are in the mediation of nation brands online.

### Anti-branding in trans-medial spaces: the challenge from without

New media have proved a double-edged sword for nation branders. Via the Internet and other ICTs, nation branders have gained access to myriad tools to present clean, simple and serious national images to the outside world. However, these benefits are often outweighed by other factors, what might be deemed 'biased' or 'empty' geographical knowledge (see Harvey 2005), which tend to be part and parcel of popular culture. While an effective tool in the right hands, online nation branding is particularly susceptible to manipulation and parody. The flat structure of cyberspace and the 'push-pull'

---

[15]Many Estonians, who are known for their reserved demeanour and quiet efficiency, took offence at the antics of the band which dresses in over-the-top costumes including those of mutant Wookiees and human-sized penises.

dynamics of the Internet privilege the voices of those who can entertain and/or offend, thus endowing satirists, comedians and other new media auteurs with enormous influence. Furthermore, digitally transmitted, visual-centric media such as video games and music videos promote and reinforce popular geographies of the region which are nearly impossible to mitigate through traditional branding efforts (on- or offline). A case in point: a search of 'Russia' on YouTube is likely to produce a top-hit for the 'We Love Russia' compilation videos from TwisterNederland, which portray staged and un-staged events including everything from overflowing toilets and drunkards in the streets to tanks on the beach and attempted suicides. Consequently, nation branders are frequently confronted with dire challenges to their efforts at positive representation, from country-specific parodies to a more generalized mediatization of the post-Second World as a decaying neverland populated by sex addicts, buffoons and madmen (Lipovetsky and Leiderman 2008). A full survey of new media influences on national brands is impossible due to space constraints, so instead the focus will be on three important areas: parodical/satirical videos, video games and Internet memes.

Perhaps the most famous of all satirists to lampoon the region is Sacha Baron Cohen. His long-running Borat shtick began as a spoof of the post-Soviet everyman, poking fun at Moldova then Albania before finally settling on Kazakhstan (Saunders 2008b). While much of Baron Cohen's work was situated in the realm of 'old media' (television, books, etc.), the British comedian was always careful to include a new media component to his Boratistan parody, most notably with his Kazakhstan-hosted web site which was eventually removed from the country's servers following Borat's appearance on the MTV Europe Music Awards in 2005. This was especially true in the lead-up to the premiere of his 2006 film *Borat: Cultural Learnings of America for Make Benefit Glorious Nation of Kazakhstan*, which saw Borat filming a YouTube video outside the Kazakhstani Embassy in Washington, DC in which he decried the activities of Roman Vassilenko, press attaché to the ambassador (and the point person for the management of Brand Kazakhstan in the United States), for lies and mistruths. In a testament to the power of Western cultural production, Baron Cohen antics eventually triggered a full-scale old and new media blitz (as well as a host of traditional public diplomacy efforts) by the Republic of Kazakhstan including the aforementioned videos about the country.

While Baron Cohen targeted a single country, the Australian comedian Santo Cilauro crafted an entirely new, but undeniably post-Second World, country to perpetrate similar acts of anti-branding. As a complement to the wildly popular faux travel guide *Molvanîa: A Land Untouched by Modern Dentistry* (2004), Cilauro created the eerily-Boratesque Zladko ('ZLAD!') Vladcik. The rocker was purportedly chosen to represent his (imaginary) country at Eurovision with the Eurotrash synth-ballad 'Elektronik Supersonik'. In addition to poking fun at the post-socialist obsession with Eurovision, the 3 minutes and 6 seconds spectacle engages in dozens of Eastern European stereotypes, including but not limited to bad English, awful haircuts, fashion crimes and an incurable obsession with the Space Race, nuclear missiles and autocratic power. In his follow-up single (and Internet video) 'I am the Anti-Pope', Vladcik returns with his signature mullet, but rather than the shiny, futuristic space-wear of his earlier video, he dons a priest's garb for his blasphemous send-up of Black Metal. In good branding fashion, both songs end with the nationalistic adage: 'Long Live Molvanîa!' More recently, the Eastern European brand has been influenced by the Serbian-British rapper and Internet sensation David Vujanic, whose satirical

'Bricka Bricka' persona embodies a host of Western European fears about the Balkan immigrant, from misogyny (his 'Run the World [Not Girls]' Beyoncé parody) to alcohol-fuelled violence ('Hard in Da Paint' [Eastern European Remix]). Bricka Bricka's unflinch-ingly racist video 'Ft. Kanyowski East – OTIS (Eastern European Remix)' displays the Soviet flag numerous times as the rapper spouts epithets about dozens of minorities, visu-ally affirming the 'uncivilized' nature of migrants for the former 'East'.

As an increasingly 'important political, social and cultural phenomenon' (Robinson 2012a, 414) and 'prevalent feature of popular culture in everyday life' (Ash and Gallacher 2011, 351), video games are perhaps the most important site for the 'empty' representation of post-Second World space via popular geographies. Whereas the above-mentioned online videos sculpt a gestalt of the new (post-)Soviet man, interac-tive—and particularly one-person shooter games—promote pervasive, pernicious ideas about place, while evincing deleterious 'political geographies nested in the aesthetics' (Vanolo 2012, 284). Perhaps the most blatant manifestation of this trend appears in the figmental country names used in video and computer games, especially those which involve a small team of well-trained commandoes infiltrating 'enemy territory'. The mythical sites of danger include pseudo-Central Asian locales like 'Adjikistan' (*SOCOM: U.S. Navy SEALs Combined Assault*), 'Aldestan' (*Command & Conquer: Generals*) and 'Serdaristan' (*Battlefield: Bad Company*). Other fabricated post-Soviet or Eastern Bloc countries go by the names of 'Chernarus' (a post-Soviet country in *ArmA 2*), 'Novistrana' (a breakaway Soviet region in *Republic: The Revolution*) and—perhaps most blatant in its nomenclature—'Soviet Unterzoegersdorf', labelled the 'last existing appanage republic of the USSR' (Surhone, Tennoe and Henssonow 2011).[16] With their interactive structures, immersive environments, addictive properties and trans-medial, cinematic practices (Norris and Strukov 2012), video games are particularly influential in the reinforcement of 'pre-existing ideas of "otherness"' (Ash and Gallacher 2011, 355). If nation branding endeavours to embed positive geographical associations in the minds of those who have never visited a given country, then there is perhaps no greater threat to this work than the 'persuasive potential' (Robinson 2012b, 513) of gaming geographies, which, in the case of post-Second World space, invariably present stilted depictions: darkened landscapes scarred by totalitarian-industrial excesses, restive populations plagued with ethnic strife, ungoverned territories rife with 'loose nukes' and other WMD and hollowed-out cities ravaged by the stillborn transition to market capi-talism. This post-Sovietesque 'ruin porn'—often echoed in traditional media representa-tions of the region (e.g. NBC Sports' coverage of the Olympics since the 1988 games in Seoul)—is a far cry from the 'reality' of the immaculate Nordic chalets of Estonia, the roaring hearths and happy halls of southern Bohemia, the pristine forests of the Russian Far East or the convivial seaside resorts of Crimea.

The last category of new media generated by Western cultural producers is that of the Internet meme, generally defined as any piece of highly contagious bit of culture that spreads through digital networks contributing to the international mediatization of

---

[16]Created by Ukrainian software developer GSC Game World, the popular personal computer game *S.T.A.L.K.E.R.: Shadow of Chernobyl* uses a real place in the region, i.e. Chernobyl (and in a subsequent ver-sion, Pripyat), as a 'zone' where greed prevails and mutants roam. While this game was generated locally, it has proved extremely popular on the international market through its English-language version.

places, spaces and peoples. Such units of culture are typically somewhat silly or frivolous in nature, though not always. During the 2004 US presidential election campaign, the visage of Vladimir Putin unexpectedly became an Internet sensation through the meta-meme 'Putin Sees Alaska'. The Putin meme evolved after Republican vice presidential candidate Sarah Palin, attempting to establish her foreign policy *bona fides* in a widely viewed interview with then-CBS news anchor Katie Couric, commented: 'As Putin rears his head and comes into the air space of the United States of America, where—where do they (sic) go? It's Alaska. It's just right over the border'. The meme included a variety of images, including one with Putin leering over the Arctic with his eyes set on Alaska (accompanied by a McCain–Palin logo) and another with Putin using ridiculously large binoculars (ostensibly to see the Alaskan shoreline). Coming when it did, this new media-fuelled phenomenon served to paint Russia both as a belligerently resurgent world power *and* the butt of American jokes, adding further validity to the notion of the 'new Cold War' (Ciută and Klinke 2010).[17] The Palin–Putin meme interestingly echoed the 'You Forgot Poland!' meme of the previous election, which grew out of President George Bush's retort to his Democratic challenger in a debate over the Iraq War.[18] The subsequent flurry of attention was capitalized by a wide variety of netizens to either make flippant references to Poland or to link other entities to the US's (easily forgotten) Eastern European ally.

In addition to these overtly political memes, the Internet has also spawned countless cultural memes associated with the post-Second World. The aforementioned Borat is responsible for many of these, including spoofs of Kazakhstan as the '#1 Exporter of Potassium' and dozens of other digital jibes at the Central Asian republic. The capacity of the Web to preserve older media has also had an impact on shaping the image of the post-Soviet nations, especially when the Internet gives new life to parodic American television commercials from the late Soviet era, e.g. the Wendy's 'Soviet Fashion Show' or Ukrainian comedian Yakov Smirnoff's Miller Lite advertisements.

### Parting thoughts: a new (media) world order?

While all nations face manifold challenges as they seek to manage their images in the digital world, the countries of the post-Second World have experienced particularly acute difficulties in the past two decades. Deeply embedded stereotypes and the forces of neoliberalism have combined to make online and other forms of digital public diplomacy a rather problematic affair. Smaller countries with well-developed digital infrastructures and more open/pluralistic political systems (Czech Republic, Latvia, Slovenia) have fared better than larger, less 'wired' states, particularly those with lower levels of political freedom (Russia, Ukraine, Serbia). That being stated, in nearly every country across the region, a substantive effort has been made to colonize some portion of cyberspace with

---

[17]Putin's own predilection for being photographed with large game and/or shirtless has also generated a flurry of memes across cyberspace.

[18]In the 30 September 2004 debate at the University of Miami, Bush chided his opponent Senator John Kerry for not including Poland in the list of the so-called 'Coalition of the Willing' that joined the US in its 2003 invasion of Iraq.

the aim of positively mediating the national brand. This serves as a testament to power of new media in new Europe-Asia. However, the steady elision of the border between digital and popular culture in the age of YouTube and Twitter is likely the beginning of what portends to be a perennial battle for control of national image on the Web. Undoubtedly, this 'new world order' of communication and information exchange (Deibert 1997) will have a direct impact on notions of citizenship and civic engagement, as well as continue to rewrite the rules of politics and politicking. With the conditions of the global information ecosystem in a state of almost constant flux, it is important to identify emergent trends in the increasingly interlinked relationship between new media and national image production/maintenance and—importantly—the position of the individual in relation to the imaginary of the (digital) nation. What is clear is that the deterritorialized, decentred, horizontal nature of cyberspace creates conditions where older forms of political propaganda cannot function with effectiveness, thus prompting states which once relied on comparatively closed information systems to adapt quickly to the new realities. Time will tell if government-sanctioned web sites, tweets and Facebook pages will emerge as the dominant forces in mediating the nation in cyberspace or if activists, satirists and hucksters will prevail ultimately turning the states of new Europe-Asia into mediatized effigies, caricatures and memes, devoid of substance and quiddity.

### Motion pictures

*Borat: Cultural Learnings of America for Make Benefit Glorious Nation of Kazakhstan.* 2006. Directed by Larry Charles. 20th Century Fox. 84 min.

### Video games

*ArmA 2.* Developed and distributed by Bohemia Interactive. Designed by Ivan Buchta. Released 17 June 2009.
*Battlefield: Bad Company.* Developed and distributed by EA Digital Illusions CE. Designed by Patrick Bach and Tobias Falk. Released 23 June 2008.
*Command & Conquer: Generals.* Developed and distributed by EA Games. Designed by Dustin Browder. Released 10 February 2003.
*Republic: The Revolution.* Developed by Elixir Studios. Designed by Demis Hassabis. Released 27 August 2003.
*SOCOM: U.S. Navy SEALs Combined Assault.* Developed by Zipper Interactive. Distributed by Sony Computer Entertainment. Released 7 November 2006.
*Soviet Unterzoegersdorf: Sector I.* Developed and designed by monochrom. Released 1 March 2009.

### Internet videos

'Borat – White House, Washington'. Published 18 October 2007. YouTube. http://www.youtube.com/watch?v=RuNjzh2p68g (last accessed 1 February 2014).
'BRICKA BRICKA – Hard In Da Paint (Eastern European Re-Mix)'. Published 14 April 2011. YouTube. http://www.youtube.com/watch?v=D9mM61pYCQs (last accessed 1 February 2014).
'EESTI LAUL 2013: Winny Puhh – Meiecundimees üks Korsakov läks eile Lätti'. Published 2 March 2013. YouTube. http://www.youtube.com/watch?v=2dllo85ZSUk (last accessed 1 February 2014).
'Elektronik Supersonik – Zlad'. Published 25 February 2006. YouTube. http://www.youtube.com/watch?v=lp_PIjc2ga4 (last accessed 1 February 2014).
'Ft. Kanyowski East – OTIS (Eastern European Remix)'. Published 11 September 2011. YouTube. http://www.youtube.com/watch?v=6hNDHXjsFWA (last accessed 1 February 2014).
'I Am The Antipope Zlad!' Published 11 October 2012. YouTube. http://www.youtube.com/watch?v=FxxhAS16vB0 (last accessed 1 February 2014).

'Kosovo – The Young Europeans'. Published 26 October 2009. YouTube. http://www.youtube.com/watch?v=dQRGHAdQjR0 (last accessed 1 February 2014).

'Pussy Riot – Как в красной тюрьме / Like a Red Prison'. Published 16 July 2013. YouTube. http://www.youtube.com/watch?v=qOM_3QH3bBw (last access 1 February 2014).

'We Don't Want a Put In'. Published 3 April 2009. YouTube. http://www.youtube.com/watch?v=LV1_s73f1-U (last accessed 1 February 2014).

'We Love Russia Compilation, August 2012 (TNL)'. Published 3 September 2012. YouTube. http://www.youtube.com/watch?v=WD2Dhnu3tP0 (last accessed 1 February 2014).

'Wendy's Commercial – Soviet Fashion Show'. Published 3 September 2006. YouTube. http://www.youtube.com/watch?v=5CaMUfxVJVQ (last accessed 1 February 2014).

'Yakov Smirnoff Miller Lite Commercial' (1985). Published 11 November 2007. YouTube. http://www.youtube.com/watch?v=AbP1DVeJCT0 (last accessed 1 February 2014).

## Sources

Alexander, Colin R. 2013. 'Review of *Branding Post-Communist Nations: Marketizing National Identities in the "New" Europe*'. *Digital Icons: Studies in Russian, Eurasian and Central European New Media* 9:135–139.

Anderson, Benedict. 1991. *Imagined Communities: Reflections on the Origin and Spread of Nationalism*. London: Verso.

Anholt, Simon. 2002. 'Nation-Brands and the Value of Provenance'. In *Destination Branding: Creating the Unique Destination Proposition*, edited by Nigel Morgan, Annette Pritchard and Roger Pride, 26–39. Burlington, MA: Butterworth-Heinemann.

——. 2007. *Competitive Identity: The New Brand Management for Nations, Cities and Regions*. Houndsmill, UK: Palgrave Macmillan.

Annasoltan. 2010. 'State of Ambivalence: Turkmenistan in the Digital Age'. *Digital Icons: Studies in Russian, Eurasian and Central European New Media* 3:1–13.

Aronczyk, Melissa. 2013. *Branding the Nation: The Global Business of National Identity*. Oxford and New York: Oxford University Press.

Ash, James, and Lesley Anne Gallacher. 2011. 'Cultural Geography and Videogames'. *Geography Compass* 5 (6):351–368.

Bachmann, Veit, and James D. Sidaway. 2008. '*Zivilmacht Europa*: A Critical Geopolitics of the European Union as a Global Power'. *Transactions of the Institute of British Geographers* 34:94–109.

Barr, Michael. 2012. 'Nation Branding as Nation Building: China's Image Campaign'. *East Asia: An International Quarterly* 29 (1):81–94.

Bernal, Victoria. 2006. 'Diaspora, Cyberspace and Political Imagination: The Eritrean Diaspora Online'. *Global Networks* 6 (2):161–179.

Bhabha, Homi K. 2001. *Narrating the Nation*. Edited by Vincent P. Pecor, *Nations and Identities: Classic Readings*. Malden, MA: Blackwell Publishers.

Billig, Michael. 1995. *Banal Nationalism*. London: Sage Publications.

Bolter, Jay David, and Richard Grusin. 2000. *Remediation: Understanding New Media*. Cambridge, MA: MIT Press.

Bowles, Anna. 2005. 'ReNut: A Cyberian Adventure'. *Russian Life* 48 (2):41–47.

Brinkerhoff, Jennifer M. 2009. *Digital Diasporas: Identity and Transnational Engagement*. Cambridge: Cambridge University Press.

Candan, Menderes, and Uwe Hungar. 2008. 'Nation Building Online: A Case Study of Kurdish Migrants in Germany'. *German Policy Studies* 4 (4):125–153.

Carducci, Vince. 2006. 'Culture Jamming: A Sociological Perspective'. *Journal of Consumer Culture* 6 (1):116–138.

Christensen, Neil Blair. 2003. *Inuit in Cyberspace: Embedding Offline, Identities Online*. Copenhagen: Museum Tusculanum Press.

Cilauro, Santo, Tom Gleisner and Rob Sitch. 2004. *Molvanîa: A Land Untouched by Modern Dentistry*. Woodstock and New York: The Overlook Press.

Ciută, Felix, and Ian Klinke. 2010. 'Lost in Conceptualization: Reading the "New Cold War" with Critical Geopolitics'. *Political Geography* 29:323–332.

Deibert, Ronald J. 1997. *Parchment, Printing and Hypermedia: Communication in World Order Transformation*. New York: Columbia University Press.

Dinnie, Keith. 2008. *Nation Branding: Concepts, Issues, Practice*. Oxford: Butterworth-Heinemann.

Dittmer, Jason. 2007. 'The Tyranny of the Serial: Popular Geopolitics, the Nation, and Comic Book Discourse'. *Antipode* 39 (247–268).

Dittmer, Jason, and Klaus Dodds. 2008. 'Popular Geopolitics Past and Future: Fandom, Identities and Audiences'. *Geopolitics* 13 (3):437–445.

Dodds, Klaus. 2008. 'Hollywood and the Popular Geopolitics of the War on Terror'. *Third World Quarterly* 29 (8):1621–1637.

Downing, John D. H. 2013. ' "Geopolitics" and "the Popular": An Exploration'. *Popular Communication* 11 (1):7–16.

Dunnett, Oliver. 2009. 'Identity and Geopolitics in Hergé's *Adventures of Tintin*'. *Social & Cultural Geography* 10 (5):583–599.

Duvall, Raymond, and Latha Varadarajan. 2003. 'On the Practical Significance of Critical International Relations Theory'. *Asian Journal of Political Science* 11 (2):75–88.

Eastland, Terry. 2005. 'Starting Over'. *Wilson Quarterly* 29 (2):40–47.

eBiz. 2013. *Top 15 Most Popular Search Engines | December 2013* 2013 [cited 26 December 2013]. Available from http://www.ebizmba.com/articles/search-engines.

Economist. 2012. 'Europe's Great Exception'. 19 May, 69–70.

Emory, Margot, and Benjamin J. Bates. 2001. 'Creating New Relations: The Internet in Central and Eastern Europe'. In *Cyberimperialism?: Global Relations in the New Electronic Frontier*, edited by Bosah Ebo, 93–109. Westport, CT: Praeger.

Eriksen, Thomas Hylland. 2007. 'Nationalism and the Internet'. *Nations & Nationalism* 13 (1):1–17.

Erll, Astrid, and Ann Rigney. 2009. 'Introduction: Cultural Memory and Its Dynamics'. In *Mediation, Remediation, and the Dynamics of Cultural Memory*, edited by Astrid Erll and Ann Rigney, 1–13. Berlin: Walter de Gruyter.

Fan, Ying. 2008. 'Soft Power: Power of Attraction or Confusion?' *Place Branding and Public Diplomacy* 4:147–158.

Fuchs, Christian. 2012. 'A Contribution to the Critique of the Political Economy of Google'. *Fast Capitalism* 8 (1): no pp. Available online at http://www.uta.edu/huma/agger/fastcapitalism/8_1/fuchs8_1.html.

Gaither, Thomas Kenneth. 2007. *Building a Nation's Image on the World Wide Web: A Study of the Head of State Web Sites of Developing Countries*. Amherst, NY: Cambria Press.

Gilmore, Fiona. 2002. 'Spain—A Success Story of Country Branding'. *Brand Management* 9 (4–5): 281–293.

Gorham, Michael. 2011. 'Virtual Rusophonia: Language Policy as "Soft Power" in the New Media Age'. *Digital Icons: Studies in Russian, Eurasian and Central European New Media* 5:23–48.

Grayson, Kyle, Matt Davies and Simon Philpott. 2009. 'Pop Goes IR? Researching the Popular Culture-World Politics Continuum'. *Politics* 29 (3):155–163.

Gurevitch, Michael, Stephen Coleman and Jay G. Blumler. 2009. 'Political Communication—Old and New Media Relationships'. *Annals of the American Academy of Political and Social Science* 625:164–181.

Guzzini, Stefano, and Anna Leander. 2006. *Constructivism and International Relations: Alexander Wendt and His Critics*. New York: Routledge.

Harvey, David. 2005. 'The Sociological and Geographical Imaginations'. *International Journal of Politics, Culture, and Society* 18:211–255.

Herron, Erik S. 1999. 'Democratization and the Development of Information Regimes'. *Problems of Post-Communism* 46 (4):56–68.

Highfield, Tim, Stephen Harrington and Axel Bruns. 2013. 'Twitter as a Technology for Audiencing and Fandom: The #Eurovision Phenomenon'. *Information, Communication & Society* 16 (3):315–339.

Hitchcock, Peter. 2003. *Imaginary States: Studies in Cultural Transnationalism*. Champaign, IL: University of Illinois Press.

Hopf, Ted. 1998. 'The Promise of Constructivism in International Relations Theory'. *International Security* 23 (1):171–200.

Imre, Anikó. 2009. *Identity Games: Globalization and the Transformation of Media Cultures in the New Europe*. Cambridge, MA: MIT Press.

Innis, Harold A. 1972. *Empire and Communications*. Toronto: University of Toronto Press.

Jansen, Sue Curry. 2008. 'Designer Nations: Neo-liberal Nation Branding – Brand Estonia'. *Social Identities* 14 (1):121–142.

——. 2012. 'Redesigning a Nation: Welcome to E-stonia, 2001–2018'. In *Branding Post-Communist Nations: Marketizing National Identities in the 'New' Europe*, edited by Nadia Kaneva, 79–98. New York and London: Routledge.

Jebril, Nael, Erik Albæk and Claes H. de Vreese. 2013. 'Infotainment, Cynicism and Democracy: The Effects of Privatization vs Personalization in the News'. *European Journal of Communication* 28 (2):105–121.

Johnson, Erica, and Beth E. Kolko. 2010. 'E-government and Transparency in Authoritarian Regimes'. *Digital Icons: Studies in Russian, Eurasian and Central European New Media* 3:15–48.

Kalathil, Shanthi, and Taylor C. Boas. 2003. *Open Networks, Closed Regimes: The Impact of the Internet on Authoritarian Rule*. Washington: Carnegie Endowment for International Peace.

Kaneva, Nadia. 2012. 'Nation Branding in Post-Communist Europe: Identities, Market, and Democracy'. In *Branding Post-Communist Nations: Marketizing National Identities in the 'New' Europe*, edited by Nadia Kaneva, 3–22. New York and London: Routledge.

Kitchin, Robert M. 1998. 'Towards Geographies of Cyberspace'. *Progress in Human Geography* 22 (3): 385–406.

Lewis, Nick. 2011. 'Packaging Political Projects in Geographical Imaginaries: The Rise of Nation Branding'. In *Brands and Branding Geographies*, edited by Andy Pike, 264–287. Cheltenham, UK and Northampton, MA: Edward Elgar Publishing.

Lipovetsky, Mark, and Daniil Leiderman. 2008. 'Angel, Avenger, or Trickster? The "Second-World Man" as the Other and the Self'. In *Russia and Its Other(s) on Film: Screening Intercultural Dialogue*, 199–219. Basingstoke: Palgrave Macmillan.

Lomov, Anton. 2013. 'Isolated Turkmenistan Bows to Internet Age'. *Agence France Presse*, 25 July.

Lundby, Knut. 2009. *Mediatization: Concept, Changes, Consequences*. Bern: Peter Lang.

Mansel, Tim. 2013. 'How Estonia Became E-stonia'. *BBC News*, 16 May.

Marat, Erica. 2009. 'Nation Branding in Central Asia: A New Campaign to Present Ideas about the State and the Nation'. *Europe-Asia Studies* 61 (7):1123–1136.

Mauranen, Anna. 2005. 'English as Lingua Franca: An Unknown Language?' In *Identity, Community, Discourse: English in Intercultural Settings*, edited by Giuseppina Cortese and Anna Duszak, 269–293. Bern: Peter Lang.

McChesney, Robert W. 2012. 'Farewell to Journalism? A Time for a Rethinking'. *Journalism Studies* 13 (5–6):682–694.

McLuhan, Marshall. 1960. 'Effects of the Improvements of Communication Media'. *Journal of Economic History* 20 (4):566–575.

Miazhevich, Galina. 2012. 'Ukrainian Nation Branding Off-line and Online: Verka Serduchka at the Eurovision Song Contest'. *Europe-Asia Studies* 64 (8):1505–1520.

Morgan, Nigel, Annette Pritchard and Rachel Morgan. 2002. 'New Zealand, 100% Pure. The Creation of a Powerful Niche Destination Brand'. *Brand Management* 9 (4/5):335–354.

Morozov, Evgeny. 2009. 'Moldova's Twitter Revolution'. *Foreign Policy.com*: http://neteffect.foreignpolicy.com/posts/2009/04/07/moldovas_twitter_revolution.

——. 2011. *The Net Delusion: How Not to Liberate the World*. London: Allen Lane.

Mungiu-Pippidi, Alina, and Igor Munteanu. 2009. 'Moldova's "Twitter Revolution" '. *Journal of Democracy* 20 (3):136–142.

Naim, Moises. 2007. 'The YouTube Effect'. *Foreign Policy* 158:103–104.

Nikiporets-Takigawa, Galina. 2013. 'Tweeting the Russian Protests'. *Digital Icons: Studies in Russian, Eurasian and Central European New Media* 9:1–25.

Norris, Stephen M., and Vlad Strukov. 2012. 'Editorial: Cinegames: Convergent Media and the Aesthetic Turn'. *Digital Icons: Studies in Russian, Eurasian and Central European New Media* 8:no pp.

Nune, Alfred. 2003. 'Internet and Albania: a Paradoxical Ambivalence'. In *New Media in Southeast Europe*, edited by Orlin Spassov and Christo Todorov, 31–49. Sofia: Southeast European Media Centre.

Ó Tuathail, Gearóid. 1996. *Critical Geopolitics: The Politics of Writing Global Space*. Minneapolis: University of Minnesota Press.

Olins, Wally. 2002. 'Branding the Nation—The Historical Context'. *Brand Management* 9 (4/5):241–248.

Papadopoulos, Nicholas, and Louise Heslop. 2002. 'Country Equity and Country Branding: Problems and Prospects'. *Brand Management* 9 (4/5):294–314.

Plate, S. Brent. 2012. 'Introduction: The Mediation of Meaning, or Re-Mediating McLuhan'. *Cross Currents* 62 (2):156–161.

Power, Marcus, and David Campbell. 2010. 'The State of Critical Geopolitics'. *Political Geography* 29:243–246.

Price, Richard, and Christian Reus-Smit. 1998. 'Dangerous Liaisons? Critical International Theory and Constructivism'. *European Journal of International Relations* 4 (3):259–294.

Purcell, Darren, Melissa Scott Brown and Mahmut Gokmen. 2010. 'Achmed the Dead Terrorist and Humor in Popular Geopolitics'. *Geoforum* 75:373–385.

Ranganathan, Maya. 2011. *Eelam Online: The Tamil Diaspora and War in Sri Lanka*. Newcastle: Cambridge Scholars Publishing.

Risen, Clay. 2005. Branding Nations. *New York Times Magazine*, 11 December, 61.

Robinson, Nick. 2012a. 'Video Games and Violence: Legislating on the "Politics of Confusion" '. *Political Quarterly* 83 (2):414–423.

——. 2012b. 'Videogames, Persuasion and the War on Terror: Escaping or Embedding the Military-Entertainment Complex?' *Political Studies* 60 (3):504–522.

Rusciano, Frank Louis. 2003. 'The Construction of National Identity—A 23-Nation Study'. *Political Research Quarterly* 56 (3):361–366.

Sandlin, Jennifer A., and Jennifer L. Milam. 2008. ' "Mixing Pop (Culture) and Politics": Cultural Resistance, Culture Jamming, and Anti-Consumption Activism as Critical Public Pedagogy'. *Curriculum Inquiry* 38 (3):323–350.

Santianni, Michael. 2003. 'The Movement for a Free Tibet: Cyberspace and the Ambivalence of Cultural Translation'. In *The Media of Diaspora*, edited by Karim H. Karim, 189–202. London: Routledge.

Saunders, Robert A. 2005. 'Virtual Irredentism? The Redemption and Reification of the Albanian Nation in Cyberspace'. *Albanian Journal of Politics* 1 (2):137–165.

——. 2008a. 'Buying into Brand Borat: Kazakhstan's Cautious Embrace of Its Unwanted "Son," '. *Slavic Review* 67 (1):63–80.

——. 2008b. *The Many Faces of Sacha Baron Cohen: Politics, Parody, and the Battle over Borat*. Lanham, MD: Lexington Books.

——. 2009. 'Wiring the Second World: The Geopolitics of Information and Communications Technology in Post-Totalitarian Eurasia'. *Digital Icons: Studies in Russian, Eurasian and Central European New Media* 1:1–24.

——. 2010. *Ethnopolitics in Cyberspace: The Internet, Minority Nationalism, and the Web of Identity*. Lanham, MD: Lexington Books.

——. 2012. 'Brand Interrupted: The Impact of Alternative Narrators on Nation Branding in the Former Second World'. In *Branding Post-Communist Nations: Marketizing National Identities in the 'New' Europe*, edited by Nadia Kaneva, 49–78. New York and London: Routledge.

Schmidt, Henrike, and Katy Teubener. 2006. ' "Our RuNet"?: Cultural Identity and Media Usage'. In *Control + Shift: Public and Private Uses of the Russian Internet*, edited by Henrike Schmidt, Katy Teubener and Natalja Konradov, 14–21. Norderstedt: Books on Demand.

Semetko, Holli A., and Natalya Krasnoboka. 2003. 'The Political Role of the Internet in Societies in Transition: Russia and Ukraine Compared'. *Party Politics* 9 (1):77–104.

Sharp, Joanne P. 2000. *Condensing the Cold War: Reader's Digest and American Identity*. Minneapolis: University of Minnesota Press.

Sinha Roy, Ishita. 2007. 'Worlds Apart: Nation-branding on the National Geographic Channel'. *Media, Culture & Society* 29 (4):569–592.

Starr, Paul. 2004. *The Creation of the Media: The Political Origins of Modern Communications*. New York: Basic Books.

Strömbäck, Jesper. 2008. 'Four Phases of Mediatization: An Analysis of the Mediatization of Politics'. *International Journal of Press/Politics* 13 (3):228–246.

Strukov, Vlad. 2009. 'Russia's Internet Media Policies: Open Space and Ideological Closure'. In *The Post-Soviet Russian Media: Conflicting Signals*, edited by Birgit Beumers, Stephen Hutchings and Natalya Rulyova, 208–221. London and New York: Routledge.

——. 2013. 'Pussy Riot: From Local Appropriation to Global Documentation, or Contesting the Media System'. *Digital Icons: Studies in Russian, Eurasian and Central European New Media* 9:87–97.

Surhone, Lambert M., Mariam T. Tennoe and Susan F. Henssonow. 2011. *Soviet Unterzoegersdorf*. Beau Bassin, Mauritius: Betascript Publishing.

Sussman, Gerald. 2012. 'Systemic Propaganda and State Branding in Post-Soviet Eastern Europe'. In *Branding Post-Communist Nations: Marketizing National Identities in the 'New' Europe*, edited by Nadia Kaneva, 23–48. New York and London: Routledge.

Szondi, György. 2007. 'The Role and Challenges of Country Branding in Transition Countries: The Central and Eastern European Experience'. *Place Branding and Public Diplomacy* 3 (1):8–20.

Trandafoiu, Ruxandra. 2013. *Diaspora Online: Identity Politics and Romanian Migrants*. Oxford and New York: Berghahn Books.

Uffelmann, Dirk. 2011. 'Post-Russian Eurasia and the Proto-Eurasian Usage of the Runet in Kazakhstan: A Plea for a Cyberlinguistic Turn in Area Studies'. *Journal of Eurasian Studies* 2 (2):172–183.

van Ham, Peter. 2001. 'The Rise of the Brand State: The Postmodern Politics of Image and Reputation'. *Foreign Affairs* 80 (5):2–7.

——. 2005. 'Branding European Power'. *Place Branding* 1 (2):122.

——. 2010. *Social Power in International Politics*. Abingdon, UK and New York: Routledge.

Vanolo, Alberto. 2012. 'The Political Geographies of Liberty City'. *City* 16 (3):284–298.

Voronina, Olga. 2013. 'Pussy Riot Steal the Stage in the Moscow Cathedral of Christ the Saviour: Punk Prayer on Trial Online and in Court'. *Digital Icons: Studies in Russian, Eurasian and Central European New Media* 9:69–85.

Warf, Barney. 2009. 'The Rapidly Evolving Geographies of the Eurasian Internet'. *Eurasian Geography & Economics* 50 (5):564–580.

Warner, Jamie. 2007. 'Political Culture Jamming: The Dissident Humor of "The Daily Show with Jon Stewart" '. *Popular Communication* 5 (1):17–36.

Zehfuss, Maja. 2003. *Constructivism in International Relations: The Politics of Reality*. Cambridge and New York: Cambridge University Press.

Zhurzhenko, Tatiana. 2013. *Yulia Tymoshenko's Two Bodies*. Eurozine 2013. Available from http://www.eurozine.com/articles/2013-06-25-zhurzhenko-en.html.

# Contesting Bulgaria's Past Through New Media: Latin, Cyrillic and Politics

## ORLIN SPASSOV

### Abstract

This essay investigates the conflict between two different ways of writing in Bulgarian on the web: using the standard Cyrillic alphabet and using the Latin script. Initially, the reason for using the Latin script was purely technical: the absence of appropriate software for decoding Cyrillic fonts. However, the Latin script remained popular even after the encoding problems were solved, acquiring new ideological meanings and provoking political controversies. This essay discusses the subcultural, cultural and political consequences of these developments.

THE WIDESPREAD USE OF COMPUTER-MEDIATED communication, facilitated by the growing availability of new media on the global scale, raises a number of issues about the place of various languages in national and transnational cyberspaces and about the influence of the internet on language use. There has been extensive research on the way the technological revolution has been accompanied by important social, cultural and political changes (Castells 2000, 2001, 2009; Poster 2001; Jenkins 2006; Shirky 2008). Relatively less attention, however, has been paid to the role of the internet with respect to language in a socio-cultural perspective, and more specifically in relation to discursive practices in individual countries. Some of the important studies in this particular field focus on the 'linguistic revolution', caused by the expansion of the internet, and follow the effect of 'netspeak' on local languages (Crystal 2001, 2011); others are dedicated to computer-mediated communication on the multilingual internet, to the use of English as a *lingua franca* and to the specific experience in online communication in countries with languages other than English (Danet & Herring 2003). Some scholars study the interaction between technology and language, as well as the developments allowing the internet and mobile phones to change profoundly interpersonal relationships (Baron 2010).

In this essay, interest is focused on a more specific problem: the influence of the internet and mobile phones on the use of traditional Cyrillic script in Bulgaria. The process involves intensive parallel use of Latin script in online communication, with all the ensuing cultural and political consequences. Of course, such a problem is not typical of Bulgaria alone. In the early years of computer development, the established

167

American Standard Code for Information Interchange (ASCII) unsurprisingly adopted the English alphabet as a base for the character-encoding scheme (Yates 1996, p. 113). Major world languages such as Arabic, Chinese and Japanese initially were not supported by computer technology at all (Yates 1996, p. 114). Later, the rapid development of the internet and the influx of numerous languages in it led to the urgent need for the encoding of characters and symbols outside the English alphabet in the new medium. Thus, for quite a considerable period of time, at least until 2007, when ASCII was surpassed by the UTF-8 (Unicode Transformation Format) standard, many languages using non-Roman script encountered difficulties when reproducing their graphics online. The problem more specifically concerned communication in Bulletin Board Systems, email correspondence, communication in internet relay chat (IRC) and in online forums. That is why in this period internet users in many countries were frequently forced to 'Romanise' the computer-mediated communication they used. The problem concerned mostly countries with a script totally differing from the English one such as, for example, Greece and Russia. Although in many cases there were more or less successful adaptations of ASCII to local writing systems, such specific character sets rarely resolved the problems in general. They began to be discarded with the widespread application of the Unicode standard, which provided a comprehensive presentation of most of the world's writing systems on the internet for the first time.

In fact, the use of Roman script by languages that were in possession of their own writing system was a practice familiar long before the emergence of the internet. Transliteration was necessitated for various cultural, political and commercial reasons and was adopted as a linguistic instrument all over the world. However, the Romanisation of a written language has frequently encountered considerable resistance and has been perceived as a threat to local cultural identity. This ideological controversy systematically reoccurs throughout the history of communication with the invention of every new communication media. Usually, the language of the country in which a certain technology is first developed gains exclusive privileges in the early period of its use. Such is the story of the Morse code, of the telegraph and the telex, which were controlled by English-speaking countries for a long time and began to offer support to other languages only decades later (Yates 1996, p. 114). On the basis of analogous examples related to the use of the internet, some authors speculate on the idea of 'typographic imperialism' (Danet & Herring 2003), deriving it from the more widely discussed concept of 'linguistic imperialism' (Phillipson 1992; Phillipson & Skutnabb-Kangas 2001). The forms of 'ASCII-isation' of languages other than English give grounds to conclude that such standards privilege certain speakers, communities and countries and hamper others in terms of the use of new media (Yates 1996; Palfreyman & Al-Khalil 2003; Warschauer *et al.* 2002).

Although, as we have seen, the technological problems have generally been overcome, the use of Latin script is still frequently encountered in online communication in many countries where this is not a standard script. The reasons vary: in some cases, the QWERTY keyboard layout is more familiar and easier to use; in others, youth subcultures find in this discursive practice an appropriate means of presenting their identity, or simply an easier (because of the lack of institutional control) way of written expression. As a result, numerous systems for automatic transliteration between Romanised versions and original fonts have appeared. In most cases, the use of Latin characters as a substitute for the original local writing system is limited to short texts exchanged in the course of online communication, or in the form of mobile telephone text messages. Typically these are exclusively informal types of communication. Although they are widespread, these practices are rarely considered as neutral. They are a direct expression of the tension between local and global identities, and are often criticised from conservative and nationalistic positions, most frequently along the lines of resistance to the processes of Westernisation or Americanisation. At the same time, in many cases the Latinisation of computer-mediated communication is encouraged by the wide use of Romanised texts offline in advertisement campaigns and in public inscriptions of institutional names (for example, in the names of banks, bars and restaurants). The cultural homogenisation imposed through popular culture and mass consumption is also an important factor in the increasingly easier 'interchangeability' of fonts in many parts of the world.

The next section of the essay explores how these common tendencies affect the group of Slavonic languages, to which Bulgarian belongs. The West Slavonic languages (Polish, Czech and Slovak) use Latin script. From the South Slavonic group, Slovenian and Croatian also adhere to this practice. In Serbia, Cyrillic and Latin have an equal status as official writing systems. Bosnian and Montenegrin also allow usage of both alphabets, although Cyrillic is gradually being ousted by Latin (Hansen 2011; Hunt 2011; Lowen 2010). This is because after the disintegration of Yugoslavia, preferences for the Latin script have been frequently charged with political meanings. They are used to emphasise the difference between the members of the former federation and to distinguish the independence of languages like Croatian and Montenegrin. In East Slavonic languages (Russian, Ukrainian and Belarusian), as well as in Bulgarian and Macedonian, only Cyrillic is used traditionally. It is precisely in this last group that the parallel use of Latin script in online communication is the subject of heated debates and tension revolving around radically opposed arguments 'for' and 'against'. The only exception is Macedonia, where in parallel to the official use of Cyrillic, the bi-scriptal system inherited from the time of pre-1990s Yugoslavia is used widely in online communication.

Again on the territory of South-Eastern Europe, the case of Greek is particularly symptomatic. As a European language with its own unique graphemic system, Greek suffered severely from the problem of computer encoding of its specific characters in the early years of the internet. To avoid difficulties, it was frequently necessary to resort to transliteration with the help of the Latin script. This practice is known as 'Greeklish' (Greek + English), and sometimes as 'ASCII Greek'. In spite of the fact that the software problems were resolved later, the use of Greeklish remained, albeit on a smaller scale. This is largely due to the lack of adherence to normative transliteration and the resulting 'facilitation' in the use of the written language: the dispensation of the need for strict adherence to orthography or grammar (Translatum 2011). Like similar cases in other countries, Greeklish is not perceived as a problem simply related to an alternative way of writing. The subject has been broadly commented on as a 'wider socio-cultural and ideological phenomenon', and has become an important part of the discussions about the role of language as a key element of national identity in present-day Greece (Koutsogiannis & Mitsikopoulou 2003).

### Latin compared to Cyrillic: the Bulgarian experience

In the past 22 years the issues concerning the use of the Latin alphabet for Bulgarian have been the subject of an intense and highly politicised debate in Bulgaria, and they clearly reflect some of the key language policies adapted and practised in the country. This essay will look briefly at the main developments in this debate, as well as at the effects of the clash between Cyrillic and Latin.

Three dimensions to this problem have been identified: technological, cultural and political. These three dimensions are in fact intertwined, and each one of them problematises the relation between language and national identity. Before we go into details, it is important to note the most general developments against the backdrop of which the dispute about the use of the Latin script has unfolded.

Our consideration of the technological context reveals that after the advent of email and the internet in Bulgaria in 1991 and 1992, respectively, problems arose with writing in Cyrillic in this new electronic medium. No computer applications had been developed to support Cyrillic characters for email correspondence. The problem was a software one, but it did not concern popular word processing programmes like *Microsoft Word*, which were easily adaptable to Cyrillic fonts, but not to internet-related software. Thus, use of the Latin script became the standard in email communication, but not in standard computer word processing. As the practice of writing Bulgarian with Latin characters became increasingly widespread with the growth of access to the internet, experts began looking for solutions to deal with the problem of character encoding.

The second aspect is related to the cultural dimensions of the overall linguistic situation in Bulgaria after 1989. A veritable linguistic revolution quickly followed the political changes in the country. Everyday speech made its way into the media and the official sphere, political discourse quickly fell under the influence of the media, while styles close to the tabloid press began to be used widely in literature. An arbitrary, indiscriminate attitude towards the Bulgarian language became the norm—in sharp contrast with the previous *status quo*, when there was strong ideological control not only over the content but also over the language of the Bulgarian media. Bulgarian society received this development in two diametrically opposed ways. Whereas most of the people involved in media and politics hailed it as an expression of a desired linguistic liberalisation, intellectuals and linguistic theoreticians tended to see its extremes as a potential threat to the purity of the Bulgarian language. A large part of the public, however, obviously showed solidarity with the dominant trend, eagerly buying newspapers that used such more liberal Bulgarian. Ultimately, this led to a deficit of serious, high-quality language discourse—in the media, in politics and in the public sphere in general (Spassov 2004). The influx of the Latin script—both on and off the internet—was easily interpreted as part of the crisis of linguistic standards and quickly began to stimulate that very crisis.

The third essential component of the debate is political. Rethinking the legacy of socialism proved to be especially difficult for Bulgarian society. A lack of consensus and a shortage of rational arguments prevailed, typifying the course of the discussions. The emotional reactions of total condemnation or unreserved praise of the socialist past found no points of agreement, often dividing the public debate into two radically opposed camps (Spassov 2000). In turn, the process of European integration, which became especially dynamic after the mid-1990s, encouraged discussion of what values Bulgaria would be required to become part of the European family. Thus, it was precisely the 'accession to Europe' that proved to be the active backdrop against which the problem of the alphabet was introduced in the political context. The key question, as formulated at the time, can be summed up as follows: should the Cyrillic alphabet and the notions, myths and public rituals related to it be regarded as part of the communist legacy? It is clear that an affirmative answer to this question would provide more cause for the imposition of the Latin script. Thus, the discussion revolved around two opposing viewpoints. For some people, the Cyrillic script was a key element of Bulgarian culture, while for others it was just an aspect of the communist past, a convenient instrument for retaining Bulgaria in the Soviet orbit.

Each one of the three above-mentioned factors calls into question, albeit in a different manner, the elements of national identity that are constructed on the basis of language. It is this that makes the Bulgarian public so hypersensitive to the issue of the use and possible official introduction of the Latin alphabet. Such sensitivity in matters of the alphabets has roots further back in time, in fact long before the period between 1944 and 1989 when the country fell within the orbit of Soviet influence. In the nineteenth century language played a key role in the Bulgarian nation-building process within the Ottoman Empire. Paisii of Chilandar, one of the earliest figures of the Bulgarian National Revival and author of *The Slav–Bulgarian History* written in 1762, introduced the idea of a direct relation between nation and language: 'You, Bulgarian people, do not allow yourselves to be deceived, know your kin and learn your language!' (Paisii 1980, p. 21). Paisii underscored the key role of Cyril and Methodius who 'invented the letters and [wrote] books in Slavonic', dedicating to them a special chapter titled 'About the Slavonic Teachers' in his history (Paisii 1980, p. 161). Thus, gradually the cult of Cyril and Methodius gained a foothold even before the formation of an autonomous Bulgarian state in 1878: many Bulgarian schools were named after them; the Church feast day at which the two brothers are commemorated began to be celebrated as a 'national festival of education and Slavonic letters'; and their work was integrated in the political discourse of the struggles for the liberation of Bulgaria and was widely used as an 'appeal for Slavic rapprochement' (Petkanova 1983a). After independence, the heritage of Cyril and Methodius continued to enjoy exceptional popularity. Celebration of them was quickly included in the system of official holidays of the new Bulgarian state and was used to simultaneously emphasise the longevity and continuity of state tradition, relating it to the heritage of the medieval Bulgarian kingdom. Systematic research of their work began, government grants in their name were awarded and in 1885 the millennial anniversary of the death of Methodius was solemnly celebrated at state level (point-of-view.org 2007). In 1909 the Order of Saint Cyril and Saint Methodius was established as the highest state distinction and as a dynastic order of Bulgarian kings, preserving this status until 1946, when the monarchy was substituted by a republic in a referendum (kingsimeon.bg 2011). The figures of Cyril and Methodius also proved particularly appropriate for the maintenance of state tradition because they allow it to be related directly to Orthodoxy. Thus, in the long run, their work—and particularly those aspects related to the Slavonic alphabet—was finally transformed into the national symbolic capital and encoded as part of the canon of Bulgarian culture. The period of socialism inherited and enriched this legacy further, including it in the official ideological discourse.

The debate about Latin script (or perhaps a more appropriate word here would be 'scandal') began against this general background in 1998 and culminated in two key moments. The first was an article in the elitist *Kultura* weekly, a front-page feature on the occasion of 24 May, the Day of the Slavonic Alphabet and Culture (*Den na slavyanskata azbuka i kultura*). The article was titled 'Latin Bulgarian—No Kidding' (Popova 1998). The author, Diana Popova, was an editor of *Kultura* and an art historian specialising in contemporary Bulgarian art. She wrote:

> The country is no longer isolated, there is talk of accession to Europe, and meanwhile words making international communication easier are constantly being introduced in our language .... That is why these days, with the influx of computers and the internet into our lives, the idea of introducing the Latin alphabet in Bulgaria seems increasingly acceptable to me. (Popova 1998, p. 1)

Then, reproducing part of an email written with Latin characters from a contributor to the weekly, Popova commented as follows:

> Tough to read, isn't it? Things would have been very different if we had become accustomed to Latin-script Bulgarian as children. And I think that considering that we have already adopted one alphabet that was adapted to our language, there's nothing to prevent us from adapting another with a view to our present and future needs .... After all, we pride ourselves on being a practical nation, don't we? (Popova 1998, p. 1)

In addition to the purely technical arguments, the author introduced several other arguments in favour of the idea of adopting the Latin alphabet. These included the fact that Cyril and Methodius never visited Bulgaria and they did not invent the Cyrillic alphabet; Popova maintained that 24 May was 'yet another desperate quest for firm foundations of Bulgarian identity' which only 'reinforces Bulgarian [inferiority] complexes'. The author also gave several examples of previous language reforms and mentioned the fact that the Serbs write 'in both alphabets', while the Turks and 'the Slavs in Europe' use the Latin script (Popova 1998, p. 1).

In 1998 the theses advanced in her article sounded like heresy. Knowing this, *Kultura* published the article in an attempt to provoke a debate. Contrary to their expectations, the newspaper received just one response. Although the response was angry, it did not lead to a serious discussion. It was written by Elka Mircheva from the Institute for Bulgarian Language (*Institut za balgarski ezik*) at the Bulgarian Academy of Sciences (*Balgarska akademia na naukite*). Mircheva attacked Popova in an indirect way: 'The article is objectionable less because of the author's idea of changing the [Bulgarian] writing system than because of her glaring incompetence in the field of the science of our language which she ventures into with enviable ease' (Mircheva 1998, p. 1). Mircheva concluded her criticism of Popova by expressing her firm belief that 'our past deserves respect' and that 'the ignorance of part of our present intelligentsia is flagrant' (Mircheva 1998, p. 1). She expressed indignation at the very fact that Popova's article had been published in *Kultura*. Ultimately, although Popova's article was much discussed in the informal intellectual circle around *Kultura*, the editorial question 'does the Cyrillic alphabet shape us in a particular way?' remained without a serious public response.[1]

A first serious attempt at an answer was made only two years later, when the Austrian Professor of Bulgarian Studies Otto Kronsteiner again raised the subject of adopting the Latin script. In a series of publications and interviews in the Bulgarian press in 2000, Kronsteiner launched the idea that 'when Bulgaria accedes to the European Union, introducing a parallel spelling in Latin should also be considered because the Cyrillic script causes many difficulties in cultural contacts' (Kronsteiner 2000, p. 6). The arguments in support of this suggestion varied in character and included the peculiarities of the internet medium where, in the words of the author, 'everything is in the Latin script' (Kronsteiner 2000, p. 6). The dominant arguments, however, were political. As Kronsteiner himself pointed out, 'the alphabet has a major role in politics' (Kronsteiner 2000, p. 6). According to him, 'Cyrillic has a bad name in Europe' and 'many think of it as a communist script and that the Bulgarians are actually Russians' (Kronsteiner 2000, p. 6). In Kronsteiner's analysis, 'Cyrillic letters are something infinitely alien to West Europeans' and, what is more, they are the reason for and symbol of Europe's division. That is why, for Bulgaria, writing in Latin script would mean final liberation from Russia's influence (Kronsteiner 2000, p. 6). According to the author, it was time for the Bulgarians to say goodbye to several myths: that the Russians are their liberators, that the Russians acknowledge the Bulgarian authorship of the Cyrillic alphabet and that the Cyrillic alphabet itself was invented by Cyril (Kronsteiner 2000, p. 6). The official Day of the Slavonic Alphabet and Culture on 24 May was also criticised indirectly. In Kronsteiner's opinion, 'communist propaganda intentionally emphasised the brothers from Salonika, making a connection with Russia and describing them as almost Proto-Slavonic communists' (Kronsteiner 2000, p. 6).

Meanwhile, after a series of reciprocal conciliatory gestures, in August 2000 Bulgaria's then President Petar Stoyanov conferred on Otto Kronsteiner the highest Bulgarian decoration, the Order of Stara Planina, First Class—a gesture that was immediately interpreted as a sign of political support for the Austrian professor's theses. As it happens Kronsteiner also holds other high Bulgarian awards, including the Order of Cyril and Methodius, which he received in 1985. However, because of his attitude to the Cyrillic alphabet, in 2001, the St Cyril and St Methodius University of Veliko Tarnovo decided to strip him of the honorary title of *Doctor Honoris Causa* awarded to him in 1990. However, before the university had a chance to do so,

---

[1] *Kultura*, 5 June 1998.

Kronsteiner himself returned the distinction in protest against the character of the discussion concerning his award of the title.

The publications of the Austrian scholar caused uproar. Whereas Diana Popova, as noted above, was criticised for her article in *Kultura* above all on the grounds that she lacked the competence to speak out on the issue, the effect of Kronsteiner's actions was great, firstly, because he was a respected specialist with many distinctions and, secondly, because of the fact that he was an 'external' spokesperson from the West. The reactions can be summarised as follows. The professional community of experts in Bulgarian Studies put the emphasis on Kronsteiner's politicisation of the issue, accusing him of basing them on 'grossly politicised presumptions' (Katedra 'Balgarska literatura' 2001), and declared that his professional arguments were unsubstantiated: 'our opinion as scholars is that there are no grounds for changes whatsoever', stated the Bulgarian Academy of Sciences' Institute for Bulgarian Language (Informatsionen byuletin na BAN 2001). Some media outlets latched on to the sensational aspect of Kronsteiner's proposal, while others quickly turned it into an instrument for their own political ends—as did the more openly partisan publications. Some dailies organised discussions where an attempt was made at depoliticising the debate. The *Sega* daily, for example, not only published opinions in defence of the Cyrillic script, but it was also one of the few publications that also gave supporters of the Latin alphabet the opportunity to express their opinion. A case in point was Anita Dimitrova's article 'Is the Latin Script a Monster to Be Feared?' (Dimitrova 2001). Once again, in such cases reference was made to the new media context or the issue of the influence of globalisation on national culture. According to Dimitrova, 'as long as we continue using different alphabets, the communication barrier will remain in place', but multilingualism 'can set the world back'. She supported her arguments by claiming that 'writing with Latin letters does not mean at all that we will give up our national alphabet and identity provided that [use of] the Cyrillic remains compulsory in textbooks, newspapers, official documents, literature' (Dimitrova 2001, p. 18).

Another peculiarity of the media coverage of Kronsteiner's proposal was parody, as a means of underscoring the unrealistic prospect for chances of the Latin script. Some newspapers published quotes from interviews with Kronsteiner in the Bulgarian press, transliterating them into Latin: '*No az vyarvam v Bulgaria i v neynoto myasto v Evropa i smyatam, che e doshlo vremeto tya okonchatelno da se otkusne ot vliyanieto na Rusia...*' ('But I believe in Bulgaria and in its place in Europe and think it is high time Bulgaria rid itself of Russia's influence once and for all ...').[2] An editorial in *Sega*, also transliterated into Latin, pointed out the following: '*poradi fundamentalnoto znachenie na predlozhenieto na avstriyskia profesor za nashata latinizirana budushtnost, prepechatvame go s podobavashta azbuka*' ('owing to the fundamental importance of the Austrian professor's proposal for our Latinised future, we are reprinting it in the appropriate alphabet').[3] In other cases, the media published transliterated excerpts from classic works of Bulgarian literature in order to illustrate 'the absurdity' of their transliteration into Latin characters and the

[2] *Sega*, 2 September 2000.
[3] *Sega*, 2 September 2000.

assumed losses of content that might result from such a transliteration. The most popular example in this respect was the 1919 poem 'I am Bulgarian' by the classic Bulgarian writer Ivan Vazov (1850–1921), 'translated' as '*Az sum bulgar4e*' (Vagalinska & Tsaneva 2004). Studied by generations of children at school, this poem is seen as emblematic of Bulgarian national identity. That is why its 'conversion' into Latin is regarded as an eloquent illustration of why it is impossible to adopt the Latin script. Other media published excerpts from English texts transliterated into Cyrillic characters for greater comic effect in order to reinforce the point that the idea of writing Bulgarian with Latin characters was unacceptable. In addition, some writers began to include sentences in English transliterated into Cyrillic in their works published in Bulgarian: 'Но ол ай уона ду ис хев съм фън ен ай хев дъ фийлин айм нот ди онли уан' ('Now all I wanna do is have some fun and I have the feeling I am not the only one') (Ivanov 2001).

The case also gave rise to political reactions. Extreme nationalist circles condemned 'the insane idea of the Jew Otto Kronsteiner' (Grancharov 2002) but, on the whole, their reaction was more cautious than anticipated. That was because Kronsteiner was respected in Bulgarian nationalist circles on the grounds of his thesis of the non-existence of the Macedonian language.[4] President Petar Stoyanov, in turn, was forced to distance himself publicly from the ideas of the Austrian professor he had decorated only the previous year, declaring, 'I have never defended Kronsteiner's thesis' (Stoyanov 2001, p. 7). This development was prompted by the upcoming presidential elections in 2001, on the eve of which Stoyanov once again declared that 'Prof. Dr. Otto Kronsteiner's suggestion that "a parallel spelling in Latin should also be considered" is, after all, a personal suggestion of his, and no institution in Bulgaria, to say nothing of the Presidency, has ever defended such a thesis' (Stoyanov 2001, p. 7). The president supported the opinion that 'in addition to being belated, the noise around the problem is also an overreaction: no one can question [and belittle] another's spiritual achievement' (Stoyanov 2001, p. 7). Nevertheless, Stoyanov's failure to win a second term in office was later attributed partly to his loss of authority because of his involvement in the Kronsteiner affair.

Gradually, after one or two other minor public scandals, the debate died down, but the politicised interpretations of the choice of scripts remained in circulation for a long time. They proved surprisingly close to theories of the new ethnogenetic studies that have become fashionable in recent years and that aim to prove that the Bulgarians are not Slavs and that their language is not Slavic.[5] In the context of this development,

[4]The Bulgarian academic and political authorities treat Macedonian not as a separate language but as a Bulgarian dialect, which has acquired different linguistic norms due to historical reasons. Kronsteiner supports this view. According to him, the Macedonian language was artificially separated from the Bulgarian for political reasons in the communist era: 'Every dialectologist is well aware of the fact that there is no dialect boundary between Bulgaria and Macedonia ... and typical Macedonian linguistic features are to be found in Bulgaria, as well. Therefore what we have here is an emblematic case of Stalinist disinformation, which successfully misled even the "critical" Slavistic in the West' (Kronsteiner 1993).

[5]See, for example, Tsvetkov (1998). The mass production of non-Slavic ethnogenetic versions and their huge popularity in the last two decades is criticised by Daskalov (2011). The formation of the Bulgarian historical canon and the political uses of different ethnogenetic versions are explored by Lilova (2003).

which meanwhile acquired academic popularity, the 'Slavonicisation' of the Bulgarians was interpreted as part of Russia's imperial projects in the nineteenth and twentieth centuries (Mitev 1998). The political distancing from Russia was concomitant with an attempt to trace the supposed ethnogenesis of the Bulgarians to much earlier times and much more distant lands. The thesis was: if the Bulgarians are sufficiently ancient, then they cannot be Slavs. In the context of this ethno-political discourse, the supporting arguments coincided completely with Kronsteiner's logic. Perhaps it was not by chance that Kronsteiner himself recalled that 'in their official interpretation of Bulgarian history as Slavic history, historians forgot that you were first part of the Roman Empire' (Kronsteiner 2000). As one might expect, the picture was complemented by the other extreme—the deconstructors of all that is national, who also proved to be in thematic proximity to Kronsteiner. To them, Cyrillic—and, hence, the work of Cyril and Methodius—often continued to be attributes of Bulgarian nationalism used for ideological purposes in the context of totalitarianism, and were therefore liable to criticism. Alexander Andreev, for example, mused as follows:

> Is the work of the brothers from Salonika truly a reason for special pride or self-esteem? . . . Ascribing metaphysical value to a script and to a cultural-political choice made eleven centuries ago is becoming unacceptable for more and more people who have grown up in a world without territorial communication frontiers. (Grancharov 2002)

In other cases the focus of criticism shifted onto Orthodox Christianity, conceived of in the Bulgarian context primarily in terms of its connection with Russia. As Evgeny Daynov noted, 'If we don't give up Slavism and Orthodox Christianity, we won't be accepted in Europe' (Grancharov 2002).

In a generalising commentary on the debate on the Latin script, Milena Kirova stressed the following point: 'The Kronsteiner scandal is political. It has nothing to do with Bulgarian culture' (Kirova 2002, p. 4). The political character of Kronsteiner's intervention becomes particularly clear against the background that he totally ignored the similar situation with the script used in neighbouring Greece in the debates he initiated. Although it is unique and different from Latin, the significance of the Greek alphabet is not questioned by anyone in the European Union, nor would Greece accept seriously any proposals for its substitution with Latin. Thus, it was precisely the lack of a Russian and Soviet connection in the use of the Greek graphic system that proved among the important reasons for Kronsteiner not to consider it relevant to the discussion. He simply remained silent on the issue of Greece and its alphabet. Instead he concentrated his ideological criticism on the Bulgarian use of Cyrillic reducing the complexity of its historical tradition to a black-and-white political choice.

As has been noted, his thesis drew enormous public attention, which was enhanced by Kronsteiner's flirting with Bulgaria's governing elite and by his attempt to present himself as the speaker of a seemingly generally shared Western point of view. Still, his project gained no support from local audiences. Kronsteiner remained unsupported in his actions and it was precisely this political and discursive isolation that led to the quick decline of his proposal.

The idea that the dispute about the Latin alphabet is purely political can also immediately find confirmation at the technological level. Here, too, there was significant discrepancy between the alleged problems with using Cyrillic on the internet and the actual state of affairs. In 1998, when Diana Popova wrote her article for *Kultura*, software applications for using Cyrillic in electronic mail were already available. In fact, the attempts to solve the problem date back to 1995, when Cyrillic-encoded email packages circulated on the Bulgarian market. The first Bulgarian websites, which appeared in 1995–1996, were also in Cyrillic. All this clearly shows that the debate on the Latin script shifted away from the real technological problems and became wholly politicised. A review of the development of the debate supports the view that the politicisation of scripts had detached itself completely from the practicalities of reality.

In the early period of the internet, from 1993 to around 1998, Bulgarian fonts were rarely used in email not because they were not available, but mainly for the reasons of inconvenience: the need to install additional applications; the cost of applications; the requirement that both parties in the communication process had the necessary software; the fact that many Bulgarians living abroad did not have access to such software. Software specialists, however, did not remain indifferent to the problem of what had to be done to ensure that 'our culture will be represented in all its national aspects, i.e. that it will be present through the Cyrillic script, on the internet' (Kolchev 1998, p. 546). The idea that 'the presence of the Cyrillic script—and, moreover, of the Bulgarian Cyrillic script—on the internet constitutes an important zone in cyberspace which will eventually contribute to the popularisation of Bulgaria' was formulated precisely in this technocratic milieu (Kolchev 1998, p. 546). The relevant strategies for achieving this objective were developed. Gradually, more and more solutions were offered to facilitate the use of Bulgarian in online communication: along with the necessary software, automatic two-way conversion applications between Cyrillic and Latin became available and platforms specially adapted to Cyrillic were popularised. At the same time, moderators of many websites where the participants wrote mostly in the Latin script obliged the content authors to convert their writing into Cyrillic. Numerous initiatives on the internet began to encourage writing in Cyrillic and grassroots campaigns were launched for restricting the use of the Latin script. Many of these campaigns turned the issue into an ethical one, underscoring the unpatriotic conduct of 'Latin-script lovers'. A telling example in this respect is the campaign conducted in 2010 by the nationalist party *Vatreshna Makedonska Revolyutsionna Organizatsia* (VMRO, Internal Macedonian Revolutionary Organisation) on the occasion of the Day of Slavonic Alphabet and Culture. Their slogan was 'Be Proud. Write in Cyrillic' (vmro.org 2010). A number of amateur videos in defence of Cyrillic were posted on Vbox7, the most popular Bulgarian video social networking site. Typical appeals here were 'O Bulgarian, write in Cyrillic!' and 'It's up to you! It's never too late to break your bad habits. Write in Cyrillic!' (vladkopicha 2010). The Bulgarian internet was flooded with user-made banners in support of Cyrillic, such as that shown in Figure 1.

Today the problem of using the Cyrillic script on the internet has been resolved entirely at the technical level. Yet even so, the practice of writing Bulgarian with Latin characters remains widespread in internet chat rooms, on a number of social

FIGURE 1. 'YOU!!! DO YOU WRITE IN CYRILLIC?'
*Note*: the man is depicted against the background of the Bulgarian national flag and he is dressed in the colours of the flag. *Source*: 'Blog na programista ot planinata', available at: http://blog.pvalley.info/?p = 429, accessed 26 June 2012.

networking sites and in social media such as Vbox7 and YouTube, in email correspondence and on many online forums. On all these platforms the Latin scripts coexists with the Cyrillic script. Why is this? There are several explanations.

First of all, the Latin script has been in use for a long time during which, as we saw, appropriate software solutions were sought. Hence, it is now commonly used. If in the beginning transliteration was encouraged by the existence of technical problems with writing in Cyrillic, Latin script usage eventually became a cultural practice in its own right, imperceptibly turning into a subcultural phenomenon. For example, on the Bulgarian internet 'some veteran chatters, even when they are sure that there will be no problem with readability, do so [write with Latin characters] out of habit and by tradition, while others do so in order to identify and present themselves as "veteran" chatters or simply as chatters' (Kirova 2001a). In this way the use of Latin script began to lend a measure of historicity to the Bulgarian internet, and also, and more importantly, it provided new opportunities for playing with online identities. On the other hand, for many ordinary users, the Latin script simply became an automatic choice, maintaining their—already practically superfluous—psychological comfort that the recipient will definitely be able to read the message.

Secondly, as a subcultural practice, writing with Latin characters became a form of opposition to imposed cultural standards, and above all to the codifications regulating the use of language. Despite the existence of standards for transcription between Latin and Cyrillic since 1956 (Kirova 2001b), arbitrary forms of using the Latin script have been established. For example, curious combinations of Latin characters and Arabic numerals came to be used to render specific Bulgarian phonemes, for which the Latin script does not have corresponding signs (for example: *'da se sre6nem v 6 na 4adyrite'* 'let's meet at 6 at the sunshades'). The failure to abide by certain rules is also characteristic of ordinary users, but the internet generation and network subcultures have found in this a new niche for counter-cultural behaviour. Thus, the Latin script has also proved to be a conscious alternative to the Cyrillic, frequently perceived as a symbol of the dominant official culture and hence representing certain restrictions. As Alexander Kiossev notes, the use of this 'non-standard Bulgarian' can be regarded under certain conditions as a 'form of democracy and a display of freedom *vis-à-vis* the binding codes' imposed by the educational system (Kiossev 2008). At the same time, however, in most cases such cultural practices are not accompanied by the development of sufficient competences in language use offline. The resistance at the linguistic level is against the wide range of values imposed by institutions such as schools and the media. That is why writing with Latin characters has gradually become symptomatic of the crisis in linguistic norms I noted at the beginning of this essay.

What is at issue, however, is not only subcultural differences and freedom. On the Bulgarian internet the use of the Latin script continues to be a widespread practice for yet another, quite prosaic, reason: because it is a convenient means of concealing illiteracy. The lack of an online linguistic standard and orthographic rules are precisely what guarantee this opportunity. The increasingly spreading social phenomenon of illiteracy is most common precisely among the young generation. As internet use is no longer determined by educational differences, the bearers of this new illiteracy freely enter the internet with its users. Imperceptibly, the lack of accepted standards for Latin transliteration has also affected the way Cyrillic is used on the internet, and more generally, the overall linguistic strategy of many online communities. The low culture of usage of internet communication channels is becoming the norm and not infrequently puts off the more serious and committed participants. Ultimately, the boundary between subcultural behaviour and illiteracy quite frequently proves to be conveniently obliterated by the use of the Latin script.

The long-unresolved problem of the standardisation of computer keyboards has also contributed to this process. While the disputes over whether and how phonetic Cyrillic should be declared a Bulgarian state standard have continued for years (the question remains open), younger users have quickly adopted this manner of writing without waiting for any externally imposed regulations.[6] For the generation learning to work not on typewriters but directly on computers, phonetic Cyrillic has proved to be much easier, not least because of the very absence of strictly fixed spelling rules. However, writing with Latin characters has proved to be even easier, where the keyboard is used on a 'what-you-see-is-what-you-get' basis and rules are almost

[6]In the Bulgarian context, the term 'phonetic Cyrillic' refers to the use of the standard Latin (QWERTY) keyboard for writing in Bulgarian. This means that the computer keyboard itself does not necessarily have to be Cyrillic-encoded. In this case, only the specific Bulgarian phonemes are rendered by special combinations of keys; in all other cases the standard QWERTY layout is used. Phonetic Cyrillic has become popular thanks to computer use. It differs from the standard Cyrillic keyboard layout which was developed for the 10-finger typing system and was standardised at the time when typewriters were in common use. Today this system continues to be widely used under the name BDS (the acronym for '*Balgarski darzhaven standart*' (Bulgarian state standard)), but it is preferred mostly by the older generations and in professional word processing.

becoming extinct. This explains the opinion, widespread on the internet, that it is much easier to write with Latin characters: 'I write in Latin because it is easier'; 'people write in Latin because they are lazy' and so on.

In turn, the long absence of clear rules for transliteration between Cyrillic and Latin has contributed to linguistic chaos both online and offline. The Transliteration Act became effective only in March 2009 and could hardly influence the ingrained habits related to the use of the Latin script online.[7] The official variant of transliteration itself is not obligatory for citizens as authors of 'private texts' (Ognyanova 2009). The advantages of the Transliteration Act can be seen in the standardised transliteration into Latin of the names of settlements and cultural and historical sites on signs, websites and maps, but not in subcultural practices, which by definition seek to oppose official rules. The recent proposals to include transliteration classes in Bulgarian language curricula in schools are undoubtedly beneficial because they can improve literacy, but will hardly be taken in earnest by network subcultures precisely because they are institutionally imposed. Today the dominant attitudes of young internet users in Bulgaria are related to the lack of respect for institutions and political elites which impose the rules of the game (Ditchev & Spassov 2009). That is why practically any initiative coming from above encounters some resistance on the various online social networks, inhabited mostly by the younger generation. These acts of disagreement seek their own forms of expression and frequently find them in the aestheticisation of the gesture of protest, and in the opportunities to play with different cultural resources and identities (Ditchev & Spassov 2011). The use of Latin script easily becomes part of such genres of micro-resistance against established standards and hierarchies in the post-socialist world.

Yet another significant factor working in favour of 'Latinisation' has emerged in parallel with the spread of mobile phones: the use of a Short Messaging Service (SMS). The situation strongly resembles the early problem with the use of Cyrillic in electronic mail: there is the near-absolute dominance of the Latin script in the exchange of text messages. This is because of the same cycle of technological problems, caused by the lack of appropriate software familiar from the dawn of the internet, as well as by difficulties related to the lack of Cyrillic keyboard layout on phones. The existing confusion around transliteration also contributes to the problems. Meanwhile, developments here are fast-moving and solutions for writing SMS texts in Cyrillic have already been proposed. Practically all recent-generation mobile phones support Cyrillic codes and are sold with Cyrillic characters integrated in the keyboard (in addition to Latin ones).

Against the backdrop of the solution of these problems, one unexpected obstacle to SMS messages in Cyrillic has become apparent. It has to do with the higher rates charged for SMS messages sent in Cyrillic that Bulgarian Global System for Mobile Communications (GSM) operators charge their users. While the maximum permissible length for an SMS message written using the Latin script is 160 characters, in Cyrillic the length is merely 70 characters for the same price: therefore, an untransliterated SMS costs twice as much. The reason is in the specifics of encoding, for which Cyrillic 'consumes' more bytes—the very process of encoding

[7]For details, see Econ.bg (2009).

'eats' characters which remain free for content in the case of Latin letters. Thus, the financial pressure on users in fact acts in favour of the Latin script. At this point, however, the initiatives calling for eliminating 'the discrimination against Cyrillic with respect to Latin letters when sending SMS messages' have not found particular support (Institut po matematika 2009).[8]

Obviously, mobile phone operators in Bulgaria are not doing much to encourage the use of Cyrillic. The messages the companies send to users (account information, various promotions) are always in the Latin script. What is more, even an SMS text written in Cyrillic and sent from a computer to a mobile phone is automatically 'translated' into Latin, rendering the effort of the sender meaningless. Therefore, Cyrillic has long remained on the periphery of mobile phone communications.

In turn, the dominant position of the Latin script in the widely popular mobile communications has an active impact on the character of writing on the internet. This is because of the increasing overlap of mobile phone and internet technologies. In Bulgaria you can send texts from a GSM not only to other mobile phones, but also to emails, social websites, chats, personal pages and blogs. The multimedia opportunities of the mobile phone permit direct access to the internet, participation in games, and so on. The use of the Latin script is dominant in practically all of these new activities. They affect young users mostly, but older generations are also involved.

In the last few years, beyond the sphere of mobile phones, Cyrillic has been regaining its position everywhere, and without any pressure from 'above'. Appropriate solutions at the technological level, on the one hand, and the emerging consensus among the majority of internet users, who use Cyrillic, that pressure must be put on those using the Latin script to stop doing so, on the other, lie at the heart of this process. As noted above, automatic applications converting postings written with Latin characters into Cyrillic are being installed on many websites. More and more moderators of online forums and chat rooms are prohibiting the use of the Latin script. As a whole, the process has become irreversible; but if the victory of Cyrillic seems certain today, the same cannot be said of the victory of literacy and linguistic culture. However, it would be too naïve to attribute the responsibility of this to the Latin script: the main responsibility lies with the educational system.

Ultimately, the fears of the expansion of the Latin script have proved to be exaggerated. They were part of the greater national anxiety at the time of transition after 1989, which coincided with the rise of globalisation. The sense of a crisis relating to everything connected to the concept of nation had a powerful impact on the attitude towards the Latin script. The process of Bulgaria's integration into the EU additionally fed concerns related to the loss of various forms of sovereignty. However, as I have demonstrated in the essay, the use of the Latin script has little to do with cultural imperialism because it is largely caused by temporary technical problems, or it is a matter of conscious choice of expression at the subcultural level. The effect is rather the opposite: it was precisely the unexpected boom in the use of the Latin script that fuelled a debate about the importance of Cyrillic. According to the logic by which globalisation actually lends new meaning to and mobilises national cultural capital,

---

[8]In 2009, the Member of the European Parliament Kristian Vigenin conducted an active campaign against the discriminating rates for sending SMS messages in Cyrillic.

the Latin script 'terrified' the defenders of Cyrillic and generally generated what we can call a discourse of defence. The conservative position proved to be the winner in the long run. This is precisely the 'Kronsteiner effect' as a whole: provoking defence, consolidation of national values, discussions of a restrictive law on language use,[9] persistent affirmation of the 'uniqueness' of the Cyrillic alphabet, and so on. The ensuing 'identity panic' (to borrow the phrase of Etienne Balibar) strongly limited the possibilities for a more rational approach toward the Latin script (Balibar & Wallerstein 1991).

Against this background, positive signals have come from an unexpected direction. Previous practice invariably shows that the global communications and software industries sooner or later will provide a solution to the problem of Bulgarian Cyrillic. Apple made its debut with Cyrillic-encoded software on the Bulgarian market back in 1992. In 2001, Google came out with a Bulgarian version of the search engine. In 2002, Microsoft issued the long-awaited localisation in Bulgarian for its *Windows XP* and *Office XP* basic products. In 2004, the Skype instant-messaging network was translated into Bulgarian and, in 2008, Bulgarian was added to the set of supported languages on Facebook. Without waiting for Bulgaria to resolve the problems with phonetic Cyrillic, Microsoft offered its users free software for phonetic keyboard layout, allowing adaptation according to individual preferences. This pre-emptive move by the software giant made further efforts involving standardisation of the keyboard on the part of the Bulgarian state meaningless. In this case, too, the move towards Cyrillic came from outside while national institutions wallowed in extremely complicated and fruitless discussions. The new opportunities for using the Cyrillic script upon registering the names of internet domains are an additional example of solutions coming from outside the national sphere of influence and practically contribute to promoting the cause of 'the Slavic script'.

In the long run, global economic logic works in the direction of supporting smaller national languages and their traditional writing systems. I expect in the near future that Bulgarian mobile phone operators will become receptive to this idea. This is simply yet another economic opportunity as it will signify an expansion of the market by reaching out to the users who do not speak English and cannot use the Latin script freely. Furthermore, for large global companies the national is an important constitutive part of a successful brand. Thus, the people 'guilty' of dependence on the Latin script surprisingly proved to be at the basis of the renaissance of Cyrillic, and the debate about which was the 'right' script ended with an unexpected economically-determined *dénouement*.

### Conclusion: new times, old priorities

Some scholars hope that the tensions between 'small' and 'great' languages—and respectively between the various systems of writing—will disappear thanks to the use

---

[9]A Bulgarian Language Bill was submitted to the National Assembly in 2004 but it was not passed. The reasons given in support of the bill explicitly underscored that 'Cyrillic is the writing system of the [Bulgarian] language'. For details, see http://www.lex.bg/forum/viewtopic.php?f = 3&t = 31938, accessed 30 August 2009.

of more sophisticated machine translation (Climent *et al.* 2003); however, this view makes the whole debate appear to be largely techno-utopian. While specialists search for solutions, discussions about the place of national languages continue to be very much alive in present-day politics. Bulgaria's accession to the EU is discussed actively in relation to the contributions the country has or could make to the Community. If the confidence of Bulgarians as members of the EU is frequently low because of the usually low ranking of the country in various studies concerning the standard of living and the degree of economic development, the Cyrillic script invariably continues to be considered as important symbolic capital convertible on the international scene. Thus, the intra-Bulgarian arguments about the significance of Cyrillic are easily transformed into notions about its role in the new international context for the country. Today the old, familiar discourses about the contribution to Slavdom, are usually substituted by discussion of the significance of Cyrillic for the EU:

> Old Bulgarian Cyrillic, Old Bulgarian books and the literature written in Bulgarian are undoubtedly unique contributions to European culture with which we identify ourselves in the EU .... Without ... generating notions of uniqueness and exceptionality of our centuries-old past and for the superiority of the Bulgarian language over the others ..., it is enough to remember that deep in the roots of the contemporary Euro-Atlantic civilization stands the heroic deed of the Holy Brothers [Cyril and Methodius], because they were the first to rise against the trilingual dogma[10] in the Europe of the day and declared the right of each nation to have its own writing and language. (Boyadzhiev 2008)

The easy transformation of the Cyril and Methodius cult into 'proof' of the Bulgarian cultural contribution to European civilisation is significant. It implies a lack of real change in the construction of Bulgarian national identity after accession to the EU. The historical canon has been simply adjusted to the new political context and the old instruments for overcoming the national complex were once again reinvented. This is a populist response to the crisis of the changing Bulgarian society and it sadly misses the chance to acknowledge Cyril and Methodius for what they really were—not national heroes but mediators who belong simultaneously to nobody and to everyone.

*University of Sofia 'St. Kliment Ohridski'*

## References

Balibar, E. & Wallerstein, I. (1991) *Race, Nation, Class: Ambiguous Identities* (London, Verso).

Baron, N. (2010) *Always On: Language in an Online and Mobile World* (Oxford, Oxford University Press).

Boyadzhiev, T. (2008) 'Ezikovata situatsia u nas v istoricheski i savremenen plan i evropeyskata ezikova politika', *balgarskiezik.org*, available at: http://www.balgarskiezik.org/3-2008/T_Boyadzhiev.pdf, accessed 18 September 2011.

Castells, M. (2000) *The Rise of the Network Society, The Information Age: Economy, Society and Culture*, Vol. I. (Cambridge, MA & Oxford, Blackwell).

[10]Cyril and Methodius translated some of the most important liturgical books and broke the 'trilingual dogma' of sermon in only the three 'sacred' languages: Hebrew, Greek and Latin. The Slavonic language was formally approved as an official liturgical language by Pope Adrian II in 867 (Petkanova 1983b).

Castells, M. (2001) *The Internet Galaxy: Reflections on the Internet, Business and Society* (Oxford, Oxford University Press).

Castells, M. (2009) *Communication Power* (Oxford & New York, Oxford University Press).

Climent, S., Moré, J., Oliver, A., Salvatierra, M., Sànchez, I., Taulé, M. & Vallmanya, L. (2003) 'Bilingual Newsgroups in Catalonia: A Challenge for Machine Translation', *Journal of Computer-Mediated Communication*, 9, 1, November.

Crystal, D. (2001) *Language and the Internet* (Cambridge, Cambridge University Press).

Crystal, D. (2011) *Internet Linguistics* (London, Routledge).

Danet, B. & Herring, S. C. (2003) 'Introduction: The Multilingual Internet', *Journal of Computer-Mediated Communication*, 9, 1, November.

Daskalov, R. (2011) *Chudniyat svyat na drevnite balgari* (Sofia, IK Gutenberg).

Dimitrova, A. (2001) 'Lamia li e latinitsata?', *Sega*, 20 December.

Ditchev, I. & Spassov, O. (eds) (2009) *Novite mladi i novite medii* (Sofia, Institut Otvoreno obshtestvo).

Ditchev, I. & Spassov, O. (eds) (2011) *Novi medii—novi mobilizatsii* (Sofia, Institut Otvoreno obshtestvo).

Econ.bg (2010) 'Zakon za transliteratsiyata', available at: http://www.econ.bg/law86416/enactments/article159883/zakon_za_transliteraciyata, accessed 20 October 2010.

Grancharov, A. (2002) 'Robi na svobodata', *Tangra.org*, available at: http://www.bulgaria88.narod.ru/IwanMladenow.htm, accessed 21 August 2010.

Hansen, B. (2011) 'Slavonic Languages', in Kortmann, B. & Auwera, J. (eds) (2011) *The Languages and Linguistics of Europe: A Comprehensive Guide* (Berlin & Boston, MA, Walter de Gruyter GmBH & Co. KG).

Hunt, S. (2011) *Worlds Apart: Bosnian Lessons for Global Security* (Durham, NC, Duke University Press).

Informatsionen byuletin na BAN (2001) 'Stanovishte na Institut za balgarski ezik pri BAN otnosno predlozheniyata za zamyana ka kirilitsata s latinitsa', *bas.bg*, 5, available at: http://www.bas.bg/fce/001/0145/files/bul50.pdf, accessed 23 July 2010.

Institut po matematika i informatika i Institut za balgarski ezik, BAN (2009) 'Initsiativa za vavezhdane na standarti za transliteratsia mezhdy kirilitsa i latinitsa i za fonetichna klaviatura', *Metodii.com*, available at: http://www.metodii.com/bg_Osnovania.html#tar2, accessed 29 June 2010.

Ivanov, S. (2001) 'Neshto predi kraya na sveta', *cult.bg*, available at: http://cult.bg/ind_lit_proza_full.php?id=710&PHPSESSID=b709562, accessed 28 November 2011.

Jenkins, H. (2006) *Convergence Culture: Where Old and New Media Collide* (New York, New York University Press).

Katedra 'Balgarska literatura' na VTU 'Sv. Sv. Kiril I Metodii' (2001) 'Otkrito pismo do Dr. Otto Kronsteiner, Pocheten doktor na Universitetite v Sofia i Veliko Turnovo, Balgaria', *Liternet*, available at: http://liternet.bg/vtu/doc/oto_mail.htm, accessed 20 August 2010.

kingsimeon.bg (2011) 'Orden "Sv. Sv. Ravnoapostoli Kiril i Metodii"', available at: http://www.kingsimeon.bg/pages/show/id/93, accessed 18 September 2011.

Kiossev, A. (2008) 'Otteglyaneto ot cheteneto ima sotsialna tsena', *temanews.com*, available at: http://www.temanews.com/index.php?p=tema&iid=341&aid=8203, accessed 26 November 2010.

Kirova, L. (2001a) 'Bilingvizum i digrafia v rechta na balgarskite geimuri', *Liternet*, available at: http://liternet.bg/publish3/lkirova/gamers.htm, accessed 16 July 2010.

Kirova, L. (2001b) 'Digrafiyata v pismenata praktika na balgarskite potrebiteli na internet', *Liternet*, available at: http://liternet.bg/publish3/lkirova/digraphia.htm, accessed 16 October 2010.

Kirova, M. (2002) 'Farsat Kronstainer', *Kultura*, 5 April.

Kolchev, Zh. (1998) 'Kirilitsata v internet', in Milev, R. (ed.) (1998) *Balgarsko Mediaznanie*, Vol. 2 (Sofia, Balkanmedia).

Koutsogiannis, D. & Mitsikopoulou, B. (2003) 'Greeklish and Greekness: Trends and Discourses of "Glocalness"', *Journal of Computer-Mediated Communication*, 9 1, November.

Kronsteiner, O. (1993) 'Razpadaneto na Yugoslavia i badeshteto na makedonskia knizhoven ezik (kasen sluchai na glototomia)', in Kochev, I., Alexandrov, I. & Kronsteiner, O. (eds) (1993) *Sachinyavaneto na t. nar. Makedonski knizhoven ezik*, available at: http://www.scribd.com/GreeceLAOS/d/30790793, accessed 18 January 2012.

Kronsteiner, O. (2000) 'Kirilitsata razdeli navremeto Evropa na Iztochna i Zapadna—intervyu na Toni Nikolov s Otto Kronsteiner', *Demokratsia*, 1 September.

Lex.bg (2008) 'Proekt na Zakon za balgarskia ezik', available at: http://www.lex.bg/forum/viewtopic.php?f=3&t=31938, accessed 20 October 2009.

Lilova, D. (2003) *Vazrozhdenskite znachenia na natsionalnoto ime* (Sofia, Prosveta).

Lowen, M. (2010) 'Montenegro Embroiled in Language Row', *BBC News*, 19 February, available at: http://news.bbc.co.uk/2/hi/8520466.stm, accessed 18 January 2012.

Mircheva, E. (1998) 'Za balgarskia ezik—maytapsiz', *Kultura*, 5 June.

Mitev, P. (1998) 'Za "slavyanizatsiyata" na vazrozhdenskite balgari', in Kolev, P. & Dimitrov, D. M. (eds) (1998) *Istoria na balgarite. Potrebnost ot nov podhod. Preotsenki* (Sofia, Tangra TanNakRa).

Ognyanova, N. (2009) 'Zakon za transliteratsiyata', Blog Posting, Mediyno pravo, 13 March, available at: http://nellyo.wordpress.com/2009/03/13/2-2/, accessed 12 June 2010.

Paisii (1980) *Slavyano-Balgarska Istoria* (Sofia, izdatelstvo Balgarski pisatel).

Palfreyman, D. & Al Khalil, M. (2003) '"A Funky Language for Teenzz to Use": Representing Gulf Arabic in Instant Messaging', *Journal of Computer-Mediated Communication*, 9, 1, November.

Petkanova, D. (1983a) 'Konstantin Kiril—Dennitsa na slavyanskia rod; chast 11: Kiril i Metodii i balgarskoto Vazrazhdane', *kroraina.com*, available at: http://www.kroraina.com/knigi/dp/dp_11.htm, accessed 18 September 2011.

Petkanova, D. (1983b) 'Konstantin Kiril—Dennitsa na slavyanskia rod; chast 7: Triumfat v Rim', *kroraina.com*, available at: http://www.kroraina.com/knigi/dp/dp_7.htm, accessed 18 September 2011.

Phillipson, R. (1992) *Linguistic Imperialism* (Oxford, Oxford University Press).

Phillipson, R. & Skutnabb-Kangas, T. (2001) 'Linguistic Imperialism', in Mesthrie, R. (ed.) (2001) *Concise Encyclopaedia of Sociolinguistics* (Oxford, Elsevier Science).

point-of-view.org (2007) '*Kak Balgaria praznuvashe predi 1944 g.*', available at: http://www.point-of-view.org/?p=332, accessed 18 September 2011.

Popova, D. (1998) 'Latinski balgarski—bez maytap', *Kultura*, 22 May.

Poster, M. (2001) *What's the Matter with the Internet?* (Minneapolis, MN & London, University of Minnesota Press).

Shirky, C. (2008) *Here Comes Everybody: The Power of Organizing Without Organizations* (New York, Penguin Books).

Spassov, O. (2000) *Prehodat i mediite: politiki na reprezentatsiya (Balgaria 1989–2000)* (Sofia, Universitetsko izdatelstvo Sv. Kliment Ohridski).

Spassov, O. (2004) 'Serious Press, Tabloid Context and the Qualities of the Public Sphere', in Spassov, O. (ed.) (2004) *Quality Press in Southeast Europe* (Sofia, European University 'Viadrina', Frankfurt (Oder) and Sofia University 'St. Kliment Ohridski').

Stoyanov, P. (2001) 'Nikoga ne sum zashtitaval tezata na Kronsteiner', *Sega*, 29 October.

Translatum (2011) 'Greeklish Converter—Greek Characters to Latin and Latin to Greek', available at: http://www.translatum.gr/converter/greeklish-converter.htm, accessed 18 September 2011.

Tsvetkov, P. (1998) *Slavyani li sa balgarite* (Sofia, Tangra TanNakRa).

Vagalinska, I. & Tsaneva, M. (2004) 'Prakti4na azbuka', *Tema*, 41, 18–24 October.

vladkopicha (2010) 'Balgarino, Pishi na Kirilitsa!!!', Online Video, *Vbox7*, 28 July, available at: http://vbox7.com/play:2a78aee1, accessed 21 November 2010.

vmro.org (2010) 'Nachinanie "Pishi na kirilitsa" provedoha mladezhite na VMRO v Plovdiv', available at: http://www.vmro.org/index.php?option=com_content&view=article&id=329:2010-05-25-07-46-56&catid=6:actual&Itemid=25, accessed 16 November 2010.

Warschauer, M., El Said, G. R. & Zohry, A. (2002) 'Language Choice Online: Globalization and Identity in Egypt', *Journal of Computer-Mediated Communication*, 7, 4, July.

Yates, S. (1996) 'English in Cyberspace', in Goodman, S. & Graddol, D. (eds) (1996) *Redesigning English: New Texts, New Identities* (London, Routledge).

# Ukrainian Nation Branding Off-line and Online: Verka Serduchka at the Eurovision Song Contest

## GALINA MIAZHEVICH

### Abstract

This essay looks at the role of the Eurovision Song Contest—an annual regularised European media event—in fostering the reconceptualisation of national selves of participating states. It draws attention to the representations of 'new' post-Soviet national brands at Eurovision in their interspatial (national, international and transnational) and inter-temporal (Soviet and post-Soviet) dimensions using the case of Ukraine. The essay argues that, through the manipulation of gendered, sexual and ethnic stereotypes, and by exploiting a kitsch idiom, the Ukrainian entry by Verka Serduchka in 2007, as well as the online responses to the show, aimed to articulate a vision of European nationhood, which simultaneously staked a position among other states of the former Soviet Union and reconfigured relationships with the shared Soviet past.

THE TELEVISION PHENOMENON OF THE EUROVISION SONG CONTEST was initiated in 1956 by the European Broadcasting Union (EBU) as a means to foster pan-European identity in post-war Europe. It grew from the core Western European countries and now includes participants from Eastern Europe, which are also members of the EBU. This recent 'eastward enlargement' of Eurovision, amplifying the tensions in defining Europeanness (Baker 2008, p. 174), has evoked a more pronounced inter-cultural exchange between the Western European and other European participating countries. As a result, the Eurovision Song Contest, in which various national narratives intersect, constitutes a fruitful ground for investigation of the flow and counter-flow of nation-branding discourse. Here nation branding is understood as a 'product' not only to be sold for external consumption, to a foreign audience, but to be actively produced in discourse around the competition online.

Recent media convergence has radically transformed the traditional televised Eurovision format 'with its centrally controlled production transmitted to passively

Supplementary material for this essay can be found in the online version at: http://www.tandfonline.com/loi/ceas20 (Figures S1, S2 and S3). This article was written with the generous support of the Gorbachev Media Fellowship. I would also like to thank Stephen Hutchings for his valuable comments on earlier versions and very useful feedback by Jeremy Morris and Vlad Strukov.

conceived national audiences' (Coleman 2008, p. 137). In the digital age, European and non-European audiences can watch (and discuss online) previous years' entries, and monitor their current national selection rounds and those of other participants on an almost daily basis.[1] As Jenkins (2006, p. 11) puts it, 'new media technologies enabled the same content to flow through many different channels and assume many different forms at the point of reception'. It is important to recognise how the variety of media and the types of access to the same content in a network society influences the perception and promotion of a national brand (Castells 2001).

The post-broadcast era demystifies the role of Eurovision as a television spectacle, as it turns the viewer into a user capable not only of consuming, but of actively engaging with this technologically mediated event. This essay will not discuss the participatory nature of the contest's voting procedure (Bolin 2006; Baker 2008), but will focus instead on online reactions towards the televised final. I adopt a cultural studies position where the online audience is seen as able to reflect on 'televisual epistemes of reality/fiction' and co-produce the meaning of the shows 'through the sharing of their reading' (Rajagopalan 2010, p. 88). The online mediascape, such as the Eurovision-related forums, which simultaneously involve various local, diasporic and other communities, is informative for understanding the construction of nationhood and the image (brand) of various nations. It is especially instructive when one deals with such articulations within the blurred borders of the vast post-Soviet space. The focus is on the former Soviet Union Republic of Ukraine, whose intermediate position between Russia and the rest of Europe makes for a particularly acute sense of the cultural polemics with both Western Europe and Russia.

In order to understand the nation branding within the post-broadcast Eurovision spectacle, this essay examines the notion of taste. It has been convincingly argued that kitsch as a primary form of excess has assumed the role of the Eurovision Song Contest's 'governing aesthetics and imaginary' (Allatson 2007, p. 87). As a result, participating nations seek to invoke kitsch in the form of ever-increasing sexual and aesthetic excess to comply with the philosophy of the contest. In this case, the notion of bad taste is treated as sexual and aesthetic performative excess. I interpret the notion of 'excess' as involving the deliberate breaching of the norms of 'good taste', and 'kitsch aesthetics' as precluding single 'readings' or definitions of the cultural phenomenon it mimics. The essay looks at contest entries and the online reactions to them (using Ukraine as a case study) to underscore what the post-broadcast engagement with the European cultural phenomena of kitsch or bad taste can tell us about the emerging nation-branding strategies.

It is argued that post-Soviet attempts to engage with the culture of sexual and aesthetic excess characteristic of Eurovision are twofold. On the one hand, they involve intra-cultural dialogue.[2] The Eurovision performances bring together rigid,

---

[1]The event involves a slow build-up starting with the national selection of each country's representative and then the actual contest when, over several days, the semi-finalists from different countries compete for a place in the final. Some of the countries (Britain, France, Germany and Spain) are guaranteed a place in the final and are exempt from the semi-finals. Eligibility to participate is not determined by geographic inclusion within mainland Europe or the European Union. (Israel, Turkey and, more recently, three former Soviet republics in the Caucasus take part in the competition.)

[2]When talking about intra-cultural and inter-cultural dialogue I ground my argument within Lotman's model (1990) of inter-cultural dialogue.

official conceptions of the national self and alternative non-mainstream identities. On the other hand, I look at the inter-cultural and transnational aspects of Eurovision. Here the post-Soviet performances, with their excessive aesthetics and aberrant sexuality, can be read in terms of an implicit dialogue with West European constructions of 'bad taste'. Bearing this close interrelationship in mind, the essay—while focusing on the multidirectional and complex process of inter-cultural dialogue—must also account for the applicable aspects of the intra-cultural dimension of national cultural production. However, it is the manifestation of the nation-branding function in the case of the Soviet legacy, which involves both inter-cultural and intra-cultural elements, which is of interest. As the 'post-Soviet' phenomenon is explored from predominantly the inter-cultural perspective (rather than the intra-cultural one), Ukrainian entries are viewed as only one version of 'post-Sovietness', in a sea of several competing and overlapping post-Soviet selves.

This essay starts by clarifying the notions of kitsch and the role of new media in propagating certain aspects of taste. After the methodology is outlined, the essay looks at the Ukrainian performances at Eurovision. The analysis focuses on the show by a famous cross-dresser Andrei Danilko (and his character of Verka Serduchka) in 2007 along with the online public perception of his Eurovision brand. Danilko enjoyed overwhelming popularity in Ukraine (and in Russia) before going on to compete at the Eurovision Song Contest. It will be demonstrated that his achieved celebrity status represents a combination of increased opportunities for social mobility in the post-Soviet realm and the transformation of media systems. The 2007 Eurovision show is of particular interest, as it deviates from an established format adopted by other Ukrainian Eurovision Song Contest shows. After the necessary contextualisation of the show against the background of other Ukrainian entries (2004–2009), the essay provides an in-depth analysis of the type of imagery foregrounded in Serduchka's show and cross-references it with the online forum responses. It is argued that the debates about the language of the song, its aesthetics, the appeal to ethnic symbols and the choice of sexual persona for the performance are all linked to (re)conceptions of Euro-taste and the place of Ukraine in New Europe. The uncovered features of post-broadcast contestation of taste and nation branding at the Eurovision Song Contest conclude the essay.

*Eurovision shows as a kitsch expression of national identity*

The role of regularly occurring 'media events' is highlighted by Dayan and Katz (1992), who describe their function in the construction of 'imagined communities' in an age of the disappearance of genuinely communal ways of living. Similarly, Coleman, discussing the Eurovision Song Contest (2008), articulates the need for different ways of binding more and more volatile citizens together in a common cause. He talks about the contest as a 'cultural embarrassment' (Coleman 2008, p. 127) and claims that at the moment 'national symbolism can only be rendered fit for popular consumption by ironic distancing' (Coleman 2008, p. 130). It is precisely the ever-increasing 'kitschification' of the Eurovision Song Contest (chipping away at the conventions of 'good' high cultural taste) that makes the production of competitive national symbolic commodities possible. It provides the philosophy and the emotive appeal of the Eurovision Song Contest.

However, in order to explicate kitsch as a framework for the Eurovision performances, one first needs to turn to issues surrounding the construction of taste (and its arbitrariness). As Bourdieu (1984) argued, the aesthetic concept of 'taste' is defined by those in power and is closely linked to the dynamic of social relations (class taste). However, advanced industrialisation has made this diffusion problematic, as it obscures class differences and weakens the correlation between social class and taste. The impossibility for a clear-cut definition of a particular style and aesthetics, and their promotion as legitimate in the age of mass production, was reflected by artists in their mocking imitation of high cultural forms (kitsch). Although kitsch was originally seen as an alternative construct to high culture, authenticity and uniqueness, this perception gradually changed. Now kitsch is regarded as a phenomenon with no single point of 'reading' or defining; 'a style which, admittedly, in keeping with the structure of the underlying society, [is] looser, more disparate and richer in contrast than earlier styles' (Goudsblom & Mennell 1998, pp. 32–33).

The digital environment provides a new platform for the expression of taste. The online networked sphere, which brings together various groups with diverging 'taste cultures' (Gans 1999), is incompatible with the rigid ordering of taste. Very often this intersection between popular culture and active use of new media results in the production of extremely popular kitsch artefacts. Sometimes it happens accidently when, for instance, a video or a misspelled word spreads virally online becoming a new internet meme.[3] Whether it is an ironical engagement with contemporary fashion, a subversion of the mainstream discourse or a response to 'technologisation' of human beings, new media expose the elusive nature of aesthetic value with its now countless mutations. Although there is a tendency to see it as subsumed by lowbrow culture or accessible, entertaining popular taste, the same grassroots and self-organising tendencies of the new media go beyond the ideology of mere consumption, and provide virtual space for a skilled worker or a citizen engaged in online 'deliberation' (Mutz 2006). In any case it enables a more transparent platform for self-expression and offers opportunities for 'ordinary' people to become media producers (Couldry 2000).

Concurrently, there is an emerging 'tendency towards the establishment of hierarchies, figures of authority, and the realms of resistance' online (Strukov 2010, p. 166). Even though the real socioeconomic status of online users might not be known, the 'new forms of decentralized dialogue' (Poster 2001, p. 182) among temporal and voluntary organised individuals may reflect their belonging to a social group, preferences and lifestyles. Hierarchy can be established on the basis of the longevity of participation (for example, in the forum) or deliberately displayed by the features of the electronically mediated discourse (such as jargon). The post-Soviet online sphere is particularly known for the 'infocratia' (Strukov 2010, p. 158) or web hierarchies (for example usage of

---

[3]The examples of post-Soviet memes include humorous videos of political figures on YouTube (for example, 'Yanukovich and the wreath', available at: http://www.youtube.com/watch?v = Qqvm5Djy 5wI&feature = related, accessed 20 July 2011) and various phrases from the so-called *padonki* language (e.g. *'preved medved'*, see http://knowyourmeme.com/memes/preved-medved-%D0%BF% D1%80%D0%B5%D0%B2%D0%B5%D0%B4-%D0%BC%D0%B5%D0%B4%D0%B2%D0% B5%D0%B4%D1%83/edits/52045, accessed 20 July 2011).

digital technologies to achieve notoriety and/or to link to the 'selected' circles or elite in *Runet*). Moreover, the loose online networks often attract people with compatible views in line with the like-mindedness and reinforcement model (Mutz 2006). Thus, by circulating or recycling only certain discourses online and clustering around certain taste groups, the users tend to demonstrate an awareness of a 'legitimate' culture (Bourdieu 1984) and their cultural competence, thus marking their social distinction in the digital mediascape.

Before engaging with the online reconstruction of Eurovision tastes by the post-Soviet post-broadcast audience it is important to understand ongoing cultural discourses around the Eurovision Song Contest. The conventions of taste displayed at the Eurovision Song Contest fall within a certain cultural time and space.[4] They can be compared to those adopted by the 'public taste' TV series *Eurotrash* and its presenter Antoine de Caunes, whose manners are a parody manifested in a skilful utilisation of the British stereotypical representation of a 'stupid' Frenchman. Eurovision's propensity to exceed the bounds of 'good taste' can be contextualised within the framework of 'Euro-trash'. It includes European cultural phenomena masquerading as *avant-garde* high art; a type of entertainment associated with continental Europe emulating high cultural forms, but perceived to be of low quality.

As a result, the culture of 'bad' taste provides a recognisable idiom within which participating states can position themselves with regards to the rest of Europe, whether aligning themselves with it, or distinguishing themselves from it. In other words, Euro-trash is a tool with which to delineate the imaginary boundaries of 'Europe', both online and offline. The matters of taste can prove useful not only in drawing boundaries in a now borderless European space, but also in detecting (non)existing boundaries of 'national internets'. By promoting a certain mode of excess, Western European entries strive to maintain a coherent version of a privileged European space, preventing 'others' from entry. In particular, Euro-trash can be deployed as a gesture of condescension towards the 'Other' Europe, which, in turn is mastering the language of Euro-trash in order to 'catch up' with and mimic its 'elder brothers'.[5]

---

[4]The culture of bad taste ('banality', 'cliché', 'kitsch') can mean different things in different national contexts and ideological regimes. For instance, the concept of camp, which is closely related to kitsch, has its origins in French slang. Now camp, or things with camp appeal, have acquired global connotations and can be described as exaggerated, theatrical or pertaining to homosexuals, as a mode of performance and enjoyment (Sontag 1964). In turn, the refamiliarisation of the oppressive Soviet past via kitsch (Allatson 2007, p. 95) involves such culturally specific aspects as *steb*. Following Yurchak, *steb* is a 'form of irony that differed from sarcasm, cynicism, derision, or any of the more familiar genres of absurd humor' (2006, p. 250). What makes *steb* slightly different from sarcasm is the ambivalence of irony displayed, as one is left to wonder whether the ironic aesthetic practice is a support, ridicule or mixture of the two. It is similar to Bakhtinian 'carnival' (1981), with one exception—this form of resistance is not a carnivalesque parody, as it combines resistance to authoritative symbols with a feeling of affinity towards them.

[5]The cult of Euro-trash has been recently supplemented with a broader, and growing, West European tendency towards distanced critique of the Eurovision Song Contest in which the new European participants now set the trend (as for example in Terry Wogan's ironic commentary, as discussed by Coleman (2008)).

*Methodology*

The study grounds itself within a reflexive performance-based cultural studies agenda.[6] Its methodology presupposes contextualisation, in-depth (and alternative) reading of individual cases or texts. In the case of the Eurovision Song Contest it is a close reading of individual shows. The entries under analysis either possess the typicality that make them representative of the whole corpus (during 2004–2009) or a singularity highlighting the background they come from (Serduchka's Eurovision entry in 2007). By focusing on the exceptionality of the 2007 show, which breaches norms and confounds trends (of other shows), I highlight the difference and draw attention to the unspoken assumptions and absences (such as distinctive post-Soviet attitudes towards homosexuality). I draw attention to representations of a post-Soviet national brand in both its interspatial (national, international and transnational) and inter-temporal (Soviet and post-Soviet) dimensions. The analysis touches upon the following issues: varying perspectives on what constitutes 'kitsch' within the act (dress, performance, sexual demeanour, music or appeal to ethnocultural symbols); how the act is presented (whether it is a self-conscious celebration, deprecation or *steb*). I also trace the usage of Soviet imagery (which symbols are used and in which ways) as well as the exploitation of '*estrada*' formats (the combination of pop culture and high art).[7]

Although a close reading of the 2007 show is provided, the focus is on the discursive place of online Eurovision Song Contest forums. First of all, the slow build up of the Eurovision annual televised final (somewhat similar to a major sporting tournament) makes it particularly pertinent in terms of online reactions. Then, in line with a cultural studies framework, the audience is viewed as able to process media texts in a variety of ways and to 'resist' the dominant codes and manipulation by elites and producers (Hall 1980). The new media (with the wide-ranging spectrum of opinions, fluidity and scope for creativity) enhance the co-creation of meanings and sharing of the 'readings' of the various (television) spectacles. At present the Eurovision Song Contest elicits wide online participation through, for instance, its official website where anyone can sign up for updates, create a profile and access the Eurovision blog entries of others or relevant YouTube videos. The online user can easily navigate relevant content and engage (anonymously if desired) in various discussions.

The web forums analysed here include several specialised online forums (devoted solely to Eurovision, such as the contest's official website[8]) and several online debates triggered either by publications (such as an announcement of the possible protest by radio Europe-FM against Serduchka's Eurovision participation) or real life events

---

[6]Interdisciplinarity of cultural studies makes the discipline incommensurable with the unified theory (White & Schwoch 2006, pp. 2–3). Despite its own 'blind spots' and contradictions (such as subjectivism or arbitrariness), this paradigm satisfies claims for uncovering 'truths', validity and generalisability of the findings.

[7]*Estrada* is a peculiar Soviet phenomenon and is defined as 'a wide-ranging term that includes pop music but also applies to modern dance, comedy, circus art, and many other performance not on the "big," classical stage' (MacFadyen 2002, p. 3).

[8]Eurovision Song Contest Official Website, available at: http://www.eurovision.tv/page/home, accessed 5 September 2010.

(demonstrations by nationalist groups (Fawkes 2007)).[9] Online debates on the Eurovision Song Contest expose first, the meaning-making and construction of national brand (inter-cultural dimension) and second, the dynamic of how the official culture is challenged by popular culture (intra-cultural dimension). The provided snapshot of online attitudes represents the first (and thus limited) inquiry of this sort. However, it is the combination with a cultural studies inflected analysis of the shows which allows a certain generalisability of the findings.

## Ukrainian Eurovision shows: the spectacle and spectators

### Ukrainian Eurovision entries: 2004–2009

A very brief overview of the Ukrainian Eurovision entries of 2004–2009 shows that most of the performances are constructed in line with a traditional gender model: they feature a lead female singer (with the one exception of an all-male hip-hop band in 2005). Also, Ukraine is like many other smaller countries, in particular recent accessions to the EU, which 'capitalise on the stereotypes that are usually attached to their homelands' (Le Guern 2000) and build performances around 'a folkloristic musical style' (Björnberg 2007, pp. 21–22).

The 2004 winner—Ruslana—with her 'Wild Dances' utilises a 'primeval' or 'tribal' sexuality, which was displayed by the performer's leather attire, wildly floating long hair, foot stomping and the introduction of the trembita, the Hutsul's musical instrument.[10] The 2006 entrant, Tina Karol, in her tiny 'kitschified' folk dress looked as though she had just emerged from a traditional Ukrainian festival. Her dance team wore costumes alluding to Cossack culture and used tambourines—a 'seductive instrument of cultural exoticism' (Boym 1994, p. 119)—strengthening this impression. In both cases the conscious self-orientalisation strategy was employed to enable 'easier consumption' by Western audiences. In recent years (2008–2009) the artists have moved away from this over-essentialisation of national traditions. However, they still tend to create a sexual and aesthetic overload (within a traditional gender model) to address a foreign gaze which exoticises differences.[11]

These carefully orchestrated performances construct the Ukrainian brand not only for the Western public but also for regional audiences. In 2005, when the country hosted the contest, its song—allegedly an 'anthem' of the Orange Revolution (Kuzio 2005, p. 34)—was performed in Ukrainian by the hip-hop band GreenJolly and

---

[9]I do not deal here with the notions of fan communities and fan culture as it goes beyond the scope of this essay. Similarly, I omit the debate about the impact of new media on sociability and community creation (the so-called cyber-idealists compared to cyber-critics debate triggered by Rheingold (1993)).

[10]Hutsuly are the dwellers of the Ukrainian Carpathian mountains. Here they represent an authentic, unique culture with an ancient, rich and timeless history.

[11]For example, in 2009 Svetlana Loboda, who is believed to have enhanced her looks with the help of plastic surgery, dressed in a revealing red basque and high black leather boots ('cyber-courtesan') while performing pole-dancing moves and suggestively licking her lips during the performance. Her song also re-positioned Ukraine on the European map drawing on transnational narratives, as the alternative title for the song 'Anti-crises girl' addressed the global issue of the day.

accompanied by peasant-like dancers freeing themselves from the chains of 'Big Brother'. The entry also appropriated elements of Western culture of bad taste, as the lead singer wore a hoodie with '69' on the back. The same cultural symbol, albeit in a different context, was used in 2007.

### The Ukrainian 2007 entry: Verka Serduchka

The entry by Serduchka seemed to master a different format of excessive aesthetics and sexuality, and by doing so (in a self-ironical way), accentuated the elements of the national brand. While many saw this entry either as an excessive vulgar (camp) performativity or a manifestation of Sovietness and post-Sovietness, thereby undermining Ukrainian independent nationhood, the show's role in rethinking sexual freedom after the fall of the USSR and the geopolitical status of contemporary Ukraine should not be underestimated.[12] However, before contextualising this statement, I will provide a brief account of the professional path of the performer.

Although Andrei Danilko started his music career under his real name, it was his stage persona of Verka Serduchka that brought him celebrity status. Serduchka is a flamboyant middle-aged woman, alluding to an image of a typical female train attendant, who openly explores (by using deviant excess sexuality) 'taboo' issues.[13] Verka was one of the characters of the Danilko Theatre troupe, which toured the cities of Russia and Ukraine in the late 1990s. Later this character became the host of a talk-show called *SV-show* ('SV' stands for the Russian abbreviation of '*Spalnyi Vagon*' or 'Sleeping Car'), which from 1997 until 2002 was broadcast in Ukraine and Russia (Crescente 2007, p. 4). In his comic performances Danilko manoeuvres between various cultural forms (stand-up comedy, quasi-comic chat show and narrative music video) and appropriates the framework of polysemy, as his act does not adhere to 'a single cultural, national or linguistic paradigm' (Morris 2010, p. 207). Talking to various celebrities, Verka's use (and popularisation) of '*surzhyk*'—a mixture of Ukrainian and Russian language associated with low educational status and provinciality—among other things, conveys reflexive irony towards Ukrainian nationhood. However, in his cross-dressing career the artist cannot be labelled as a drag queen in its Western understanding. Danilko's act constitutes 'the first act approximating drag' on mainstream Ukrainian and Russian TV or 'the grotesquely exaggerated nature of cross-dressing' (Morris 2010, pp. 196, 206).

Danilko's successful career to a large extent is based on a networked logic enabled by new communication technologies (Jenkins 2006) and processes of celebritisation through new media. Verka's brand is a post-broadcast brand. It ranges from televised songs, the *SV-show* (and its online archive) to his YouTube music videos constructed as a series of clips to be viewed in succession, thus creating an online television channel dedicated to Verka Serduchka. The various media platforms (blogs, Twitter,

---

[12]See Fawkes (2007), Morris (2010) and 'Odna iz Kievskikh radiostantsii ob'yavila voinu Verke Serdyuchke', *Korrespondent.net*, 19 February 2007, available at: http://korrespondent.net/showbiz/179870, accessed 10 September 2010.

[13]One of these issues is homosexuality. Attitudes towards homosexuality in Ukraine should be related to the Soviet practice of criminalising homosexuality (since 1933) and the inherited tendency to treat it as non-normative (*nenormativnaya*) sexuality.

Facebook, etc.) create an illusion of accessibility and the 'real' presence of Verka.[14] They are the closest way for the fans to 'interact' with Verka and the most accessible way to validate and reconstruct their 'imaginary' of this celebrity.

Next, the persona of Verka is intriguing. It bridges ordinary and high profile, familiar and strange. In its function, this fabricated character resembles an avatar used in computer games (Meadow 2008) or in online communication in blogs, etc. As a rule, Danilko does not want to be seen without his 'three-dimensional stage mask' (even in his off-stage interviews). This 'mask' of a transgendered person reverberates with the post-Soviet online culture where users often tend to express themselves with the help of exaggerated 'sexually encoded avatar[s]' (Strukov 2010, p. 148) and by actively playing with class, gender and other identities. Moreover, Verka's image echoes the logic of the virtual space with its anonymity and fantasy: the construction of a fake female character is one of the most common patterns of male online users (Nakamura 2002).

Finally, Danilko's popularity is, to a large extent, determined by a particular set of historic conditions (combined with the growth of new media in former Soviet states). Verka bridges not only the broadcast and post-broadcast divide, but also plays on the Soviet and post-Soviet, totalitarian and post-totalitarian metamorphosis. The 'extra-territorial qualities of cyberspace' (Strukov 2010, p. 165) mean that Verka's brand mediates different geographical and temporal spaces. Playing with the controversial themes of post-Soviet identity and searching for new post-Soviet aesthetics, Serduchka's act signals the transformation of cultural memories and 'talks' to a large former Soviet Union space.[15] His hybrid insider-as-outsider position and rapidly acquired celebrity status allow Danilko to act as a post-Soviet 'cultural intermediary' (Bourdieu 1984). Finally, Verka's brand highlights contemporary values (for example his mockery of mainstream pop culture and at the same time capitalisation on it) and legitimises emerging particular taste cultures. The subsequent analysis of his 2007 show and the online attitudes provide further contextualisation of this.

The 2007 performance is constructed within the paradigm of the carnival but it goes beyond appropriation of elements of visual folk kitsch by embracing transgressive and absurd qualities of carnivalesque parody. In line with Serduchka's brand, the 2007 Eurovision entry conveys several, at times, subversive messages and operates within a number of ironic modalities. The references to the Soviet past are apparent from the beginning as the first line of the song '*sieben sieben ai lyu-lyu*' is vocalised. It is a slightly altered phrase representing a mixture of German and a made up 'mumbo jumbo' echoing a famous sentence from the Soviet film *Diamond Arm*.[16] The ridicule of a failed Soviet utopia unfolding on the stage is crowned by a Silver Star shining from the singer's

---

[14]For instance, see Verka's twitter account (http://twitter.com/#!/SerduchkaVerka), the presence of mobile ringtones (Verka's Eurovision tune became one of the most popular mobile ringtone downloads); and Verka's fan sites in the former Soviet Union (http://serduchka.org, http://serduchka-club.ru and http://www.liveinternet.ru/community/serduchka, accessed 12 December 2011).

[15]As a result, Verka is categorised by some (Crescente 2007; Morris 2010) as post-Soviet rather than Ukrainian.

[16]In the film this phrase is uttered by a Western prostitute luring the main character, an honest Soviet worker on (a party approved) trip abroad, to her 'boudoir'.

head.[17] This is combined with an obvious rebellion against Russian dominance through the refrain '*lasha tumbai*' (a made-up phrase resembling 'Russia goodbye') and the prevalence of the Ukrainian and English language in the song. This aspect of the song seems to reflect a particularly tense phase of Russian–Ukrainian diplomatic relations. As the frenzy of the show progresses, fostered by an up-tempo music, the spectator is plunged into an even more absurd space. However, a post-Soviet Russian-speaking viewer, for whom the 'Soviet-ness' constitutes a 'common place' (Boym 1994), can easily find a way to 'translate' numerous references and contextualise meanings.

The stage outfit of the dance team resembles either a Soviet Pioneer uniform or a soldier's or nurse's outfit. The fact that the start of the song is in German strengthens the reference to World War II (The Great Patriotic War). This mocking of political correctness (merging 'camp' with the cultural memory of World War II) is a post-Soviet phenomenon, which echoes the legacy of a negative attitude toward all things Western (including Western liberal values). The framework of self-irony or *steb* allows the singer to 'get away' with ridiculing the sacralised World War II legacy. Furthermore, the tune, the use of an accordion and dance (with choreography reminiscent of folk *polka* at a rural wedding or an old fashioned Soviet disco) brings us back to the tradition of Soviet village festivals. By the end of the show, Serduchka's celebration of sexual liberation turns into a mockery and an ironic protest against sexual colonisation (the camp show inevitably links to pornography and the West, which is 'rotten').[18] Here the artist—dressed up in a short silver dress with the name 'Serduchka' and '69' on the back of it—runs around the stage 'pestering' the dance team slapping only female support singers on their bottoms.

As Ukrainian Eurovision performances described above showed, they can be treated as a 'double voiced' act which self-consciously parody Western imaginings of an exotic, yet sexually promiscuous East (in many cases signalled by the inclusion of ethnic burlesque alongside sexual excess). However, Verka's masculinity, which is quite obvious beneath a purposefully chosen, glamorous yet unflattering, female dress, turns the same message into a bizarre and absurd point. The show's grotesque representation of Eastern European 'exoticism' as an attractive difference (for the Western audience) disrupts a predictable associative chain of fixed sexual identities and exotic sexual availability. At the same time, the show's carnivalesque orientation renders various evoked 'others' less 'frightening' teasing the audience with a glimpse into an 'utterly unknown' with its alterity disarmed (Allatson 2007, p. 94).

As Serduchka strives to combine locally distinct and globally recognisable intertextual images in a highly eclectic show, it elicits confusing or conflicting meanings (e.g. subversive references to World War II). Even the performer may find it difficult to draw a line between the associated meanings, shifting in 'unexpected and absurd directions' (Yurchak 2006, p. 252) during this carnival. Moreover, an insight into the career progression of Serduchka's 'mediated personality' (Morris 2010, p. 196) demonstrates that the artist is unable to geopolitically locate his stage persona or at

---

[17]This star resembles the one at the clock tower of the Kremlin—the centre of the former Soviet empire.

[18]'*Zagnivayushchii Zapad*' is an (ironic) USSR cliché. This is a common phrase, which was used to describe 'all things Western' and is translated as 'rotten West'.

times to separate himself from it. In other words, he tends to combine 'resistance to authoritative symbols [with] a feeling of affinity or warmth towards them' (Yurchak 2006, p. 252). This makes him stand closer to *steb* rather than a kitsch framework.

Clearly, it is difficult to make any extrapolations and generalisations on the basis of one show. However, it can be stated that by his '*teatralizatsiya*' and moving in a multidirectional plane '[making] generic references... back and forth using Soviet Estrada tradition' (MacFadyen 2002, p. 35) the artist communicates to various communities, especially to the large diasporic post-Soviet one. In order to provide further insights into Serduchka's Eurovision entry, I will turn to the online forums where the performance was extensively discussed.

## Online discussions: the 2007 entry by Serduchka

As the analysis above demonstrates, the Eurovision Song Contest in general, and Serduchka's entry in particular, constitute part of a carnival where various messages are transmitted and the lack or excess of taste is implemented as a conscious strategy. How does the Ukrainian public perceive the show and make sense of the representation of their 'national brand'? How do online users respond to this combination of counter-cultural aspects of Westernisation with a nostalgic 'Soviet-ness'? I am interested in how pop culture is seen and used to articulate and remediate what it means to be part of a national–cultural collective by fragmented diasporic transnational communities. First, I will outline general trends in the online posts and then explore the identified common topics within the theme of nation branding. The quotes from online posts are usually given without exact reference to the website, unless the whole sentence (rather than a word or a phrase) is quoted; the nickname of the user is given if it is available on the site. Although in many cases bilingualism in Ukraine is nonreciprocal (Bilaniuk 2005 ), it is necessary to account for forums in both languages, Ukrainian and Russian.[19]

The online discussions are characterised by an emotive charge resulting from ambivalence of attitudes (such as 'I hate Serduchka, but') and contradictory comments posted by the same user. Sometimes the users who actively contribute to these forums are neither ardent supporters nor fervent adversaries of the contest:

> All these people who are complaining here and protest; can you tell me, are you all Eurovision fans? Or do you seriously believe that political or economic leaders watch it? I do not like Verka Serduchka—I do not listen to his songs; I also do not like Eurovision—I have not watched it ever. Let him go. At least we have something to talk about ... not ... like GreenJolly. Verka—is big progress.[20]

---

[19]The language of the comments are Russian, Ukrainian or a mix of the two; some of the comments are typed using Latin letters (by users from abroad or by people without access to a Cyrillic keyboard). It is difficult to judge the location of the users or their national belonging or gender. One possible way to conceptualise this communication is to use the notion of diasporic public spheres (Appadurai 1997) enabled by new media.

[20]Di_D, Comment, 'Odna iz Kievskikh radiostantsii ob'yavila voinu Verke Serdyuchke', *Korrespondent.net*, 19 February 2007, available at: http://korrespondent.net/showbiz/179870, accessed 10 September 2010.

In terms of the content, there is a trend towards generalisation, as the public picks up on several elements such as language or type of imagery (rather than going into a detailed analysis of the show). Slang words or adaptations of foreign words (such as 'Serduchka rulez') are quite common. They denote certain symbolic capital and the user's belonging to a 'select community' of 'media-savvy' people. References to popular online memes perform a similar function (Figure 1).

The identified sub-themes are taste and the nature of the contest; the stage persona of the performer, the post-Soviet reflexivity and redefinition of sexuality and homosexuality; and the stance of the country in relation to the rest of Europe. The issue of taste is one of the most frequently mentioned. It is characterised by the multiplicity of meanings: the show is described as trash or non-trash; as clownishness or *steb* or taken at face value. Occasionally users post their reflections about the stages in the development of culture, postmodernism and the parody, thereby revealing a well informed and critical stance.[21] Typically, there is a discordance, as the country is forced to succeed in the 'spectacle of a very low quality' (*nizkoprobnoe zrelishche*) to present its 'competitive' national brand: 'Disappointments all around... extremely over-politicised, over-discriminated (possibly) and forgive me idiotic show... disappointing... if Danilko would have come first, it would not be shameful... only a handful of miserable songs worth paying attention to'.[22] On the one hand, the online participation appears to be more socially inclusive. On the other hand, it exposes the co-presence of various groups with diverging taste cultures and levels of 'cultural competence' (Bourdieu 1984).

Sometimes confusion over the arbitrariness of taste is triggered by nostalgia for high art, purity and clear-cut forms and the desire to delineate high and low culture inherited as part of the pre-Soviet and Soviet literary culture. As a result, such utterances implying that only 'true art' can withstand the test of time, are mixed together with the parallels drawn between the artist and global literary figures:[23] 'I consider Danilko to be a genius actor, who created a national hero comparable to Sancho Panza or Švejk'.

The online audience is more concerned with Danilko's stage persona than with his Eurovision performance, which might be determined by his popularity in Ukraine and the former Soviet Union and his established profile as a versatile performer. The posts display different degrees of intimacy ('Andryusha' or 'Serdyun'ka') and confusion between his onstage and offstage persona, as he is equally often referred to as Serduchka, Verka, Andrei, Danilko and in mixed format as 'Verka—he'. The positive labels are mixed up with overtly negative: 'actor', 'clown', 'comic', 'vile creature'

[21]Supplementary material for this essay can be found in the online version at: http://www.tandfonline.com/loi/ceas20 (Figure S1).

[22]Cured, Comment, 'Verka Serduchka na Eurobachanni', *LitForum*, available at: http://www.litforum.net.ua/showthread.php?t = 1348, accessed 18 September 2010.

[23]For example: 'The lyrics in the song is no, to put it politely, Pushkin :-) To pick up on the lyrics—is somebody's order and paranoia. Believe me if we are going to translate all foreign songs, which are on the radio, we will find out even more about ourselves. One should not translate hits, you will be hugely disappointed, trust me—a translator' (Panialanna, Comment, 'Verka Serduchka: Prosti menia Rossia', *TopPop*, 20 June 2007, available at: http://www.toppop.ru/press/verka_serdyuchka__prosti_menya__rossiya/, accessed 20 September 2010).

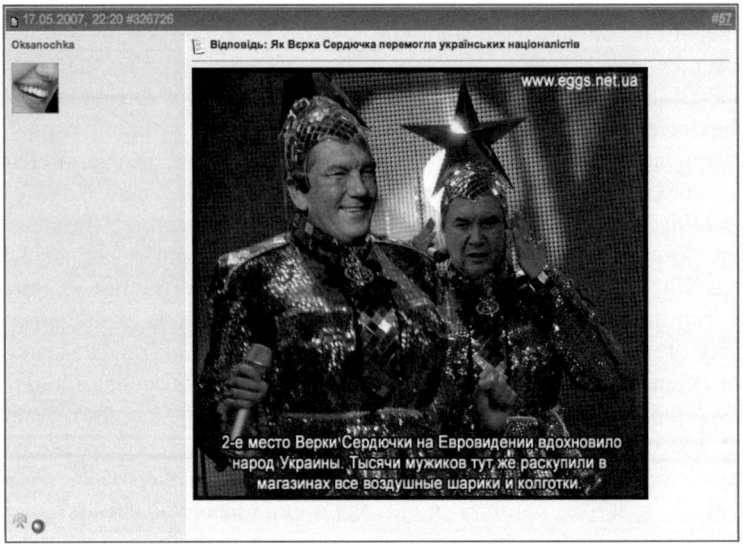

FIGURE 1. THE 'RECYCLED' SERDUCHKA'S STAGE OUTFIT, WHICH IS INCORPORATED IN THE FORUM
DISCUSSION, DEPICTS TWO KEY UKRAINIAN POLITICAL FIGURES—V. YUSHCHENKO AND V.
YANUKOVYCH.
*Source*: 'Yak Verka Serduchka peramogla Ukrainskikh natsionalistiv', *Forum Fainogo Mista*, 2007,
available at: http://forum.te.ua/showthread.php?p=337737, accessed 17 September 2010.

(*gadina*), 'talent' and 'cultural haemorrhoid'. This is linked with the issues of taste:
'The only people who like Serduchka are those who would "gobble anything". It's
idiocy, Serduchka is the shame of Ukraine. This character is good only for primitive
men and women';[24] and 'Clowns and comics were at all times, but S(m)erdyuchka is
too much'.[25] Here mass culture is often equated with low culture and primitivism (even
if the clownishness of the act is acknowledged), which casts a shadow on the
Ukrainian national brand and the country's capacity for cultural production.

An ambivalent attitude towards the performer is largely triggered by his toying with
Soviet and national legacy and 'taboo' topics (homosexuality), which I explore in turn.
Firstly, his allusion to a common pool of cultural (mostly Soviet) symbols is quite
controversial. The comments signify the division of the national public into two groups:
on the one hand those whose mental maps are largely built with reference to the USSR
and nostalgia for the Soviet past, and on the other, a more pro-Ukrainian group, which
feels that the 'Ukrainian brand' is undermined. The latter group is prolific in its negative
comments: this is the product for 'Russians who will gorge on anything';[26] the

[24]'V Kieve sozhgut chuchelo Verki Serdyuchki', *Korrespondent.net*, 7 March 2007, available at:
http://korrespondent.net/showbiz/181973, accessed 18 September 2010.

[25]'*Klouny i komiki zavzhdi buli, ale S(m)erduchka-tse perebir*'. This is a play on words; in Ukrainian
it means something which stinks ('Odna iz Kievskikh radiostantsii ob'yavila voinu Verke Serdyuchke',
*Korrespondent.net*, 19 February 2007, available at: http://korrespondent.net/showbiz/179870, accessed
10 September 2010).

[26]'V Kieve sozhgut chuchelo Verki Serdyuchki', *Korrespondent.net*, 7 March 2007, available at:
http://korrespondent.net/showbiz/181973, accessed 18 September 2010.

promotion of 'a post-imperial inferiority complex, LittleRussia' (*malorossiishchina*); 'he is insulting Ukraine'; 'he is no longer a Ukrainian artist, creating his shows with the Russian viewer in mind'.[27] The assertion of Ukraine's independent statehood is clear as the civic term to denote belonging to the Russian Federation instead of a more conventional ethnic term being used (*rossiyane* compared to *russkie*).

However, the online perception of Serduchka's show as a post-Soviet cultural text can also be more nuanced. To start with, it is seen as an interesting nation-branding phenomenon, which is constructed for the European public, and can potentially engage the former Soviet Union public ('*Sovok*' will vote').[28] In addition, some find positive aspects (and occupy a position of superiority in relation to 'Big Brother') due to the fact that Verka managed to 'get lots of money out of the CIS'. This goes together with the ideas about the cultural trauma after the fall of the USSR: 'Ukraine has been exposed to shame for a long time... plus Serduchka now... one thing more. We should be used to it'. Conversely, the performance is seen as capable of unifying the nation (irrespective of its quality): 'Let the guy go and show to Europe what a good SHOW is. Stop complaining. Go yourself and do something better'.[29]

Serduchka's entry—in line with the tradition of fools or jesters (Partan 2007)— 'articulates' the issue of alternative sexual identities in a mocking, ironical, provocative, eccentric style. This triggers online debates in the public and private spheres (discourse on 'abnormal' sexuality), which go well beyond what is permitted on air. The online analysis of Serduchka's show is linked to a pronounced homophobia towards the artist ('a creature without any voice dressed up as a transvestite'), towards the European public ('they say there are lots of gays over there') and towards the contest itself ('the show should be renamed as Euro-gay-vision' (*EuroPiderBachanne*)). Although the 'image' of the country is at stake, the price of victory should not be too high: 'we do not want to bend over (*prognut'sya*) and satisfy the needs of European deviant public'. This is combined with still homophobic but quite ironic comments such as 'If they don't give us... first place... [we] will prosecute them for the infringement of gay rights! Let Verka sing[a] duet with Elton John :)'.[30]

However, alternative identities are suppressed in this discussion, as only sporadic moderating comments, coming from (presumably) heterosexual people, are present: 'and you would not believe: most of the Eurovision viewers and voters in Europe are gays, and they like everything bright and out of the ordinary. So Serduchka might be very close to winning at such a contest'.[31] The comment also indicates the worldwide awareness and mobility of the user (his or her special social status). Thus, the online debates on nation branding play out the complexities of post-Soviet identity. Moreover,

---

[27]'Kun Naszval Verku Serdyuchku Produktom Khokhlyatskogo Masskulta', *Korrespondent.net*, 13 February 2007, available at: http://korrespondent.net/showbiz/179108, accessed 17 September 2010.

[28]'*Sovok*' (literally 'scoop') is a derogatory reference to the Soviet people or residual features of this mentality.

[29]'Odna iz Kievskikh radiostantsii ob'yavila voinu Verke Serdyuchke', *Korrespondent.net*, 19 February 2007, available at: http://korrespondent.net/showbiz/179870, accessed 10 September 2010.

[30]'Kun naszval Verku Serdyuchku produktom khokhlyatskogo masskul'ta', *Korrespondent.net*, 13 February 2007, available at: http://korrespondent.net/showbiz/179108, accessed 17 September 2010.

[31]'V Kieve sozhgut chuchelo Verki Serdyuchki', *Korrespondent.net*, 7 March 2007, available at: http://korrespondent.net/showbiz/181973, accessed 18 September 2010.

the analysis demonstrates that the forums are not necessarily politically progressive (as in the display of homophobic comments) or have the potential to transform popular opinion (people tend to express one position or visit the same forums).

Ukraine, like other countries of the former Soviet Union which need both to cater for the evolving national self and various external 'Others', struggles to balance the tensions between the two. The tension between the proximity with and distance from Europe (Le Guern 2000) is articulated in various ways. On the one hand, there is an inclination to be part of that unity (which is embodied by the TV show with a certain set of rules and linear format), hence, the anxiety-driven debates about the quality of the presented show. On the other hand, there is a desire (in an online non-normative space) to assert one's nationhood, to be accepted the way the nation is, as articulated by 'Ukrainian bumpkin':

> Comrades, is our boisterous Serduchka worse than those Finnish monsters? We are always as Pronya Prokopovna[32] trying to cover up our bumpkin nature. And why? Yes, WE ARE LIKE THIS. We do not comply with the standard, without that European veneer of glamour; but we are not Europeans in the long run. Why do we need it? We should be the way we are.[33]

There is also a feeling of inferiority ('put to shame', 'will be laughing at us' (*porzhut nad nami*), 'Europeans will not get it') and a negative outburst, sometimes echoing Cold War narratives ('their' Eurovision Song Contest, 'parody', 'degradation', 'faceless glamour-wiping singers', 'rotting West'). The last two trends are exemplified in the following quotes: 'I am particularly pleased that this creature could not get first place...; the "persona" has lost its appeal, and one should bring something newer to Europe',[34] and 'I am for Verka! She will create a stir and furore within an over-fed and bored European public'.[35]

The contradiction between the reconceptualisation of Euro-taste and aspirations for a place in a New Europe can be summarised by one user's comments: 'One needs to sing what "people are ready to gobble"; otherwise there is no point in participating in the contest'.[36] This aspect of discussion highlights the nature of organisation of individuals online, who can freely migrate within the digital mediascape and form temporarily tactical alliances if and when required (in this case, the users are temporally unified about the priority of Verka's brand over the quality of the show). Overall, the comments follow two general trends: while some agree that the strategies of sexual and aesthetic excess are employed in order to be able to 'speak' to Western Europe, others take this strategy 'to go beyond the conventions of "good" high

---

[32]A female character from a comedy *Za dvumya Zaitsami* (1961) which was partially filmed in Kiev and originally was in Ukrainian.

[33]Ukrainian bumpkin, Comment, 'Odna iz Kievskikh radiostantsii ob'yavila voinu Verke Serdyuchke', *Korrespondent.net*, 19 February 2007, available at: http://korrespondent.net/showbiz/179870, accessed 10 September 2010.

[34]Tim-Lit, Comment, 'Verka Serduchka na Eurobachanni', *LitForum*, available at: http://www.litforum.net.ua/showthread.php?t = 1348, accessed 18 September 2010.

[35]'Odna iz Kievskikh radiostantsii ob'yavila voinu Verke Serdyuchke', *Korrespondent.net*, 19 February 2007, available at: http://korrespondent.net/showbiz/179870, accessed 10 September 2010.

[36]'Verka Sedyuchka: Prosti menya Rossiya', *TopPop*, 20 June 2007, available at: http://www.toppop.ru/press/verka_serdyuchka__prosti_menya__rossiya/, accessed 20 September 2010.

cultural taste' (clownishness of Serduchka) at face value and do not see it as a distancing mechanism from the bad taste of the Eurovision Song Contest.

My analysis demonstrates that the digital interactive technologies provide a particular opportunity for the audience's involvement and deliberation. The multi-user reflection is characterised by the simultaneous existence of various layers of meaning. They range from habitual judgments to more conscious statements about the 'organisers' strategies'; from ironic comments 'in passing' to a serious deconstruction of the show; from a more self-reflexive attitude to a very judgmental remark (sometimes articulated in the same post). Moreover, the online reactions to the 2007 show indicate fragmentation of positions (such as a diverging under-standing of cultural hierarchies of taste) and simultaneous softening and strengthening of distinctions between the particular taste cultures of the users. The analysis exposes contradictory attitudes towards contemporary Western artistic (bad) taste and tension about the appropriation of contemporary (Western, Soviet and post-Soviet) fashion.

The spatial boundaries of the discussion about the 2007 Eurovision Song Contest show deserve a final note. Even if members of the forums cannot travel abroad, they are able 'virtually' to cross state borders and enter different cultural realms. This 'symbolic mobility' (Strukov 2010, p. 165), empowered by new media, is an important aspect for the post-Soviet audience, who might remember Soviet constraints on travel abroad and the lack of freedoms associated with it. As a result, a significant part of the debate refers to 'European' space. At the same time, the majority of the analysed forums are not directly linked to the Eurovision website. The discussions are mostly located at the forums of national web portals, regional blogs and (often Russified) social platforms.[37] While the televised Eurovision Song Contest entry is mostly constructed as a brand for the EU, the online claims for a place in New Europe are predominantly articulated for a post-Soviet rather than Western audience. The digital 'locality' acts as a cultural mediator of a transnational televised media spectacle. The analysed online contestation of national brands, thus occurs within a more 'national' internet space echoing the glocalisation framework (Appadurai 1997).

*Conclusion*

My analysis of the Ukrainian 2007 Eurovision 'brand' and the online responses to it generated several conclusions. Firstly, the identified 'kitschification' of the show both reinforces national and other stereotypes and, by creating an ironical distance, challenges and renders them ambiguous. The incorporation of folk motives and sexual

---

[37]Obviously, there are other factors potentially contributing to that, such as the language issue (e.g. a tendency to stay within one's own native language part of the official Eurovision Song Contest website rather than contribute to other discussions in English, which seems to be becoming the official language of the contest) and the broadband connection (e.g. difficulties downloading high-bandwidth content from the official Eurovision Song Contest website). Moreover, in 2007 the construction of fan culture around the Eurovision Song Contest site was still ongoing (as compared with the current official Eurovision presence where the website even has a section for an academic conference running in parallel with the contest).

excess in the show is a 'double voiced' act, which can be read at face value and as a self-conscious parody to address both Western assumptions about the sexual availability and/or promiscuity of the East, and Western longings for an East, purified of modern sexual ambiguities and still in touch with primeval male and female archetypes (Baer 2009, p. 14 ). It thereby challenges reductionist conceptions of East–West cultural geography.

Secondly, the instruments of sexual and aesthetic excess in Serduchka's 'deviant' camp show are also used to address the representation and misrepresentation of the post-Soviet 'Other' and represent the Ukrainian version of post-Sovietness, which is characterised by a daring manipulation of Soviet symbols. Here I should note that, in the post-Soviet context, the distinction between inter-cultural and intra-cultural is itself problematic. The fall of the Soviet Union generated a plurality of overlapping versions of the 'post-Soviet self'. For instance, former Soviet republics compete with one another to claim the *estrada* tradition for themselves. This competition is, on the temporal level, an intra-cultural struggle and, on the spatial level, an inter-cultural polemic.

Thirdly, the analysis of the show and the online discussion highlights the intertwining of kitsch with geopolitics. This is a feature of all Eurovision entries. The kitsch, self-ironic homosexual eroticism of Serduchka may offer a comforting sense of superiority to domestic fans aware of the ploy with which their 'unsuspecting' Western counterparts are 'fooled', but by that same token it is also a highly effective tactic for 'impressing' those counterparts with post-communism's newly found 'progressive modernity', and thus for securing votes.

Fourthly, the analysis of Eurovision forums explicates their role as a platform for national–cultural identification and the negotiation of one's individual stance. The discussions are built within and around certain spatial and temporal spaces, reaffirming particular 'cultures' of taste. The persona of the singer is at the core of the debates, where it is often reconstructed as the 'Other' at the heart of the self. Despite the variety of the topics covered, the core of the discussion constitutes the nation-branding issues ('us' compared to 'them'): how they see us and how we want to be perceived by them. The passionate online debates that the 2007 camp-style entry evoked help clarify meanings on such topics as taste and Euro-taste, cross-cultural differences and post-Sovietness. The public also expresses its opinion on the limits of the sexually permissible. However, the forums do not become a place where certain identities (such as homosexual ones) and alternative narratives were articulated. Instead, they reinforce normative cultural positions.

Finally, despite certain identified limitations of online forums, it is clear that the public often re-appropriates meanings away from those intended by the producers and artists, etc. and constructs its own taste cultures. The online mediascape blurs the distinction between highbrow and lowbrow culture, as it brings together groups of various taste- and fan-cultures that surf and contest each other's sociocultural realms. As a result, it becomes a place where various societal preconceptions or, as Bourdieu puts it 'the only legitimate mode of perception' (1984, p. 23), are challenged. Further insight into the deliberative potential of digital interactive technologies is especially important in the age of media convergence, which is reshaping popular culture and altering the interaction between the content, producers and the audience as stated by

Jenkins (2006). Future research might seek to identify how the post-broadcast audience (such as various subcultures and social movements) evades, subverts or resists the normative mediated meanings and practices.

*University of Oxford*

## References

Addley, E. (2008) 'Comic O'Grady Joins Dames of Honour', *The Guardian*, 14 June, available at: http://www.guardian.co.uk/uk/2008/jun/14/monarchy.britishidentity, accessed 11 October 2010.

Allatson, P. (2007) '"Antes Cursi Que Sencilla": Eurovision Song Contests and the Kitsch-Drive to Euro-Unity', *Culture, Theory and Critique*, 48, 1, pp. 87–98.

Anderson, B. (1991) *Imagined Communities: Reflections on the Origins and Spread of Nationalism* (London, Verso).

Appadurai, A. (1997) *Modernity at Large: Cultural Dimensions of Globalisation* (New Delhi, Oxford University Press).

Baer B. J. (2009) *Other Russias: Homosexuality and the Crisis of Post-Soviet Identity* (New York: Palgrave Macmillan).

Baker, C. (2008) 'Euro Visions: Culture, Identity and Politics in the Eurovision Song Contest', *Popular Communication*, 6, 3, pp. 173–89.

Bakhtin, M. M. (1981) *The Dialogic Imagination: Four Essays* [edited by M. Holquist and translated by C. Emerson & M. Holquist] (Austin, TX & London, University of Texas Press).

Bilaniuk, L. (2005) *Contested Tongues: Language Politics and Cultural Correction in Ukraine* (Ithaca NY, Cornell University Press).

Björnberg, A. (2007) 'Return to Ethnicity: The Cultural Significance of Musical Change in the Eurovision Song Contest', in Raykoff, I. & Tobin, R. D. (eds) (2007) *A Song for Europe: Popular Music and Politics in the Eurovision Song Contest* (Aldershot, Ashgate), pp. 13–23.

Bolin, G. (2006) 'Visions of Europe', *International Journal of Cultural Studies*, 9, 2, pp. 189–206.

Bourdieu, P. (1984) *Distinction: A Social Critique of the Judgement of Taste* (London, Routledge & Kegan Paul).

Boym, S. (1994) *Common Places: Mythologies of Everyday Life in Russia* (Cambridge, MA, Harvard University Press).

Castells, M. (2001) *The Internet Galaxy: Reflections on the Internet, Business and Society* (Oxford, Oxford University Press).

Coleman, S. (2008) 'Why is the Eurovision Song Contest Ridiculous? Exploring a Spectacle of Embarrassment, Irony and Identity', *Popular Communication*, 6, 3, pp. 127–40.

Couldry, N. (2000) *The Place of Media Power: Pilgrims and Witnesses of the Media Age* (New York & London, Routledge).

Crescente, J. (2007) 'Performing Post-Sovietness: Verka Serduchka and the Hybridization of Post-Soviet Identity in Ukraine', *Ab Imperio*, 2, available at: http://abimperio.net/cgi-bin/aishow.pl?state=showa&idart=1911&idlang=1&Code=, accessed 11 September 2010.

Dayan, D. & Katz, E. (1992) *Media Events: The Live Broadcasting of History* (Cambridge, MA, Harvard University Press).

Fawkes, H. (2007) 'Angry Ukrainian Nationalists Have Held Protests Against the Selection of a Controversial Drag Queen to Represent Ukraine in the Eurovision Song Contest', *BBC News*, 2 April, available at: http://news.bbc.co.uk/1/hi/6516927.stm, accessed 12 September 2010.

Gans, H. J. (1999) *Popular Culture and High Culture: An Analysis and Evaluation of Taste* (New York, Basic Books).

Goscilo, H. & Strukov, V. (eds) (2010) *Celebrity and Glamour in Contemporary Russia: Shocking Chic* (London, Routledge).

Goudsblom, J. & Mennell, S. (eds) (1998) *The Norbert Elias Reader: A Biographical Selection* (Oxford, Blackwell).

Hall, S. (1980) 'Encoding/Decoding', in Hall, S., Hobson, D., Lowe, A. & Willis, P. (eds) (1980) *Culture, Media, Language* (London, Hutchinson), pp. 128–38.

Jenkins, H. (2006) *Convergence Culture: Where Old and New Media Collide* (New York, New York University Press).

Kuzio, T. (2005) 'From Kuchma to Yushchenko: Ukraine's 2004 Presidential Elections and the Orange Revolution', *Problems of Post-Communism*, 52, 2, pp. 29–44.

Le Guern, P. (2000) 'From National Pride to Global Kitsch: The Eurovision Song Contest', *The Web Journal of French Media Studies*, 3, 1, October, available at: http://wjfms.ncl.ac.uk/leguWJ.htm, accessed 10 September 2010.

Lotman, Y. (1990) *Universe of the Mind: A Semiotic Theory of Culture* (Bloomington & Indianapolis, IN, Indiana University Press).

MacFadyen, D. (2002) *Estrada?!: Grand Narratives and the Philosophy of the Russian Popular Song since Perestroika* (Montreal, McGill-Queen's University Press).

Meadows, M. S. (2008) *I, Avatar: The Culture and Consequences of Having a Second Life* (Peachpit, New Riders).

Morris, J. (2010) 'Elevating Verka Serduchka: A Star-Study in Excess Performativity', in Goscilo, H. & Strukov, V. (eds) (2010), pp. 195–218.

Mutz, D. C. (2006) *Hearing the Other Side: Deliberative versus Participatory Democracy* (Cambridge, Cambridge University Press).

Nakamura, L. (2002) *Cybertypes: Race, Ethnicity, and Identity on the Internet* (Routledge, London).

Partan, O. (2007) 'Alla: The Jester-Queen of Russian Pop Culture', *The Russian Review*, 66, 3, pp. 483–500.

Poster, M. (2001) *What's the Matter with the Internet?* (Minneapolis, MN, University of Minnesota).

Rajagopalan, S. (2010) 'How to be a Well-groomed Russian: Cultural Citizenship in the Television/ New Media Interface', *Digital Icons: Studies in Russian, Eurasian and Central European New Media*, 3, pp. 87–101, available at: http://www.digitalicons.org/wp-content/uploads/2010/07/Rajagopalan-3.5.pdf, accessed 10 June 2011.

Rheingold, H. (1993) *The Virtual Community: Homesteading on the Electronic Frontier* (Reading, MA, Addison-Wesley).

Sontag, S. (1964) 'Notes on "Camp"', in Sontag, S. (1964) *Against Interpretation and Other Essays* (New York, Farrer Straus & Giroux), available at: http://www.math.utah.edu/~lars/Sontag:: Notes%20on%20camp.pdf, accessed 18 September 2010.

Strukov, V. (2010) 'Russian Internet Stars: Gizmos, Geeks, and Glory', in Goscilo, H. & Strukov, V. (eds) (2010), pp. 144–68.

White, M. & Schwoch, J. (eds) (2006) 'Introduction', in White, M. & Schwoch, J. (eds) (2006) *Question of Method in Cultural Studies* (Oxford, Blackwell), pp. 1–16.

Wilson, S. & Peterson, L. (2002) 'The Anthropology of Online Communities', *Annual Review of Anthropology*, 31, pp. 449–67.

Yurchak, A. (2006) *Everything was Forever, Until it was No More. The Last Soviet Generation* (Princeton, NJ, Princeton University Press).

# Blogging the Other: Construction of National Identities in the Blogosphere

## NATALIA RULYOVA & TARAS ZAGIBALOV

### Abstract

This essay examines the construction of Russian and Chinese national identities through the representation of the other in the blogosphere. Strategies that bloggers use to represent the other are analysed quantitatively and qualitatively. The quantitative examination is strengthened by a comparative perspective: the representations of Russian people by Chinese bloggers are contrasted with those of Chinese people by Russian-language bloggers. The qualitative examination of blogging posts is underpinned by a genre approach which leads the authors to the conclusion that the blogger's choice of genre, such as personal journal, futuristic fiction or 'letter into the blogosphere', is an important strategy in online identity construction and the representation of the other.

MOST OF THE RESEARCH INTO THE CONSTRUCTION of national identities in post-Soviet Russia has focused on strategies applied at the national level (Tolz 2001, 2003). However, very little is known about the ways in which citizens construct their national identities at the grassroots level.[1] In other words, we understand fairly well how the Russian state has manipulated the state-controlled mass media in order to promote its national agenda, and we know how individual viewers interpret nation-building messages disseminated by state-sponsored television channels (Mickiewicz 2008; Hutchings & Rulyova 2009), but we have a very limited understanding of how individuals engage in national identity discourses from the bottom up, not directly in response to a piece of state propaganda.[2] Social media and blogging, in particular, have enabled researchers to gain such an insight. This essay aims to analyse strategies used by bloggers to create their national identities through the representation of the other. Our work is underpinned by Stuart Hall's and Jonathan Rutherford's understandings of cultural 'identity as a "production", which is never complete, always in process, and always constituted within, not outside representation' (Hall 1990, p. 222). We also draw on Rudolf De Cillia, Martin Reisigl and Ruth Wodak in interpreting national identity as a type of cultural identity (De Cillia *et al.* 1999).

---

[1]For construction of local identities in provincial urban Russia, see Morris (in this collection).

[2]Strukov to some extent touches on how individuals are engaged in the construction of national identity in computer games (see Strukov in this collection).

One of the key roles in the construction of national identity is played by the representation of the other because it helps formulate cultural difference, which is conceived as 'the effect of the other': the otherness represents 'what is alien', 'the repository of our fears and anxieties' (Rutherford 1990a, p. 10). The other could be manifested in terms of sexual, gender and racial binarisms: male and female, black and white, heterosexual and homosexual, etc. These polarities are often part of dominant, national discourses which can lead to the perpetuation of racism, homophobia and misogyny. This essay is primarily concerned with the discourses of 'othering' in the Russian-language blogosphere. The need for this is justified by an increase in anti-immigrant, including anti-Chinese, sentiment in post-Soviet popular discourses. However, we find that researching the representation of the relationship with the other would have been less productive without a comparison. Hence, we introduce a comparative aspect to mirror the two 'nations' of bloggers—Russian and Chinese—to see what differences and similarities there are in their representations of each other. It also helps to tease out how national identity is formed in the collective consciousness of internet users.

Drawing on De Cillia *et al.* (1999) and Uri Ram (1994), our examination of the national in individual blogging entries is informed by the conceptualisation of nation or nationality as a narrative, or a story that people create to make sense of their social world. To select relevant blogging entries, we have focused on one aspect of identity construction: the representation of the other. Examining individual blogs, we found some common themes in the ways in which bloggers interpret the other and, as a result, reflect on themselves. We consider these unifying themes as features of the bloggers' collective national identities, and we also compare them with the themes identified in mass media discourses in relation to the other. This comparison shows similarities and differences in identity construction between the top-down mass media and bottom-up social media practices in the period after the 1990s national identity crisis (Tolz 2001, 2003).

Under Putin, the search for national identity accompanied the spread of right-wing extremism and the return of Eurasianism (Tolz 2001, 2003). Russians looked for self-identification through racial, ethnic and linguistic definitions. These differences become manifest in the Russians' othering of the neighbouring Chinese. Russia has had long-standing and tumultuous relations with China, including periods of shared experience, such as communism, and periods of confrontation. Currently, post-communist Russia looks with admiration at China's rocketing economic development. Russia's decreasing population, especially in the Far East and Siberia, is another reason for jealousy caused by asymmetry in population density in neighbouring areas of the two countries: 12 people per square km in Primorskii *Krai*, according to the *Krai*'s official website,[3] and between 10 and 300 people per square km in adjacent Chinese regions, depending on whether the area is urban or rural, according to the website of the Ministry of Foreign Affairs of the People's Republic of China.[4] Discrepancies in economic and demographic development in Russia and China feed

[3]See http://www.primorsky.ru/primorye, accessed 11 January 2011.
[4]The website of the Ministry of Foreign Affairs of the People's Republic of China, 'Population', available at: http://www.fmprc.gov.cn/eng/ljzg/zgjk/3580/t17857.htm, accessed 20 February 2012.

into the Russians' feeling of insecurity, especially in the context of the above-mentioned post-Soviet crisis of national identity. Russian mass media help disseminate fears and anxieties about the potential of Chinese demographic, economic and even military expansion (see the following section). A combination of these factors, along with growing xenophobia and nationalism in Russia, at times leads to the return of the 'yellow peril' rhetoric.[5]

Before turning to the analysis of the representations of the Chinese in Russian blogs, this essay will first discuss how the Chinese are represented in Russian mass media, and follow with an overview of academic and popular literature about China and the Chinese published in Russia since 2000. This helps us identify in our conclusions some overarching themes developing in both the mass media and social media. In the following section, we describe our approaches to the analysis of blogging entries, which include quantitative and qualitative analyses of selected blogs. The quantitative analysis is strengthened by a comparative perspective: the representations of Russian people by Chinese bloggers are contrasted with those of Chinese people by Russian-language bloggers. The qualitative examination of blogging posts is based on by a genre approach. The blogger's choice of genre is analysed in terms of its relevance to online identity construction and the representation of the other.

### *Representations of China and of the Chinese in Russian mass media*

This section demonstrates the need for our study. Our preliminary survey of some national and regional mass media conducted in the summer of 2008 suggests that messages about China and the Chinese are often contradictory and even confusing. Central television channels, such as Channel One and Rossiya, mostly give neutral or positive coverage of China in news bulletins and in such programmes as *Muslims* on the Rossiya 1 channel,[6] the travel programme *In Search of Adventures* on the Rossiya channel[7] and some documentaries. Critical comments on China and the Chinese people on Russian central television channels are infrequent but can still be found, for instance in documentaries, such as that on the Rossiya channel *The Planet of Orthodoxy* (*Planeta pravoslaviya*), in which China is criticised for forbidding foreign missionaries to spread the Russian Orthodox religion. On the whole, 'official' representations of China and of the Chinese that are disseminated by national, state-sponsored media are positive, such as that given in the popular newspaper *Komsomol'skaya Pravda*, where Larisa Ledeneva wrote about Khabarovsk's improving international links, in particular with nearby Chinese cities (Ledeneva 2008). The newspaper promotes China as 'the best country to visit' (Poluyaktova 2008). When some criticism appears in news sections, such as information about Chinese people smuggling Ussuriisk tigers from Siberia, the newspaper underlines the measures taken by the Chinese authorities to improve the situation (Ryabtsev 2008).

---

[5]According to Pavel Fel'gengauer, the term 'yellow peril' (*zheltaya ugroza*) was introduced by the German Emperor Wilhelm II during the anti-Western revolt in China in 1899–1901. For further discussion of the military aspects of Sino-Russian relations see Fel'gengauer (2008).

[6]*Musul'mane*, broadcast on 15 September 2009.

[7]*V poiskakh priklyuchenii*, broadcast on 15 August 2009, 7 December 2009, 2 August 2009 and 30 November 2009.

However, the representations of the Chinese outside the official state discourse in regional and independent media, in particular, are mixed and often negative. These mass media publish materials about increasing immigration from China (Filimonov 2008; Goryukhina 2005; Soinova 2005; Varsegov 2003; Klimov 2006),[8] and about supplies of cheap Chinese goods of poor quality to Russia. They spread rumours that the Chinese have adopted programmes on the development of the Far East and speculate about Chinese expansion (Groman 2002; Kez 2002). They alarm readers with stories about the Chinese threat (Sharavin 2001). They spread unconfirmed reports that Russia may lose some Siberian and Far Eastern lands to China (Tsyganyuk 2008) and that the Chinese could take control over Russian natural resources.[9] They discuss threats caused by the organised Chinese mafia in Russia (Lisanov 2008). Many of these anxieties are rooted in a long-term history of anti-Chinese sentiment in Russia. Some fears have been perpetuated by republishing old books online, such as Vladimir Arsen'ev's *The Chinese in Ussuriiskii Krai* (*Kitaitsy v Ussuriiskom krae*), which was written in the early twentieth century and in which the author warns of the potential dangers of a possible Chinese expansion into the Far East (Arsen'ev & Nansen 2004).

The relationship between Russia and China has recently become a popular research topic among Russian scholars and analysts also. Some of them see the improving relationship between Russia and China in a positive light. Among others, Yurii Peskov maintains that Russia and China have developed a 'new friendship' based on a growing trade between the two countries and on the common interests shared by the members of the Shanghai Cooperation Organisation, which was founded in 2001 and includes Russia, China and the countries of Central Asia (Peskov 2007). The sinologist Alexander Lukin, on the other hand, takes a more cautious approach and alerts Russians to the fact that China is quickly becoming richer and more influential than Russia. Lukin's discussion of the perceptions of China in Russia is based on an insightful historical overview of the issue (Lukin 2007). Another cautious view is expressed by the sinologist Vilia Gel'bras, who warns Russians of an increasing demographic imbalance between the bordering Chinese and Russian provinces (Gel'bras 2001). In addition, Sino-Russian international relations and cooperation have recently been examined among others by Konstantin Penzev and Evgenii Strigin (Penzev 2007; Strigin 2008).

Meanwhile, an increase in negative attitudes towards the Chinese has been registered by the Public Opinion Fund (*Fond Obshchestvennoe mnenie*, FOM): the number of people who have a positive view of China decreased from 43% in 2001 to 29% in 2008.[10] This trend also reflects a general rise in anti-immigrant sentiment. According to some experts, there is less anti-Chinese sentiment than that expressed

---

[8]See also 'Kitaiskikh gastarbaiterov v Rossii stalo bol'she ukrainskikh', *Lenta.Ru*, 5 June 2007, available at: http://lenta.ru/news/2007/06/05/chinese, accessed 20 February 2012.

[9]'My sami pozvolyaem kitaitsam upravlyat' Dal'nim Vostokom Rossii—eksperty', *Amur.net*, available at: http://hab-amur.net/news/index.php?option = com_news&task = view&id = 12172, accessed 28 February 2012.

[10]'Fond Obshchestvennoe mnenie', available at: http://bd.fom.ru/report/map/d083021, accessed 15 October 2010.

against the people from the Caucasus and Central Asia.[11] For example, according to the Public Opinion Fund, 35% of Russians share a view that the fact that there are many ethnicities living in the country is more harmful than beneficial. Some 25% of Russians surveyed said that there were conflicts between people of different ethnicities in their town or village in the last year. A total of 69% mentioned that in the last year they had read materials in the media about interethnic conflicts, which supports our observation about negative representations of the Chinese in the media.[12]

To summarise, the representations of the Chinese reflect two contradictory and simultaneously developing tendencies: on the one hand, the improving Russo-Chinese international relations facilitated by the Putin and Medvedev governments; and on the other, the growing anti-immigrant, xenophobic sentiment fed by domestic discourses. The positive coverage of China includes the representation of Russo-Chinese official visits and events organised within the framework of national programmes, such as the year of Russia in China in 2006[13] and the year of China in Russia in 2007.[14] Unlike in the national media, the representations of the Chinese in regional and independent media, including popularised academic literature, are to a greater extent tainted by the domestic discourses of nationalism, ethnic and religious intolerance, and xenophobia, as shown above.

### Theoretical approaches and methodology

After the survey of mass media presented above, we now turn to social media. Our analysis of the selected Russian and Chinese blogs is quantitative and qualitative. The quantitative analysis has been designed to reflect the comparative aspect of our approach to the representation of the other. We identify the most frequently used terms and phrases in both the Russian and Chinese blogs. The results have been collated in two tables and presented with commentary in the relevant sections. The majority of Russian blogs come from LiveJournal, LiveInternet, Ya.ru and blog.mail.ru, the four most popular blogging sites on the *Runet* (Etling *et al.* 2010, p. 12).[15] We made the first selection of 50 Russian blog posts in March 2010 and a further 575 between 13 June and 4 August 2010. Blog posts for this selection were retrieved using the query '*kitaitsy v rossii*' ('the Chinese in Russia') and harvested from a Yandex blog search.

As for the blogging posts in Mandarin Chinese, most come from the four most popular Chinese blogging sites: http://blog.sina.com.cn, http://hi.baidu.com, http://blog.163.com and http://blogbus.com. We selected 1,073 items over the period of 5–24

---

[11]Interview with Dmitri Trenin by the author, Carnegie Endowment, Moscow, 20 October 2009. Trenin is the author of the book *Russia's China Problem* (Trenin 1999).

[12]'Fond Obshchestvennoe mnenie', available at: http://bd.fom.ru/report/map/d072725, accessed 10 November 2010.

[13]'God Rossii v Kiate', available at: http://russian.china.org.cn/international/archive/russian/node_2222334.htm, accessed 20 February 2012.

[14]'God Kitaya v Rossii', available at: http://russian.china.org.cn/international/archive/zhongguonian/node_7014420.htm, accessed 1 October 2010.

[15]See http://livejournal.ru, http://www.liveinternet.ru, http://www.ya.ru and http://blog.mail.ru, all sites accessed 22 June 2012.

July 2010. To retrieve these blogging posts, we used the most effective Chinese search engine, Baidu (the equivalent of Google in English or of Yandex on the *Runet*), with the query 'the Russians'. The analysed Chinese blogs are written in Mandarin Chinese in simplified Chinese characters. As seen from above, the queries used in Russian and in Chinese differ because the query 'the Russians in China', which we tried first to match the Russian query, did not produce many results (we suggest reasons for this below), so we expanded our search by opening it up to 'the Russians' in general, not only those who live in China.

Our qualitative approach draws on Bell's understanding of the internet as a cultural space (Bell 2001) and on the close reading of the blog posts. To paraphrase Bell, we analyse how the Chinese and the Russians are 'storied into being at the intersection of different knowledges and metaphors' in the blogosphere (Bell 2001, p. 6). Hence, in our discussion of national identities we see them as those conceived discursively, 'by means of language and other symbolic systems, *produced, reproduced, transformed* and *destructed*' (De Cillia *et al*. 1999, p. 153, emphasis in original). A study of strategies used to represent the other illuminates our understanding of methods of self-identification. We used semiotic analysis to scrutinise the representations of the other in terms of stereotypes, myths, underlying narratives and associations used by individual bloggers for identity construction in blogs.

The present-day format of the blog dates back to 1996 when the term blog first appeared (Herring *et al*. 2004). When blogs first emerged they were described as a new genre related to the older genre of personal journal or diary. However, it soon became obvious that all blogs could not be defined as personal journals. Herring *et al*. suggest that blogs are best defined as 'a hybrid of existing genres, rendered unique by the particular features of the source genres they adapt, and by their particular technological affordances' (Herring *et al*. 2004, p. 10). Although the most common genre employed by bloggers is still the 'personal journal', amounting to 70.4% of the blogs examined by Herring *et al*. (2004, p. 6), it is far from covering all blog genres. Miller and Shepherd (2004) have come to the conclusion that 'the blog... as a genre... addresses a timeless rhetorical exigence in ways that are specific to its time'. They identify several ancestral genres which develop into the branches of the blog family, including political journalism (pamphlet, broadside, editorial), the journal and the diary. However, in a later article, Miller and Shepherd suggest that the blog 'is a technology, a medium, a constellation of affordances—and not a genre' (2004, p. 283). We draw on the now generally accepted understanding of the blog not as a genre but as a medium, or a publishing platform. Bloggers use these publishing platforms to write in a variety of genres (such as rhetorical, literary, fictional) and styles.

At the core of our understanding of the genre is Carolyn Miller's definition of genre as social action (Miller 1984). Miller states 'that a rhetorical sound definition of genre must be centered not on the substance or the form of discourse but on the action it is used to accomplish' (Miller 1984, p. 151). Socially motivated agents, or bloggers, act in similar ways in recurrent rhetorical situations by using certain genres for certain purposes. Hence, genres are defined by both form and social context, and bloggers actively construct their online discursive identities through their choice of a specific genre. We consider blog genres as strategies used by the agent, or the blogger in the process of identity construction.

An important feature of genres is that they are identified through typification. According to M. A. K. Halliday, there is a 'number of general *types* of situations, which we can describe in such terms as "players instructing novice in a game", "mother reading bedtime story to child", "customer ordering goods over the telephone", "teacher guiding pupils", "discussion of a poem", and the like' (Halliday 1978, p. 29, emphasis in original) Analogously, we have identified some types of blog postings in our selection, such as personal journal, futuristic fiction with apocalyptic undertones, 'a letter into the blogosphere' and fictionalised personal journal. Each type of blog posting is characterised by similarities in two main aspects: textual features (a narrative, themes, characters if there are any, chronotope); and the action that the agent or the blogger aims to fulfil (whether it is to express a certain feeling, to appeal to a particular group of readers, to advocate an idea or to disseminate information). Choosing a genre for a blog posting, the blogger commits to a set of genre conventions which act as a vehicle for discursive identity production. The blogger inscribes him or herself into a generic form and enters into a contract with the reader based on those conventions.

Our textual analysis of blog postings draws on by Mikhail Bakhtin's understanding of genre. Bakhtin (1986) identifies speech genre on the level of the speech utterance, analysing the thematic content, style and compositional structure of individual utterances. Among speech genres, Bakhtin recognises short rejoinders of daily dialogue, everyday narration, military command, business documents and commentary. He describes these genres as primary. Secondary speech genres, in his view, include novels and dramas. Genre in blogs can be identified on three levels: the level of speech utterances, the level of the blogging post and the level of the blog as a sequence of posts. On all levels, genre is inseparable from style and is inherently dialogic. The blogger and other users who can leave their comments on the blogging post are in dialogic exchange and actively produce meaning in response to previous utterances. Blogs often imitate colloquial speech to a greater extent than other written texts. Rutten has discussed imperfection in syntax and grammar in blogs by Russian writers (Rutten 2009). However, not all blogging posts are written in the style of colloquial exchanges. In addition to those, we have encountered a range of genres related to the family of literary genres including stories, futuristic fiction, 'letters in a bottle' and others. By choosing a genre, the blogger engages in meaning-making by relating an individual blogging post paradigmatically to other texts written in the same genre and simultaneously contributes to the formation or a mutation of this genre.

As for the specificity of the Russian-language blogosphere, it has become a large and popular self-publishing platform since the advent of '*Zhivoi Zhurnal*' or LiveJournal. Among other popular websites there are LiveInternet, Ya.ru and blog.mail.ru (Etling *et al.* 2010, p. 12). Blogging in Russia is used by the most diverse range of people (Alexanyan 2009), from President Medvedev, regional governors,[16] writers and poets (Rutten 2011), celebrities, journalists to ordinary citizens. Focusing on the blogs posted by Russian poets in her unpublished paper, Rutten develops Henrike Schmidt's description of the genre of the blog as 'literary fact': it is creative, commercial and virtual (Rutten 2011). There has not yet been a comprehensive

---

[16]For the further discussion of blogging by Medvedev and regional governors, see Yagodin and Toepfl in this collection; see also Goroshko and Zhigalina (2009).

analysis of genres in the Russian-language blogosphere and our study contributes a case study in which we examine a carefully selected group of blog postings on a particular topic in terms of genre.

However, before delving into the meaning of genre in the blogosphere, we discuss the results of our quantitative analysis first.

## The Russians in Chinese blogs

An analysis of the most frequently used terms appearing in blogs shows that Chinese bloggers are mostly interested in the Russian economy. We encountered 82 references to the Russian economy, including economic growth (see Table 1). In addition, they made 21 references to inflation and 22 to the 'average salary'. This shows interest in Russia's job market as well as the state of the domestic economy. The next big theme that interests the Chinese is Russian politics. The greatest number of references is made to the Russian president: 79. The number of references to the Russian government closely follows those to the president. Prime Minister Putin was mentioned 26 times. Russo-Chinese relations (45 references) represent the next popular topic for discussion. Chinese bloggers are most interested in treaties between Russia and China, in particular Aigun (47) and Pekin (21). The next topic of concern is the Russian military, including the Russian army (45), Russian intelligence (43), Russian weapons (23), nuclear crisis (21), Russian soldiers (20) and Russian military force (19).

TABLE 1

MAIN TOPICS THAT OCCURRED IN CHINESE BLOG POSTS WITH REFERENCE TO THE RUSSIANS

| Topics | Terms | Number |
| --- | --- | --- |
| Russian politics | Russian president | 51 |
| | Russian government | 50 |
| | Russian society | 30 |
| | US–Russian relations | 45 |
| | President Medvedev | 28 |
| | Prime Minister Putin | 26 |
| Russo–Chinese relations | Treaty of Aigun | 47 |
| | Russo-Chinese relations | 45 |
| | Treaty of Pekin | 21 |
| Russian economics | Russian economy | 63 |
| | Average salary | 22 |
| | Inflation | 21 |
| | Results of economic growth | 19 |
| Russian military | Russian army | 45 |
| | Russian intelligence | 43 |
| | Russian weapons | 23 |
| | Nuclear crisis | 21 |
| | Russian soldier(s) | 20 |
| | Russian military force | 19 |
| Russian people | Russian women | 34 |
| | Russian men | 25 |
| | Russian beauties (female) | 25 |
| | Russian girls | 20 |
| | Russian youth | 19 |

The final, but important, topic discussed in Chinese blogs is Russian people. Here the greatest number of references is made to Russian women, especially if we combine references to women, 'Russian beauties' and 'girls' (79 references). This is partly to do with stereotyping Russian women and partly it is indicative of existing demographic and social problems in Russia, such as gender misbalance caused by a gap between the average life span of men (59) and women (70) (Harding 2008). In addition, poor economic conditions and widespread alcoholism mostly among men also encourage Russian women to search for a foreign partner or husband. Marriages between Russian women and Chinese men have become more common.[17] To sum up, the three topics of greatest interest regarding Russia and the Russians for the Chinese include the Russian economy, the Russian government, the president and politics, and Russian women.

It is informative to compare our findings with the recent report entitled *How the Chinese See Russia*, published by an independent think tank (Lo 2010). The report takes an international relations perspective and draws on a number of interviews with leading Chinese experts and a body of specialist literature. There are some direct parallels between the report and our findings. According to the report, Russia is seen as an important power, declining internationally but still significant in the geopolitical context, especially as China's 'most direct—even though often unreliable—neighbour' (Lo 2010, pp. 1–3). Hence, bloggers in our research pay attention to Russo-Chinese treaties and Russia's military power. According to the report, the Chinese admire Putin as a leader but criticise his 'administration policy failures' (Lo 2010, p. 21). Our findings show that Chinese bloggers discuss the Russian president, in general, more than any particular leader, although both Medvedev and Putin feature in the blogs. The report states that the Chinese find the Russians difficult business partners. Our study reveals that despite that fact, bloggers are interested in the domestic Russian economy, discussing specific aspects of its development.

Many blogs are written by Chinese who have visited Russia and are based on personal, anecdotal interactions, thus it is possible to say that most posts are based on real life experience. However, these visits are often short and do not provide an opportunity for gaining an in-depth knowledge of the people living in the country. Therefore, the Russians are conceived as 'real' people, but often heavily stereotyped. As defined by Klapp, a stereotype is generally viewed as an inaccurate, rigid popular concept playing an important part in prejudice (Klapp 1962). For instance, drinking vodka is one of the main features that the Chinese associate with the Russians and, especially, Russian men:

> In Russia, drinking spirits is a symbol of being a man ….. Russian men drink alcohol like water ….. Quite often the Russians are associated with drinking vodka ….. The Russians have a capacity to drink significant amounts of spirits … like vodka, whisky, brandy. They also

---

[17]See the documentary *Pogranichnoe sostoyanie* (Reznichenko 2008). This documentary is a story about a woman from Blagoveshchensk who regularly travels to China. Her husband died and she is desperate to meet a Chinese man. The film is a reflection on social, economic and other aspects of life in the Russian Far East and bordering Chinese territories.

drink tasty, smooth liquors . . .. Russian people attach great importance to every holiday: they celebrate every holiday by drinking.[18]

One Chinese blogger links drinking vodka with accidents and comments on the biased Russian media:

Vodka isn't tasty, it feels like pure alcohol. I do not know why the *laomaozi* [hairy ones][19] like to drink it. When the weather is hot many drunk *laomaozi* drown [in rivers, lakes and water reservoirs]. This is what CCTV reported but we did not hear about it while we were in Moscow.[20]

As a nation, the Russians are seen as 'solid, sincere, tenacious, and unyielding'.[21] They are described as people who 'have spirit and psychology similar to Joseph Stalin's, the same temper, personality, and showing of force; it is a part of their personality'.[22] The Russians are also depicted as rough, generous, free, informal and lazy people who like having fun, and are relaxed, especially compared to the Chinese. This perception of the Russians as lazy and complacent echoes the findings of Lo's report, quoted above, which also states that Russia as a country is 'resting on its laurels as a traditional great power', which is 'wealthy only by virtue of its vast natural resources' (Lo 2010, p. 21).

One feature that the Russians and the Chinese have in common is that both the Russians and the Chinese appreciate their countries being strong and powerful:

In fact, Chinese and Russian nations are similar. The Chinese also worship power. The best example is a large military parade in Beijing, but the military parade on the Red Square in Russia means even more to the Russians . . .. The Chinese have a backbone. Last year's National Day parade was very refreshing to the people. Russia is a great nation, and a large military parade is a manifestation of Russia's self-confidence.[23]

One Chinese blogger links Russian freedom to randomness: 'I also found that some cars have their steering wheel on the right, some on the left side, which is consistent with the randomness and freedom of the Russian people. I found it very strange'.[24]

---

[18]Available at: http://blog.sina.com.cn/s/blog_681c3f8e0100kj8u.html, accessed 11 August 2010. The reader should be warned that some of the quotes from blogging posts are no longer available for viewing for various reasons. Texts in the blogosphere are permanently changeable.

[19]A derogatory term used to refer to the Russians.

[20]Yilu Xiangbei, Blog Post, available at: http://silentrainbow1000.blog.163.com/blog/static/4067324520106171176144, accessed 10 August 2010.

[21]buyaozhe1212, Blog Post, available at: http://blog.cnfol.com/buyaozhe1212/article/23145810.html, accessed 11 August 2010.

[22]This reference to Stalin can have more positive connotations than it might appear at first sight, as Stalin is interpreted as a strong leader. Also, Stalin was of course ethnically Georgian, not Russian but he is perceived as a manifestation of Russian national identity.

[23]http://blog.sina.com.cn/s/blog_537fd7410100ihd6.html, accessed 23 June 2012.

[24]Liuyi37, Blog Post, available at: http://blog.sina.com.cn/s/blog_6544617e0100jw3f.html, accessed 11 August 2010.

More negative assessments of the Russian character are concerned with the Russians having 'extremely low self-esteem and a big ego': 'They are empty and pretentious'.[25]

Another number of stereotypes about Russia are linked to its extreme winter temperatures, big expanses of land and Russia's geographical situation neither in Europe nor in Asia:

> My vision of that country is of one floating Siberian snow. 30 degrees below zero temperatures, unusually cold. Rude guys are drinking vodka and lying in the streets, in the snow …. It is neither Asia, nor Europe, not it is clear what it is and it makes people feel uncomfortable.[26]

Some bloggers are nostalgic about the Soviet period, including Soviet leaders Lenin and Stalin:

> There is a statue of Lenin in the station square, it is not very tall. Lenin's left hand is holding his hat, and his right hand is pointing forward. This is a great historical figure, the first revolutionary teacher of the proletariat, the leader of the former Soviet Revolution. But in the eyes of young people in present Russia, Lenin does not seem to be revered, and he is even forgotten. Some Russians even think that if the troublesome October Revolution led by Lenin did not occur, Russia could have been ahead of the USA in terms of economic development. Incredible.[27]

As noted above, the representations of the Russians depend on gender and age of those represented. While the stereotypical Russian man is a drinker, the stereotypical Russian girl is beautiful and the stereotypical Russian woman is fat: 'Russian girls in great shape, they are very beautiful. Eventually, they turn into the "Russian mother", looking like a penguin'.[28] Another blogger brings up a similar comparison: 'Russian girls are beautiful flowers but after they turn thirty they become like barrels'.[29]

To summarise, the majority of the analysed Chinese blogging posts represent Russians as the other by stereotyping and exaggerating some features of the Russian collective identity. The representations of Russians could be roughly separated into four groups: 'vodka-drinking men', 'beautiful girls', 'Russian mothers' and Russians as a nation. In the eyes of the onlookers, or bloggers, these stereotypical representations are linked to 'real life'; these national characteristics of Russians are constructed based on some—often limited—knowledge of Russia and/or Russian people. Some Chinese bloggers compare Russians to themselves but mostly in order to

---

[25]fengchun216, Blog Post, available at: http://hi.baidu.com/fengchun216/blog/item/df7fc4f ce2dd2d8fb801a056.html, accessed 13 August 2010.

[26]Yilu Xiangbei, Blog Post, available at: http://silentrainbow1000.blog.163.com/blog/static/ 4067324520106171176144, accessed 11 August 2010.

[27]Sima Pingbai, Blog Post, available at: http://blog.sina.com.cn/s/blog_537fd7410100ihd6.html# comment, accessed 10 August 2010.

[28]fengchun216, Blog Post, available at: http://hi.baidu.com/fengchun216/blog/item/df7fc4fce2dd2d 8fb801a056.html, accessed 13 August 2010.

[29]fengchun216, Blog Post, available at: http://hi.baidu.com/fengchun216/blog/item/df7fc4fce 2dd2d8fb801a056.html, accessed 10 August 2010.

emphasise some shared qualities, such as their appreciation of power and the strong state.

## The Chinese in Russian blogs

The quantitative analysis of popular topics in the Russian blogs presents a very different picture to the one we saw in the Chinese blogs. Unlike the Chinese, Russian bloggers are primarily interested in Russia's own domestic economy, politics and society (see Table 2). The greatest number of references in the analysed Russian blogs is given to Russian Orthodoxy (36) and Russian people themselves (28). A considerable number of blogging posts are concerned with the Russian population, its ethnic (32 references, in general, and 14 references to minorities) and religious constitution (Muslims are referenced 16 times). Among other ethnicities mentioned in blogs, Americans (24) and Vietnamese (14) are often featured. Confrontational terms such as 'distraction' (31), 'destroy' (12), 'expansion' (18), 'occupation' (16),

TABLE 2

MAIN TOPICS THAT OCCURRED IN RUSSIAN BLOG POSTS WITH REFERENCE TO THE CHINESE IN RUSSIA

| Topics | Terms | Number |
|---|---|---|
| Russian people | Orthodox | 36 |
| | Russian people (*russkie lyudi*) | 15 |
| | The Russian people (*russkii narod*) | 13 |
| Population and other nations | Population | 58 |
| | Millions | 27 |
| | Americans | 24 |
| | Number of people | 21 |
| | National | 21 |
| | Muslim | 16 |
| | Vietnamese | 14 |
| | Minority | 14 |
| Confrontation | Destruction | 31 |
| | Just | 21 |
| | Expansion | 18 |
| | Army | 17 |
| | Occupy | 16 |
| | Destroy | 14 |
| | Provocation | 13 |
| | World War II | 13 |
| | Destroy | 12 |
| | Warning | 13 |
| Politics and history | Territories | 22 |
| | Democratic | 21 |
| | Nationalist | 17 |
| | Communist | 16 |
| | Hitler | 14 |
| | Socialist | 14 |
| Economy | Money | 25 |
| | Trade-economic | 17 |
| | Earn | 15 |
| | Trade-economic cooperation | 13 |
| | Employment | 13 |
| | 'Big money' | 12 |

'provocation' (13) and 'warning' (13) are quite frequent. Historical and political issues discussed in blogs are marked by the use of such terms as 'territories' (22), 'nationalist' (21), 'democrat' (17), 'communist' (16), 'Hitler' (14) and 'socialist' (14). In addition, the discussion of territories with reference to China is often indicative of the Russians' anxieties regarding the possibility of losing more Russian territories to China, especially after the handover of the two islands Tarabarov and Bolshoi Ussuriiskii to China by the Russian government in 2004 (Svetlova & Miroshnikov 2004). Another category of frequently used terms is concerned with economy and finance including 'money' (25), 'big money'(12), 'making money' (15), 'trade-economic' (17), 'trade-economic cooperation' (13) and 'employment' (13).

A recurrent theme in the analysed Russian blogs is nostalgia for the former USSR, which is a theme similar to that expressed in some Chinese blogs. In such posts, the USSR is often interpreted as the peak of Russia's development. Some expand the theme of the collapse of the USSR to the extreme, saying that Russia will eventually break into pieces, and Siberia will belong to China.[30] A similar sentiment is shared by the blogger with an appropriately aggressive name, 'v_kalashnikov', who complains that while the Chinese youth are reading *How the Steel Was Tempered*, an iconic socialist realist novel published in 1934 by Nikolai Ostrovskii, the young Russians watch porn on their mobile phones.

In addition, the Russians renegotiate their perception of the Chinese alleviated from the position of the younger communist 'brother', as is stated in the song '*Moskva-Pekin*' to the economically superior and more successful neighbour.[31] The history of the 'older brother/younger brother' relationship led to 'antagonistic perceptions' in both nations: 'the Chinese seeing the Soviets as arrogant; the Soviets regarding the Chinese as petulant and ungrateful' (Lo 2010, p. 20).

The Russians are concerned about the Chinese only to the extent that the latter can impact on Russia's domestic situation. Partly, this is a result of our query 'the Chinese in Russia' instead of 'the Chinese'. However, the fact that there were very few references to 'the Russians in China' (see the 'Methodology' section above) and that there are so many references to 'the Chinese in Russia' may be indicative of two issues: the number of Chinese in Russia is significantly larger in proportion to Russia's population than that of the Russians in heavily-populated China; and the Chinese do not discuss the Russians as the other in the process of their self-identification. Both are true in relation to the Russians who are concerned about the number of Chinese immigrants in Russia, and they use the Chinese as a mirror in the process of self-identification.

Based on the above comparison of themes frequently referred to in Russian and Chinese blogs, it is possible to draw some preliminary conclusions. The discussion in Chinese blogs is primarily focused directly on the other; Chinese bloggers are interested in Russia and the Russians, ranging from politics and economy to Russian women. The interest in the Russians in China is insignificant. Russian bloggers, on the other hand, are primarily interested in the impact that the Chinese might have on

---

[30]Andrej Tim, Blog Post, available at: http://my.mail.ru/community/nationalsovet_rf/B414CE51 E16B02C.html?thread=2213C5EA740FD93, accessed 5 November 2010.

[31]'This is the Soviet Union walking ahead; / This is the mighty Soviet Union / A new China is walking next to it!' (music by Vano Muradeli, words by M.M. Vershinin).

Russia, but not so much in the Chinese people themselves. They use the Chinese as their mirror in their identity search. The Russian bloggers' fears and anxieties in relation to the Chinese are rooted in Russia's domestic problems and efforts to reinvent a national identity after the collapse of the Soviet Union. This is supported by the statement made by one Chinese student, saying that while she studied in Russia she was regularly asked what the Chinese make of Russia but she was rarely asked questions about China itself. She found surprising this 'narcissistic', or self-regarding, quality of the Russians.[32] The following section will test these preliminary conclusions by presenting a genre analysis of Russian and Chinese blogs.

### Genre in blogs

In the process of the qualitative analysis of Russian-language blogs with reference to China and Chinese, we were struck by the prevalence of certain genres, in particular, the genres of personal journal, futuristic fiction and 'the letter in a bottle' (in other words, a letter without an addressee or a letter that is written by the blogger without the expectation to receive a reply). We have observed a correlation between the content of the blogging posts and the form, i.e. between the themes expressed in the blog and the genre employed by the blogger. Certain genres are more suited for certain content. For example, bloggers share 'real life' experience in their personal journals. They employ futuristic fiction and 'the letter in a bottle' to express anxiety and fear. They fictionalise their 'experience' to create messages with ideological undertones.

### Personal journal: sharing 'real life' experience

The genre of personal journal is employed to share some 'real life' experience, such as meeting a Chinese person or a personal visit to China. Most positive representations in blogs can be found in personal journal entries aimed at breaking the existing negative stereotypes about the Chinese. Blogger R1, for example, is embarrassed by the Russian stereotyping of the Chinese after the blogger's meeting with a Chinese businessman in person.[33] A similar example is the post by the blogger motovetal who tells a story of a Chinese known as Gena whom he admires for his survival qualities demonstrated in Russia and Ukraine.[34] Some positive remarks are made in the blogs of the Russian-speaking bloggers who live in China. The blogger Shnonihov0b, for example, admires Chinese efficiency and productivity, commenting on the building of a superfast railway.[35] Positive stereotypes associated with the Chinese include such qualities as hardworking, efficient, productive and survivalist. Based on his first-hand experience of China, blogger Alexander Naumov explains why he finds a potential

---

[32]See the podcast with her interview available at: http://magazeta.com/podcasts/2010/08/04/laowaicast-35, accessed 15 November 2010.

[33]Newslife, Blog Post, available at: http://newslife.livejournal.com, accessed 15 February 2011.

[34]Motovetal, Blog Post, available at: http://motovetal.livejournal.com/35757.html, accessed 1 November 2010.

[35]Shnonihov0b, Blog Post, available at: http://shnonihov0b.livejournal.com/981.html, accessed 1 November 2010.

Chinese aggression unfounded.[36] Some bloggers share their plans for working with the Chinese. Another blogger publishes a comparative account of the development of Russian and Chinese territories and reveals Chinese strengths.[37] In all these representations of the Chinese, the signifier 'Chinese' is directly linked to the blogger's 'real life' experience of visiting China or meeting Chinese people.

*Futuristic fiction: apocalyptic forecasts*

The genre of futuristic fiction is employed by bloggers as a vehicle for nationalistic xenophobic discourses which project Russia's current domestic problems onto the future. For instance, the blogger Buntar' has published a piece 'We support justice and freedom', in which he contemplates a possible war between Russia and China, which would take place on Russian territory in the future.[38] The blogger Boris Dorokhov posts a similar futuristic story about wars between Russia, China and the USA.[39] The blogger support-man develops the nationalistic narrative by adding that Russia would use a nuclear bomb to protect itself against China and the USA.[40] In the interpretation of the blogger Farit Fatkullin, the USA aims to destroy China with the help of the Russians; the world would then be run by the World Mafia and Russia would be ruled by a Mafia Committee. Ordinary Russians would have microchips inserted into their brains and be thrown onto the Russo-Chinese border to fight the Chinese.[41]

Another representative example of such futuristic fiction has been posted in the journal of the blogger kolibri_2006. In *Moscow—the Province Khun'Zian'*, the author ironically predicts the creation of the New Russian language: one third of Russian speech will be English in Ryazan transcription with American shortenings, one third will be in the classic Prison jargon, and one third will be in the Chinese language. The extract from this text cited below reflects a range of anxieties that are circulated in the Russian media:

> The [Chinese] population was not decreasing! So, they decided to give away a part of Russia! The Russians kept oil fields, the city of Moscow known as 'Rublevskii Federal district', according to the 2008 Constitution, but soon these areas became Chinese too. The country is now called Khun'Zian' Province. The Chinese are everywhere. They never die …. There are two types of the currency: the dollar made in USA and the dollar made in China. The latter currency is made in sweatshops and has the same portraits on them as USA dollars (though the presidents have narrower eyes but the Chinese dollars are valued higher as they are harder to forge).[42]

---

[36]Alexander Naumov, Blog Post, available at: alexandernaumov.ru, accessed 1 August 2010.

[37]Persephona, Blog Post, available at: http://blogs.mail.ru/mail/persephona.74/764A437 B45118064.html, accessed 15 November 2010.

[38]Buntar', Blog Post, available at: http://my.mail.ru/community/www.buntbest.ru/4A1E86062F F971BC.html, accessed 1 August 2010.

[39]Boris Dorokhov, Blog Post, available at: http://my.mail.ru/community/writers_club/382404D D741CBB66.html, accessed 1 August 2010.

[40]support-man, Blog Post, available at: http://support-man.livejournal.com/3474.html, accessed 15 August 2010.

[41]Farit Fatkillin, Blog Post, available at: http://blogs.mail.ru/mail/fatfarit, accessed 15 August 2010.

[42]kolibri_2006, available at http://kolibri-2006.livejournal.com/256303.htm, accessed 3 November 2010.

This kind of xenophobic fantasy reveals anxieties among Russian people that are rooted in their own insecurities: inability to overcome the demographic crisis, especially in the provinces, and the slow development of the Russian economy, compared, in particular, with China. These insecurities manifest themselves in antagonistic and nationalistic discourses in relation to the nations which are perceived to be more successful: the Chinese and the Americans. These futuristic narratives often refer to the narratives disseminated in the mass media—the Chinese eating dog's meat, the poor quality products imported from China, counterfeit Chinese products, lazy Russians being taken advantage of by efficient Chinese workers, cheap labour provided by Chinese immigrants, high Chinese birth rate as opposed to low Russian birth rate. Among the most frequently circulated narratives are those of fear: there is a potential occupation of Russia by the Chinese and Americans, a loss of Russian natural resources and of the Russian language. Russian bloggers project their anxieties onto the future; with the help of black humour and irony they exaggerate existing problems, and transform them into the imaginary and take them to the level of the absurd.[43]

### Steb *as a popular device*

In addition to black humour, some blog posts use *steb*, a device characteristic of late Soviet and post-Soviet writing and viewing, which blurs the boundary between the serious and the ironic or sarcastic (Hutchings & Rulyova 2009). *Steb* requires 'a degree of overidentification with the object, person or idea at which [it] was directed that it was often impossible to tell whether it was a form of sincere support, subtle ridicule, or a peculiar mixture of the two' (Yurchak 2006, pp. 249–50). It allows the writer, the reader, or the viewer to hint at ridiculing something without actually ridiculing it. Inadvertently, it often leads to uncritical confusion of irony with its opposite. There are many examples of blog posts written with elements of *steb*. For instance, the blogger sapojnik[44] post contains 'a recipe for gentrifying the Russian *sovkomassa*' by having NATO military bases built in Russia to protect the latter from the Chinese and protect Russian democracy', which, he continues, would work only if Russia is run by the European Parliament.[45] This statement, which may appear absurd at first sight, divulges several major discourses that preoccupy Russian bloggers' imagination: the Westernisation of Russian, Russia's loss of status as a world power, Russia's unique path which differs from the East and from the West, Russia's fear of Chinese invasion.

---

[43]It is worth noting that this genre has recently been employed in pro-Putin videos produced as part of his 2012 presidential election campaign, which testifies that the genre is seen as one with a widespread appeal to the masses. See the video which manipulates oppositional discourses and portrays the future of Russia without Putin as apocalypse: 'Rossiya bez Putina? Apokalipsis zavtra!', available at: http://picgur.com/apokalipsis-zavtra, accessed 20 February 2012.

[44]Here and henceforth the names of bloggers are spelled as they appear in blogs. Hence, their spelling is irregular.

[45]sapojnik, posted on 12 July 2010, available at: http://sapojnik.livejournal.com, accessed 20 February 2012.

The blogger lisiy_hvost posts a request in a similar *steb*-fashion, addressing the Chinese:

> [Chinese are invited] to conquer Russia and build cafes everywhere, so that there are enough places where one can eat well. Chinese cafes are the most authentic in Russia because they are owned by the Chinese, run for the Chinese and are supplied with Chinese products. And prices are laughable.[46]

The generally positive comment on Chinese food is subverted by the sarcastic tone of the message, which creates a confusing effect. Apart from anxieties concerned with military and cultural occupation of Russia by the Chinese, there are linguistic and religious insecurities. For instance, the blogger kosarex spreads conspiracy theories including one about Chinese collaboration with the West against Russia.[47]

*Fictionalised personal journal*

Yet another genre of blog posts can be distinguished when bloggers present their 'real' or online experience as a piece of fictionalised narrative. In this case, the boundary between the real and fiction is blurred even further. An example of such a narrative can be found in Pfertsentakl's blog, who shares his experience of using a new service on Skype.[48] This piece of writing is self-consciously constructed and framed. At the beginning of this narrative, the blogger specifies his location: a small Ukrainian '*khutor*' (hamlet). His American female cousin recommends to him that he should check out the new Skype service. Then he describes his online meeting with a random Chinese male student. He begins with a portrait, a condescending description of the Chinese man, which draws on xenophobic stereotypes presented as 'jokes' with references to cyber-exhibitionism, porn on the internet, Chinese movies and so on. Then the blogger recites the whole of his dialogue with the Chinese man:

> So, the conversation: alas, I can't remember it word by word .... So this is 'a la conversation with the Chinese in free interpretation' (in translation from English into Russian[49]).

> I — Hi! Glad to have an opportunity to chat with you!
> C — Hi! Me too. Where are you from?
> I — From Zimbabwe. Do you know where it is?
> C — No.... In South America? (moronic Chinese)
> I — Of course not! North of Russia.
> C — Wow! I would like to visit Russia! And Zimbabwe.

[46]lisiy_hvost, Blog Post, available at: tourism_il@community.livejournal.com/tourism_il, accessed 10 August 2010.

[47]Kosarex, Blog Post, available at: http://kosarex.livejournal.com/408808.html, accessed 30 October 2010.

[48]See http://www.chatroulette.com, accessed 22 June 2012.

[49]In parentheses are the words of the blogger who is blogging in Russian but was having a conversation with the Chinese person in English. The author of this article has translated this back into English. The text has been reproduced in translation as faithfully to the source text as possible.

> I — Yes! In Russia and in Zimbabwe, we would be glad [to meet you].... By the way, where are you from?
>
> C — From China...
>
> (the most interesting bit begins now)
>
> I — Where is it?
>
> C — What? You don't know where China is? (half-surprised and half-disdained)
>
> I — Mmmm. Is it a small country which is located on the islands between Antarctic and Africa?

Then, the blogger gives a long and overblown description of the reaction of the shocked Chinese man who describes his country as a 'much respected nation in the East' with a huge population and so on. The 'punch line' of the story is that the blogger discovered a 'weapon' against the Chinese, which is that if or when they invade Russia one just needs to ask a Chinese person where China is. This dialogue is crafted as a piece of fiction but presented as a piece of 'real life' experience supported with some photos of Chinese people found on the internet to illustrate the narrative. Its only purpose is to invoke nationalistic and xenophobic feelings.[50]

### 'Letter into the blogosphere', or into the blogosphere

Some Russians use their blogs to post 'letters in a bottle', in which they express their general anxieties and fears. For example, the blogger Iurii Smolii posts a piece of advice to the Russian president, in which he shares his views on how Russian political life should be organised: through civic organisation, not party-led. He encourages mixed marriages between Chinese and Russians, and recommends opening the border between Russia and China to counteract the European Union.[51] Such letters are not expected to reach a recipient and are similar in some ways to letters in a bottle dropped into the sea in the past. In fact, they often do not have an addressee. They are used to express anxieties, frustrations and mixed-up feelings about the state of affairs in Russia. Such messages are contradictory, confused and angry; they mix up anti-Semitic, anti-Putin, racial, homophobic, xenophobic and other sentiments.

### Reposting

Bloggers often post links to, or republish, existing articles. Often these articles are of sensationalist character. For instance, there were two articles that were reposted most in July–August 2010. The first one claimed that the Chinese would take control of the Russian media, and the other one spread the rumour that the Chinese eat small children. Among other popular articles, there was a piece about imported Chinese shoes that are made in sweatshops and are bad for your feet posted by the blogger

---

[50]Pfertsentakl, available at: http://blogs.privet.ru/user/pfercegentakl-/84637706, accessed 1 November 2010.

[51]Iurii Smolii, available at: http://my.mail.ru/community/teni_vetra/377AA585044C6801.html, accessed 2 November 2010.

'who is kto'.[52] Recirculated articles are often published out of context, out of date and compiled with other articles. For instance, Kirilina Elena posted four articles from different sources published between 2003 and 2008 including 'Attention! Chinese Threat' by V Ivanov, E. Gilbo's 'Projections for the Chinasation of Russia'; 'The Chinese Will Run to Russia'; 'The Chinese Threat'.[53] Such alarmist collections appeal to emotion, perpetuate popular myths and fears.

## Conclusions

Our comparative analysis of the representations of the Chinese and the Russians in each other's blogging posts reveals national differences between the strategies that individual Chinese and Russian bloggers use to construct the other. These strategies are created bottom-up, at the grassroots level, by individual bloggers. Representations of the Chinese in the Russian blogosphere reflect Russian bloggers' fears and anxieties rooted in their domestic problems, such as decreasing population, low birth rate, nationalism, the post-Soviet crisis of national identity and xenophobia. In addition, the Chinese other is used as a mirror in the process of self-identification, complicated by the perceived loss of Russia's status as a super-power and hence injured national pride. Chinese bloggers, on the other hand, are interested in the development of the Russian economy, political situation and indeed in Russian people. The Chinese tend to write about Russia and the Russians in a matter-of-fact fashion, though their posts are often peppered with stereotypes presented as facts. The majority of blogs were written by Chinese people who had some first-hand experience of Russia or had met Russian people. We did not come across any posts written in the form of fiction, as in the Russian blogosphere. The Chinese also appear more straight-forward and less introspective in their representation of the Russians. They are more concerned with the actual state of affairs in Russia than in self-absorbed reflection inspired by anxiety which is characteristic of the Russians. The representation of the Chinese in Russian-language blogs is akin to the narratives spread in the mass media, especially in the local and independent press in which popular discourses are more spread.

Analysing Russian blogs in terms of genre has provided insight not only into the writing practices of Russian-language bloggers but also into mechanisms of meaning-making and identity construction. As a rhetorical means, genre is used 'for mediating private intentions and social exigence'; genre 'motivates by connecting the private with the public, the singular with the recurrent' (Miller 1984, p. 163). The blogger's choice of genre is determined by his or her social motive and is a strategy in identity construction. The blogger inscribes him or herself into the norms and expectations of a genre. Genres provide a framework for interpretation and meaning-making. There is a correlation between the content and the form: the blogger's topic and intentions dictate the genre. Genres are culturally specific: they must be interpreted in historical and social context. For instance, there are reasons (such as frustration with the post-Soviet political and social situation, lack of effective communication channels between

---

[52]who is kto, Blog Post, available at: http://whoiskto.livejournal.com/16429.html, accessed 3 November 2010.

[53]http://my.mail.ru/mail/a-kir/, accessed 22 June 2012.

the public and political elite) why in the post-Soviet Russian-language blogosphere the genres of futuristic fiction and 'letter in a bottle' act as vehicles to express anxiety, fear, to alarm readers or to appeal to them. Genres provide ready-made plots, structure, style and imagery which help bloggers to create texts, and make sense out of reality. These genres serve as contracts between the blogger and his or her readers by providing a set of expectations which are used as guidelines to the bloggers in their writing and readers or other users in their understanding and interpretation of the text. Genres can be described as the shared containers of meaning, which provide some DIY patterns for unprofessional writers and bloggers to construct their messages.

A genre approach to the blogs helps develop effective decoding strategies, which are necessary in the digital age when 'real life' and fictionalised stories are published next to each other and are often intertwined into one narrative. Some scholars described the online environment of texts as carnivalesque because it contains a mixture of genres, styles and discourses.[54] However, it is the genre that helps the reader and blogger intuitively find their way through texts.

*University of Birmingham*
*Brandwatch (UK)*

## References

Alexanyan, K. (2009) 'Social Networking on Runet: The View from a Moving Train', *Digital Icons: Studies in Russian, Eurasian and Central European New Media*, 1, 2, pp. 1–12, available at: http://www.digitalicons.org/wp-content/uploads/2009/11/Karina-Alexanyan-DI-2.1.pdf, accessed 28 February 2011.

Arsen'ev, V. & Nansen, F. (2004) *Kitaitsy v Ussuriiskom krae. V stranu budushchego* (Moscow, Kraft).

Bakhtin, M. (1986) *Speech Genres and Other Late Essays* [translated by V. W. McGee and edited by M. Holquist & C. Emerson] (Austin, TX, University of Texas Press).

Barthes, R. (1977) *Image, Music, Text* (London, Fontana Press).

Bell, D. (2001) *An Introduction to Cybercultures* (London & New York, Routledge).

De Cillia, R., Reisigl, M. & Wodak, R. (1999) 'The Discursive Construction of National Identities', *Discourse and Society*, 20, 2, pp. 149–73.

Etling, B., Alexanyan, K., Kelly, J., Farris, R., Palfrey, J. & Gasser, U. (2010) *Public Discourse in the Russian Blogosphere: Mapping RuNet Politics and Mobilization*, Research Publication No. 2010–11, The Berkman Centre for Internet and Society, Harvard University, October, available at: http://papers.ssrn.com/sol3/papers.cfm?abstract_id=1698344, accessed 20 February 2011.

Fel'gengauer, P. (2008) '"Kitaiskaya ugroza": voenno-tekhnicheskii aspekt', *Otechestvennye zapiski: Dykhanie Kitaya*, 3, 42, available at: http://www.strana-oz.ru/?numid=44&article=1672, accessed 20 December 2010.

Filimonov, D. (2008) 'Kitaitsy idut', *Izvestiya*, 18 June, available at: http://www.izvestia.ru/special/article3033246, accessed 1 December 2010.

Gel'bras, V. (2001) *Kitaiskaya real'nost' Rossii* (Moscow, Muravei).

Goroshko, O. & Zhigalina, E. A. (2009) 'Quo Vadis? Politicheskie w=kommunikatsii v blogosfere runeta', *Russian Cyberspace*, 1, 1, pp. 81–110, available at: http://www.digitalicons.org/issue01/pdf/issue1/Political-Interactions-in-the-Russian-Blogosphere_O-Goroshko-and-E-Zhigalina.pdf, accessed 28 February 2011.

Goryukhina, E. (2005) 'Kitaitsy v Sibiri', *Novaya gazeta*, 24 November, available at: httm://www.NovayaGazeta.Ru, accessed 22 September 2011.

Groman, S. (2002) 'Kitaitsy okkupiruyut Rossiyu', 14 January–20 October, available at: http://www.languages-study.com/demography/china-in-russia.html, accessed 3 July 2012.

Hall, S. (1990) 'Cultural Identity and Diaspora', in Rutherford, J. (ed.) (1990b), pp. 222–37.

---

[54]For a detailed discussion of the 'carnivalesque' in new media, see Bakardjieva in this collection.

Halliday, M. A. K. (1978) *Language as Social Semiotic: The Social Interpretation of Language and Meaning* (Baltimore, MD, University Park Press).

Harding, L. (2008) 'No Country for Old Men', *The Guardian*, 11 February.

Herring, S., Sheidt, L. A., Bonus, S. & Wright, E. (2004) 'Bridging the Gap: A Genre Analysis of Weblogs', *HICSS*, 4, p. 40101b, available at: http://www.computer.org/portal/web/csdl/abs/proceedings/hicss/2004/2056/04/205640101babs.htm, accessed 14 February 2011.

Hutchings, S. & Rulyova, N. (2009) *Television and Culture in Putin's Russia: Remote Control* (London, Routledge).

Kez, S. (2002) 'Kitaiskaya ekspansiya v Irkutske prinimaet neobratimyi kharakter', *Nezavisimaya gazeta*, 30 September.

Klapp, O. (1962) *Heroes, Villains and Fools* (Englewood Cliffs, NJ, Prentice-Hall).

Klimov, Y. (2006) 'Bez kitaitsev propadem', *Novye Izvestiya*, 21 November.

Ledeneva, L. (2008) 'Khabarovsk—stolitsa druzhby', *Komsomol'skaya Pravda*, 29 August, available at: http://kp.ru/daily/24154.4/369561, accessed 19 September 2010.

Lisanov, E. (2008) 'Mafiozi, klany i "vory b zakone"', *Transport Rossii*, 22 June, available at: http://avtonews.net/026/mafiozi-klany-i-vory-v-zakone?page = 0,0, accessed 14 February 2011.

Lo, B. (2010) *How the Chinese See Russia*, Report No. 6 (Paris, Russia NIS Centre).

Lukin, A. (2007) *Medved' nablyudaet za drakonom: Obraz Kitaya v Rossii v XVII–XXI vekakh* (Moscow, Vostok-Zapad: AST).

Mickiewicz, E. (2008) *Television, Power, and the Public in Russia* (Cambridge, Cambridge University Press).

Miller, C. R. (1984) 'Genre as Social Action', *Quarterly Journal of Speech*, 70, May, pp. 151–67.

Miller, C. & Shepherd, D. (2004) 'Blogging as Social Action: A Genre Analysis of the Weblog', in Gurak, L., Antonijevic, S., Johnson, L., Ratliff, C. & Reyman, J. (eds) (2004) *Into the Blogosphere: Rhetoric, Community and Culture of Weblogs*, available at: http://blog.lib.umn.edu/blogosphere/blogging_as_social_action_pf.html, accessed 1 July 2012.

Penzev, K. (2007) *Zemli Chingizkhana* (Moscow, Algoritm).

Peskov, Y. (2007) *SSSR-KNR: Ot konfrontatsii k partnerstvu* (Moscow, IDV RAN).

Poluyaktova, E. (2008) 'Edem my v Kitai bez riska -otdykh deshevo i blizko!', *Komsomolskaya Pravda*, 29 August, available at: http://kp.ru/daily/24154.4/368929, accessed 14 February 2011.

Puschmann, C. (2009) 'Lies at Wal-Mart. Style and the Subversion of Genre in the Life at Wal-Mart Blog', in Giltrow, J. & Stein, D. (eds) *Genres in the Internet* (Amsterdam, John Benjamins Publishing Company), pp. 49–84.

Ram, U. (1994) 'Narration, Erziehung und die Erfindung des jüdischen Nationalismus', *Österreichische Zeitschrift für Geschichtswissenschaft*, 5, pp. 151–77.

Rutherford, J. (1990a) 'A Place Called Home', in Rutherford, J. (ed.) (1990b), pp. 9–27.

Rutherford, J. (ed.) (1990b) *Identity: Community, Culture, Difference* (London, Lawrence and Wishart).

Rutten, E. (2009) 'From Typo to Hype. Linguistic Imperfection in Russian Literary Blogs', available at: http://www.uib.no/rg/future_r/projects/overview/from-typo-to-hype.-linguistic-imperfection-in-russian-literary-blogs, accessed 20 February 2012.

Rutten, E. (2011) 'Blog poeta kak kreatiff' (unpublished paper, available at the request of the author).

Ryabtsev, A. (2008) 'Spetsoperatsiya "Tigr"', *Komsomol'skaya Pravda*, 28 August, available at: http://kp.ru/daily/24154.4/369452, accessed 10 February 2011.

Schmidt, H. & Teubener, K. (2006) '(Counter)Public Sphere(s) on the Russian Internet', in Schmidt, H., Teubener, K. & Konradova, N. (eds) (2006) *Control+Shift: Public and Private Usages of the Russian Internet* (Norderstedt, Books on Demand GmbH), pp. 51–72, available at: http:\\www.russian-cyberspace.org, accessed 28 December 2011.

Sharavin, A. (2001) 'Tret'ya ugroza', *Nezavisimoe voennoe obozrenie*, 28 September.

Soinova, N. (2005) 'Kitaitsy v Rossii', *Sel'skaya nov'*, 1, available at: http://novsel.ru/articles/full.php?aid = 105, accessed 20 February 2012.

Strigin, E. (2008) *Era Drakona* (Moscow, Algoritm).

Svetlova, L. & Miroshnikov, I. (2004) 'Rossiya otdaet Kitayu poltora ostrova', *Komsomol'skaya Pravda*, 15 October, available at: http://www.kp.ru/daily/23383/33010, accessed 14 February 2011.

Tolz, V. (2001) 'Values and the Construction of a National Identity', in White, S., Pravda, A. & Gitelman, Z. (eds) (2001) *Developments in Russian Politics* (Basingstoke, Palgrave), pp. 269–88.

Tolz, V. (2003) 'Right-Wing Extremism in Russia: The Dynamics of the 1990s', in Merkl, H. M. & Weinberg, L. (eds) (2001) *Right-Wing Extremism in the Twenty-First Century* (London & Portland, OR, Frank Cass), pp. 251–71.

Trenin, D. (1999) *Russia's China Problem* (Moscow, Carnegie).

Tsyganyuk, A. (2008) 'Spisok ugroz rasshiryaetsya: postsovetskoe prostransto okazalos' pod pritselom', *Nezavisimaya gazeta*, 3 March.

Varsegov, N. (2003) 'Zaselyat li kitaitsy Sibir' i Dal'nii Vostok?', *Komsomol'skaya Pravda*, 11 November, available at: http://kp.ru/daily/23154/24576, accessed 25 February 2011.

Yurchak, A. (2006) *Everything Was Forever, Until It Was No More: The Last Soviet Generation* (Princeton, NJ & Oxford, Princeton University Press).

Zassoursky, I. (2007) 'Development in Russian New Media', talk given at the CEELBAS-funded workshop *Russian New Media and Civil Society*, University of Birmingham.

# Learning How to Shoot Fish on the Internet: New Media in the Russian Margins as Facilitating Immediate and Parochial Social Needs

## JEREMY MORRIS

### Abstract

This essay examines regimes of internet use, and the significance of the internet for everyday lives, in the Russian margins. The field site, a small provincial town in European Russia, was visited in 2009 and 2010. Informants were mainly families dependent on a single non-professional wage. Research materials—semi-structured interviews and participant observation—comprising an ethnography of internet use, are supplemented by survey data. Qualitative social research on new media use has critically examined technologically determinist assumptions about social effects of the internet, including the so-called 'digital divide'. The present research also seeks a contextualised understanding of new media use by considering how it is embedded in established everyday social settings and practices. The ethnographic materials and survey data collected indicate that Woolgar's rules of virtuality hold true in the Russian margins: use of new media depends on the local social context and supplements, rather than replaces real activities. Most users in the group surveyed are highly instrumentalist and have little interest in the communicative and non-grounded aspects of the media. At the same time the impetus for initial access to the internet is closely related to issues of esteem and peer recognition within a social network rather than actual need.

THE BACKGROUND NOISE IN SASHA'S HOUSEHOLD means it is almost impossible to conduct a coherent conversation. His two children are present in the living (and sleeping) room, sitting on their bunk beds and continually interrupting. The television, ignored by all, blares away on a wall-mounted steel arm, directly above the computer monitor. Two large fish tanks, brightly lit, with filters humming, sit next to the computer console. A mini hi-fi, printer, CD rack and multiple remote controls mounted vertically on the desk in charging cradles complete the techno-tableau. In the kitchen, Sasha's wife watches a different television programme on the smaller set similarly mounted on the wall above the fridge, the volume set high enough to follow the gist of the programme from the other room. Sasha sits on the only chair in the living room, pivoting towards

Supplementary data for this essay can be found in the online version at: http://www.tandfonline.com/loi/ceas20 (Figures S1, S2 and Appendix).

227

me with his back to the computer. The monitor 'wallpaper' is the popular search engine and portal Yandex. For the first hour and a half of this visit, the computer and internet, like the TV, remains ignored. Only when it is time to leave, as the children start their bedtime routine, does Tanya emerge from the kitchen to play for half an hour: a recently downloaded point and click game, the object of which is to nurture a potted plant sufficiently to produce a lasting bloom. In this setting, the internet and personal computer have been neatly integrated into daily life alongside other objects signifying that this family is 'up-to-date' without much changing it or its habits.

Writing before the advent of the World Wide Web, Neuman listed 15 sometimes contradictory propositions that had been put forward about the effects of new media on society. These propositions included, *inter alia*, the creation of an information underclass and the diminution of media conglomerates' power to shape and control the news individuals receive (Neuman 1991, pp. 5–7). To avoid an overly deterministic understanding of social change and technology,[1] he proposed an interaction model 'focusing on the issue of technology-in-use' (Neuman 1991, p. 18). For him, new media's obvious place in filling communicative gaps between one-to-one and mass communication on the one hand, and the phenomenon of instantaneous communication (conversation in the presence of others) compared to delayed communication (traditional mail) on the other, meant that in the information age there would be 'no prospect of resurrecting the technologies, life-styles, and values of the small town and rural society' (Neuman 1991, p. 9). It is not so much the 'resurrection' proposition that this essay seeks to examine with reference to the post-socialist context of the Russian small town; rather the thrust of this research is that, specifically, the internet has not subverted, replaced or fundamentally altered the existing social, communicative and cultural rhythms that predate new media use. However, it is precisely 'small towns'— the focus of this essay—that saw the fastest growth in the first decade of the century (Galitskii 2009a, p. 15). From 2002 to 2008, internet use there grew by a factor of five; by contrast in metropolitan areas it grew by a factor of three and is approaching levels similar to other European countries (Galitskii & Sidorova 2009, p. 12) at around two-thirds connectivity. Clearly, smaller urban settlements are now experiencing growth similar to that in cities about 10 years ago. There are obvious, though untested, hypotheses about the lag in diffusion to provincial areas. One is that the telephone infrastructure outside large cities is patchy and unable to support reliable broadband and the existence of local service-provider monopolies means the cost is prohibitively high (Cooper 2008).[2] Another is that Russians, as a nation living mainly in small cities and having already wholeheartedly embraced the mobile phone, have their social networking needs more than catered for by existing technologies of communication (Cooper 2008)—including the most reliable and traditional of all—dropping in on each other. This is compounded by the demography of many regions, where older people are less likely to seek access to digital networks. Finally, there is as yet little economic logic behind broadband diffusion in the regions. With an undeveloped system of electronic financial payments, e-commerce cannot be seen as a driver of

---

[1]For an overview of this tendency see Croteau and Hoynes (2003, pp. 305–7).

[2]For an extended discussion on uneven technological diffusion and the new geography of marginalisation in Russia, see Castells and Kiselyova (1998).

diffusion: the 4% of users defined as 'shoppers' (Galitskii 2009b, p. 81), new media users, still rely primarily on alternative forms of payment (such as mobile phone credit and cash-on-delivery) for their online purchases.[3]

### New media ethnography: qualitative research on actually existing internet use

Like other ethnographies of new media use, this research does not seek to test such hypotheses directly; however, in seeking to understand the actually lived experience of internet users in 'ordinary' areas—small towns in the regions—it can provide a micro check on these macro presuppositions. For example, for the vast majority, the computer-based internet as a one-to-one communication device does not compete with voice-based mobile phone use. This supports the hypothesis that existing social networks are not directly enhanced by online life—if the latter is narrowly understood as the potential for virtual communication such as through social networking sites. This is not to say, however, that internet use does not enhance or supplement network practice in other ways, the main contention of this essay. Despite being networked by relatively high-speed and reliable broadband, the working people of the small Russian town hardly ever email each other, or anyone else for that matter. Why should they when mobile or land-line communications suffice?[4] Even when contacting relatives and friends in other regions and countries, other methods of communication were preferred to email. Of course, socio-economic, cultural, as well as geographic factors explain this—most informants for this study were in manual and semi-skilled employment and therefore had never used email for professional purposes. For example, work was frequently sought online, but CVs and interviews were provided and elicited respectively in traditional non-electronic forms. Socio-economic factors are not adequate alone to explain this. Looking at a comparator grouping of well-educated businesspeople in the same town, similar dynamics of use were found. For business, email was a twice-daily-only ritual, reserved for communication with prospective partners outside the region; the internet itself was used for financial transactions with state structures such as the tax authorities.

A second imperative for ethnographies of internet use in Russia arises out of the largely qualitative sociological nature of existing materials. The ethnographic approach can supplement, enhance or challenge qualitative data when the latter adopt a functionalist approach to access and use of the internet, or where analysis adopts an implicit determinist approach (Schmidt & Teubener 2006, p. 19). The dominance of quantitative and survey-based qualitative approaches in social research on new media use in the post-socialist sphere is evident from the meagre list of area-specific literature utilising ethnographic methods. Examples include Pearce's study of social networking in Armenia (2009), Bakardjieva's study on becoming an internet user in Bulgaria (2005a), and survey and interview data on use from Kyrgystan (Driesbach *et al.* 2009).

---

[3]The situation concerning electronic financial payments was nonetheless rapidly changing over the course of two fieldwork visits as various e-cards and other systems for making payments online were becoming available to ordinary Russians, if not electronic payments linked to credit and debit cards.

[4]Consideration of the use of mobile phones is unfortunately beyond the scope of this essay, but see Lonkila and Gladarev (2008).

In Russia, survey data collected mainly by the Public Opinion Fund focus on national and regional snapshots of access and use (Galitskii & Sidorova 2009) which inevitably brush over socio-economic nuances and a more contextual relevance of place.[5] Such sociological research on the internet in Russia, while providing interesting data on, for example, the differences between types of non-users, still takes as its main heuristic tool a modernisation narrative predicated on a simple dichotomy of access or non-access. When it asks questions about actual use, these are typically formulated in generalisations such as—'to what extent has access changed your life?' (Galitskii & Sidorova 2009, p. 35). Unsurprisingly, while inhabitants of the urban centres report the internet as an integral part of their lives (41%), few rural inhabitants and those with an income of less than R4,000 a month see access as qualitatively affecting their life—only 17% and 13%, respectively. Over and above the expected uses of the medium as another leisure and entertainment source—for watching films and playing games—the opaque typology: 'reference and necessary information', at 53% of use (Lebedev & Galitskii 2009, p. 42), reveals the limitations of qualitative survey data at getting to the reality of actual use regimes of such a versatile and inherently social medium. If social research is not to avoid the trap of couching its research questions in such a way that the agency of users is assumed to take limited or prescribed forms, it must go beyond the typologies of user as the narrow 'shopper', 'financial user', and on the other hand, the unhelpfully broad category of 'socialiser' (Galitskii 2009b). Work done on new technologies' actual regimes of use is important, because in time it is ordinary users themselves who can shape the evolution and unforeseen affordances of new technology (Kline & Pincher 1996; Wyatt *et al.* 2002, p. 36).

Finally, the Birmingham and University College London workshops which generated the materials of this collection show that much research is understandably focused on the civic and political potential of new media in the post-socialist space—in particular the essays by Bakardjieva, and Gladarev and Lonkila. Invariably this entails a preoccupation with the technological elite of sophisticated consumers and producers of online material. In line with existing ethnographies of the internet I ask more mundane but equally fundamental questions about the internet in Russia such as 'how members of a specific culture attempt to make themselves a(t) home in a transforming communicative environment, how they can find themselves in this environment and at the same time try to mould it in their own image' (Miller & Slater 2000, p. 1).

Qualitative research methods are increasingly being used to question assumptions about the technologically determining social effects of new media. Scholars are building theoretical models of the interplay between technology and society to question new media's potential to replace or change existing social and cultural behaviours. This includes critically examining 'epithetised phenomena' such as activities prefixed by the word 'virtual', and the meaning of 'community' as it is variously adjectivised (Woolgar 2002a, pp. 2–3). Simultaneously, case studies using a

---

[5]An exception to an overly narrow typologisation is early research (mid-1990s) carried out by Russian behavioural psychologists which identified key motives for use as: professional, knowledge seeking, socially affiliative, corporative and collaborative, self-actualisation, recreational, communicative (Arestova *et al.* 2000).

variety of social research methods from surveys to ethnography have found that actual use of media is significantly gendered (Howard *et al.* 2001; Hargittai & Shafer 2006), differentiated on the basis of age (Söderström 2009), culture and ethnicity (Bakardjieva 2005a; Pearce 2009; Leonardi 2003; Miller & Slater 2000) and class (Crump & McIlroy 2003; Leonardi 2003; Hargittai 2010). This development of a more nuanced understanding of the relationship between social and cultural life and new media has in turn led to a reappraisal of the meaning of the 'digital divide'. Instead of thinking in terms of access only, difference in proficiencies and usage can be mapped on to existing socio-economic structures in society (Norris 2001; Hargittai 2007). Access itself continues to be strongly correlated to social background: an Australian study found the degree-educated to be six times more likely to use the internet than others (Willis & Tranter 2006, p. 53). Despite the technological *milieu* of 'pervasive communication' (Fischer *et al.* 2008, p. 529) within which social life now takes place, the internet, especially as a social capital resource and potential driver of social mobility, must be understood as a qualitatively different medium depending on the identity of the user. One response to this problem is the domestication of research and material cultures. This approach aims to move beyond adoption and use towards the meaning of services and technologies to people and how they are experienced in everyday life, especially in the home (Haddon 2006). The question of how the internet is mediated through actual use, spatial relationships, placeness, sociability and social class, is key to the research presented in this essay.

## The field site, informants and methods

The research presented in this essay derives from materials collected during two periods of ethnographic fieldwork conducted in November 2009 and July–December 2010. The fieldwork site was an urban settlement of 15,000 inhabitants in the Kaluga region of Russia with which the researcher had a longstanding connection as a source of ethnographic research material for other projects. I call this settlement 'Izluchino'.[6] The main group of informants for previous and parallel projects had been blue-collar workers; for the purposes of collecting information on internet use it was decided to focus on this same group: not only was a ready-made pool of informants available to the researcher thanks to previous research in this field site, but anecdotal evidence pointed to this group as newly connecting to the internet from around 2005 onwards.

From an existing informant base of nine families, all of whom relied on manual and semi-skilled wages for their primary household income, snowball sampling was undertaken in Izluchino during November 2009 to identify members of those household's immediate and extended social and kinship networks that had recently (in the previous three or four years) started to use the internet at home. As was to be expected in a physically and socially compact sample and field space, a limited number of genealogies of use of the internet were identified. Early-adopters (by the social and geographical standards of the field) were some of the primary informants of this research. To put this into perspective, the earliest use of the internet at home among

---

[6]For ethical reasons, mainly relating to other research carried out at this site, I adopt a pseudonym for the town, as well as anonymising material relating to individual informants.

informants was 2006. By contrast, professionals in the same field space had been observed by the researcher using dial-up as early as 1999. Some of the present informants had clearly 'skipped' the dial-up phase for reasons of cost or accessibility.[7] During fieldwork, semi-structured interviews with key informants were undertaken; however, the vast majority of materials presented in this chapter derive from participant observation and rich open-ended conversational research focusing on five domestic settings. A variation on this methodology was considered: installing software on the computer with permission from informants to track actual internet use;[8] however, for ethical reasons this was rejected. In a few cases informants gave permission to view their web browser's 'history' which was very useful.

In addition to ethnographic methods, two paper surveys (sample sizes: $n = 54$ in 2009 and $n = 88$ in 2010) on access were carried out. By way of comparison, this allowed the evaluation of the ethnographic informants in terms of the typical trajectories of access and habits of use from the survey data. The survey tried in part to replicate those administered by the sociologists Galitskii and Sidorova (2009); in addition it asked about routes into access, histories and hierarchies of use, specific information about websites visited, local websites accessed, the domestic setting of use and interests relating to the internet beyond the virtual sphere. This survey was administered by a research assistant in households selected by a random walk method using different temporal sampling intervals (Bernard 2000, p. 150) after a period of pre-testing in a local public library. As is standard practice in ethnography, survey data are not presented separately; the data are used to indicate the generalisability or otherwise of informants' use regimes.[9]

Another group of informants—local business owners—served as a further comparator. A small number of formal and semi-structured interviews were conducted with this group. Like the survey data, such materials are mentioned here only when they inform the focus on 'ordinary' use by non-technical, non-professional people. Such users have little or no vested economic interest in the technology.[10]

Before entering the field I had constructed a linear model of increasing penetration of an individual or family's use but this remained little more than an ideal. The model would have begun with arranging an informal interview, usually in the home of the informant, in which I would replicate the main points of the separately administered survey discussed above. This combination of interview and interaction would be conducted one-to-one and take about half an hour. After this, depending on the willingness of the informant, I would ask them to give me a 'tour' of their online habitat, sitting with them at the computer or from a suitable vantage point. We would visit the sites the informant most often visited, and I would prompt them about the

---

[7]It should be noted that many younger people (in their 20s) tended to rely on 3G connection technology when asymmetric digital subscriber line (ADSL) broadband was not an option, when living in rented or parental accommodation. Quite often connectivity through 3G was comparable to dial-up in terms of speed.

[8]This alternative was also suggested by an anonymous reviewer.

[9]Supplementary data for this essay can be found in the online version at http://www.tandfonline.com/loi/ceas20 (Appendix).

[10]As for example, in Latour's 'simple customer' model, quoted in Bakardjieva (2005b, p. 3).

existence of sites catering to local interests such as forums about the *oblast'*, about Izluchino and special-interest hobbies of which I had learned in the interview.

From the first interviews it was clear that the above format would be unlikely to succeed. Firstly, the majority of informants lived in 'one-room' flats[11] and in most cases it was in this single living and sleeping space that the computer was located. In some cases too, interaction with informants took place in the evening when the whole family was present. It was difficult enough to replicate the main points of the survey in this environment, let alone discuss in private the use-regimes of individuals. However, what this research setting did allow was the observation of the internet in use by families in a 'natural' setting (Schrøder 1999; Bakardjieva 2005b, p. 79). Also, researcher-effects need to be accounted for;[12] in a few cases it was clear that informants were uncomfortable about an observer watching their every click over their shoulder. Most of the time, however, the process of gaining access through informants' acquaintances via the snowballing technique meant that informants, after discussing the main reasons for the research, were put more or less at ease. In some settings, careful physical positioning of the researcher at a distance away from the computer often resulted in the informant, increasingly immersed in their own online experience, losing situational awareness, and therefore providing the researcher with periods of arguably 'candid' use.[13]

In addition, most hands-on sessions were conducted, not on initial contact, but during a second follow-up meeting sometimes lasting more than two hours, with an interval of a day or more after the semi-structured interview or initial contact had taken place. Often in the interim, or even immediately prior to the second session, less purposive interaction was sought with informants, taking part in the domestic social gatherings typical of the spatially constrained leisure activities of the small town.

A final point needs to be made about the personal computer and internet themselves as props for the ethnographer to gain trust from informants. More than once I was encouraged to connect a pen drive I had loaded with photos of my own domestic setting and life to show interested informants; often I found myself opening browser windows to show informants some of the kinds of sites I used—though this in itself could be viewed as problematic in terms of encouraging informants to read the researcher as a technologically 'learned' person (Bakardjieva 2005b, pp. 80–81). On the other hand, by limiting such researcher-led use of technology to its more banal and everyday characteristics—news sites, information and hobby-related forums, rather than revealing professional activities—this mode may be described as contributing to a

[11]Flats typically comprised a single living space of about 12–15 m², a small bathroom and kitchen, and in some cases a covered balcony. Thus a family of four persons might have only 25 m² of 'living space' available to them.

[12]These are discussed at length in this research context by Bakardjieva (2005b).

[13]'Situational awareness' relates to an individual's awareness of their physical environment and sense of time. Research into online learning has shown that the high level of cognitive load imposed by many internet experiences can be disorientating (Eveland & Dunwoody 2001, p. 56). In the present research, goal-directed concentration on a screen leading to loss of informant's awareness of the researcher was frequently exacerbated by the typical domestic setting encountered—a single family multi-use room. The effects of background 'noise'—movement of people, other conversations, other technological media resulted in a very high level of cognitive load indeed for informants.

dialogic approach to new media ethnography,[14] and redresses at least some of the power imbalance implicit in the researcher–informant relationship in front of the screen.

In the following thematic sections I present the main research materials gathered from the two fieldwork periods in Izluchino. Three dominant themes on use and reasons for access—educational, informational and leisure—combine material from personal observation, survey data and interviews. For informants, pseudonyms, age and occupation are given.

*From access and narratives of virtual arrival, to everyday pragmatic regimes of use*

The most common answer to the question of 'Why did you first get connected (at home, with a monthly broadband contract) to the internet?' was usually a slightly uncomfortable pause followed by variations on: 'I don't really know'. Informants found it very difficult to articulate a concrete need to justify their decision to invest in what was always a quite costly service.[15] At first a narrative of general usefulness was pursued: the 'information' strand discussed below. 'You really need it to keep up to date [*byt' v kurse*]—the internet means we're not living in the stone-age'.[16] When pressed on this point, Sasha found it very difficult to pinpoint any of his actual internet use in terms of generic information gathering. Like all informants and most respondents he never visited news sites. A common arrival-narrative developed which hinged on three 'reasons' for connection, often linked to the purchase of the personal computer through which access was gained. In many cases informants had not initially bought a PC to gain access to the internet. Each of these are really *post-hoc* 'reasons' and can be shown to lack explanatory power for initial access. I conclude the section with a description of the main, usually indirectly articulated, reason for connection that emerges from the user narratives—a reason which stems from a general anxiety about not having access to the latest technological gadget and the peer-recognition that comes from getting connected before others. I define this phenomenon as 'internet inertia'.

*'Without the internet at home, the children will be educationally disadvantaged'*[17]

The internet as an educational aid was proposed as a reason for access by more than half of informants, but schoolwork was rarely, if ever, carried out with the aid of either computer or the internet. In fact, in at least two families who gave education as a

---

[14]The ethical considerations of giving informants some kind of ownership over the research materials has long been an ideal in ethnographic research (Hammersley & Atkinson 1995, p. 279)—the 'technique' of bringing personal items and scraps of the researcher's own home life to the field in order to establish trust and credibility is well known in ethnography. A similar consideration is the extent the researcher should or can take on the role of contributor to the lives of informants (Hammersley & Atkinson 1995, pp. 274–75)—an example of this relates to the online guidance given by the researcher to the informant Valerii, see the section on informational affordances.

[15]Up to 5% of take-home breadwinner pay.

[16]Author's interview with Sasha, 39, manual worker, November 2009.

[17]Author's interview with Tanya, 37, primary-school teacher, November 2009.

reason for connection, a password was installed on the PC to prevent child access altogether. Parents embarrassedly mentioned the tendency of otherwise innocuous Russian sites to carry links to adult sites: 'We were looking for some pictures of dinosaurs or something for little Lena's school project, but when we clicked on some link from a site all these pictures of naked women appeared in the margin of the screen. After that we stopped using it with her for school work'. Tanya's use was itself exceptional in that she frequently used online teaching resources; she also browsed with her eight-year-old daughter for pictures to print out and illustrate school projects. However, when Tanya was pressed on the subject she revealed that the majority of her pupils did not have access (as shown by survey data and predicted by the national statistics on access[18]) and therefore there was never any expectation or even mention of the internet as a resource for school projects. Tanya's own use with her daughter appeared to have been an experiment at the beginning of their honeymoon period of access, which was quickly superseded by a ban on underage use in their household— even by their 14-year-old son. At least for this family, initial connection appears to have been prompted by the professional needs of Tanya, and is therefore unrepresentative of the sample as a whole.[19]

An overview of the current use of the internet shows the educational impetus to be a *post-hoc* rationalisation for some, and for others a genuine if unrealistic ideal prior to access, that was quickly dispelled by actual use, or abandoned in favour of other uses. Similarly, some expressed disappointment with the communicative and socially-situated potential of the internet. This is why I use the term 'honeymoon' period: initial delight and fascination of informants at the entertainment and leisure aspects of use, such as viewing and downloading films and music, rather quickly gave way to an acute sense of the paucity of reliable information and content relating to local concerns and communication. (I discuss these concerns in more detail in the following section.) I do not wish to argue that for families the internet will not in time become a useful educational resource, merely that there are two obvious barriers to this at present: the *Runet*'s overwhelmingly 'adult' focus (though this is changing rapidly), and the continuing lack of access in schools and homes.

## 'The internet? Well, it's information, isn't it?'[20]

The most common reason for initial access given (by most interview informants and over 70% of survey respondents) was that as an informational resource the internet occupies an unassailable position in contemporary lives: that without access, informants would somehow be 'out of the loop'. At first glance there is little to dispute here; and in reviewing the ethnographic material plenty of examples were observed of the utility of internet access in fulfilling informants' immediate and parochial informational needs. However, the term 'information' needs to be carefully qualified. Often, when technological change is discussed in terms of networking characteristics allowing unbounded communication and informational flows (Castells

---

[18]On this point see Galitskii and Sidorova (2009).
[19]Author's interview with Tanya, 37, primary-school teacher, November 2009.
[20]Author's interview with Dima, 35, arc-welder, November 2009.

1996, 2001), what is at least implied is the potential widening of the socio-cultural informational frame of the individual user.[21] In its lay iteration, informational affordances can be viewed as analogous to the 'informational utopics' described by Hetherington (1998), whereby reification of such utopian technological visions via networking of individuals takes place. The uses of virtual literacy in the utopian readings of new media relate to the same narratives of self-improvement and self-education to be found accompanying the rise of other communicative technologies and affordances,[22] for example public libraries.

In the Russian small town, the affordance of 'information' turned out to be epistemologically framed by informants not as utopian, nor even as particularly pragmatic, but as strongly or exclusively instrumentalist. However, the very nature of narrow instrumentalism often served to undermine the 'informational' affordance as an initial justification for access. The title of this essay refers to the best example encountered of this paradox—one informant's interest in underwater harpoon fishing. This hobby is not included in the section below on leisure use for two reasons. Firstly, Dima, the informant who liked to visit an online forum on harpoon fishing in Russia, was at pains to compartmentalise his online experience relating to fishing as 'informative'—an almost technical and intellectual affordance separate from the actual practice of fishing:

> For me it is just like reading a book on it. You're looking for the specific answer to a specific problem—like how to hold your breath under water for longer. There aren't any books so I started searching on the internet and I found this forum. I just find the information I need but I don't want to discuss it with these 'participants' [uchastniki]. I don't class myself as a 'participant'—I mean, some people just sit there posting all the time—when do they actually get a chance to do any fishing!

Dima articulated a distinction between an informational affordance relating to a personal interest and more communicative aspects which remained unexplored. A similar narrative emerged from discussions with others about interests as diverse as knitting patterns for clothes and the maintenance of diesel engines (also a 'hobby').[23]

The second reason relates to the first in a negative sense. If 'informational' use was posited as active, goal-orientated gathering of knowledge, then leisure use was usually placed on an opposite pole by informants—passively to 'kill time' together with family or friends, most typically in viewing so-called 'fun' (prikol'nye) sites comprising user-

---

[21]Lay interpretations of the socio-culturally widened communicative opportunities of the internet often include references to intercultural communication—either directly between individuals located in different countries, or indirectly in terms of the informational resources the internet provides for people to discover 'thick' descriptions of cultural knowledge pertaining to their own or other countries and cultures.

[22]After Gibson (1979), as discussed in Bakardjieva (2005b, p. 19), I use the term 'affordance' to describe what a technology offers a potential user in the context of that person's social and cultural life. See Bradner et al. (1999) on the unpredictable and user-predicated 'social affordances' of technological change.

[23]Author's interview with Dima, 35, arc-welder, November 2009.

posted jokes and images designed to amuse and/or shock, playing online computer games, and downloading films, games and music. The extreme instrumentalism observed served to undermine the 'informational' narrative of initial access in the following ways. Firstly, some interests that emerged as central to informants did not pre-date access. Therefore informants could not coherently rationalise the impetus for initial access by means of furthering their interests or hobbies by online means. Secondly, when online interests coincided with pre-existing ones, it emerged that internet use to support a hobby had developed only recently—such as in the case of Sveta, who had been knitting for years, but only in her fourth year of internet use had started downloading clothing patterns.[24]

But what of the less narrowly defined 'informational' affordances of the internet? As has already been mentioned, no one used the internet to supplement their diet of TV news, whether local, national or international. The paucity of the more generic informational online life can be partially explained by examining informants' reports of their initial use. This indicates a pre-access idealisation of the informational potential of the internet that was frustrated by initial experience, perhaps leading to a rejection of many aspects of the 'infosumptional' rationalistic ideals of use (Bakardjieva 2005b, p. 169). Sasha[25] reported that at first (in 2006) he and his wife had looked for local news information—a site with user-generated local content had appeared with commentary on issues of community interest, but quickly this had 'got broken' (*slomalsya*) and the websites on local themes (*lokalka*) were now not interesting.[26] The chat forums relating to Izluchino were used only by a narrow segment of young internet users. Sergei expressed frustration at this situation—'they're going to use the disused clay pits as a rubbish tip for Moscow but you can't find anything out about it'. He was concerned about the employment and health implications of this issue, but had given up on the *lokalka* as a source of information.[27] Finally, a telling 'informational' example that underscored the importance of approaching and understanding internet use in terms of competencies was observed as a result of interaction between researcher and informant. Valerii had heard a rumour that taxes on car owners were to rise sharply in the following year. He owned a foreign-made car with a powerful engine and was afraid that this would hit his pocket hard—but where could he find out the particular tax-rate for the Kaluga region? We tried looking on the site maintained by the regional authorities but were unable to find the information there. 'I told you it wouldn't be there—there's never anything useful there', he rebuked me, despite previously admitting he had never visited the site.[28] After a number of searches we found the information on a region-specific forum for motorists—but this slightly inconvenient operation merely confirmed for Valerii the internet's lack of utility for obtaining locally specific information—whether relating to news, official information or even the weather. In addition, like all informants, there

[24]Author's interview with Sveta, 43, shop-assistant, July 2010.

[25]Author's interview with Sasha, 39, manual worker, July 2010.

[26]The primary meaning of '*lokalka*' is 'local-area network', but informants used the word to describe online content serving the town in which they lived, which at the time of the research fieldtrip was extremely limited.

[27]Author's interview with Sergei, electrician, 28, December 2009.

[28]Author's interview with Valerii, delivery driver, 45, December 2009.

was no mention from Valerii of contributing his own user-generated content to enhance the local online space. Finally, in contrast to these findings, a review of locally generated content relating to some larger settlements (the *oblast'* capital and other towns with a population over 50,000) revealed some significant use of forums for local people to discuss local community issues—such as illegal gambling—and in some cases to apply pressure on local authorities to take action.

*'Entertainment? Well, yes, but there are more important things to do on the internet'*[29]

In contrast to the purportedly informational or educational utilities of access given, there was often an understandable initial reluctance or discomfort in admitting the potential of the internet as a leisure resource. However, the entertainment resources that the internet provided were often revealed—usually only after the informant had prioritised the utilitarian reasons given above. Partly this could be explained by the abstracted description of the use of the term 'entertainment' (*razvlechenie*). When given concrete examples of activities observed—downloading electronic games, cartoons for the children or the latest cinema release—then informants readily agreed that these activities were a significant impetus for initial connection.[30] Informants did not make the indexical leap from the umbrella term to specific practices. Nonetheless the physical (offline) exchange and sharing of downloaded material significantly contributed to social intercourse and the maintenance of personal networks. This practice supported local communication and socialisation. Such socialisation is also linked to the physical citing of internet access in the domestic space—the discussion of which follows this section.

There were, of course, a number of incidences of solitary online gaming observed. Typically these sessions involved school-aged children and young adults living at home with parents. However, this activity was not the norm. The 'entertainment/leisure' use was rooted in sociality. A new game would be downloaded by an informant prompting a visit from extended family or friends to make a CD-ROM copy or play together. Alternatively, a film or game was downloaded, a copy made of it on recordable media, and a short journey undertaken by the downloader himself to visit friends. In a number of cases, the online or downloaded game provided the focus for the only permissible use of the internet by children under direct observation by parents, or as a whole-family activity. However, in terms of the share of time spent online by families and individuals alike, gaming and leisure featured as a relatively minor activity. Characteristically it performed a function of delineating a temporal threshold between evening and bedtime. Children and adults alike would 'play for half an hour' before it was time for the former to go to bed. Gaming was also important for some workers in killing time between and after shifts when no one else was 'around' online, such as the middle of the day. Perhaps one of the most striking findings that emerged from discussions of leisure use was that informants reported few examples of interest in computer gaming prior to their connection to the internet—indicating another difficulty in comparing current use to the

---

[29]Author's interview with Vadim, 28, electrician, July 2010.

[30]Informants were generally quite sophisticated users of online sources of pirated material and the software enabling such use. However, on the other hand the researcher observed an alarming number of cases of users falling victim to online scams and computer failures due to viruses and Trojans.

purported reasons for initial access. This may be due to the fact that computer access for many did not significantly pre-date internet access.

### *Narrow and locally determined instrumentalism*

Three main affordances of internet access as understood by informants have been conceptualised and examples of them within concrete regimes of use described. What emerges from comparing narratives of the initial use-justifications for access with everyday experience is that no affordances provide adequate explanatory power for such access. In the first 'educational' justification it emerged as largely *post-hoc*. Indeed, in a number of families the internet was largely off-limits to children. Even taken as a 'naïve' ideal by pre-access individuals, an 'educational' justification can only carry a small amount of weight as a justification for a connection costing up to 5% of monthly wages for a family. The second and most dominant narrative, 'informational use', is also problematic. If we examine current use, it is dominated by practical self-educational and self-provisioning purposes; the latter includes finding the right knitting pattern for a daughter's autumn dress, or the correct order in which to dismantle parts of a diesel engine. Key informant Dima's narrowly instrumentalist use of the internet is not untypical of the case studies in this research. His use consists of reading up on the techniques of underwater harpoon fishing. He almost never checks his email, does not contribute to forums about his hobby, does not read the news online, and registers for social networking sites then never returns to them.[31] Current use for him revolves around a narrow, instrumentalist need—he is learning to shoot fish on the internet. However, this is actually an offline activity, rooted in a locally and socially specific life-world. (The field site borders a national park with excellent river fishing opportunities; fishing in company is an important male blue-collar leisure activity.) In contrast to actually encountered narrow regimes of informational use, the strong narrative of general informational utility of access reported by informants—'you can keep up to date with what's going on'—is difficult to accept at face value. A more compelling interpretation is provided if we focus on the first part of this statement—'keeping up to date'.

Access, as one might expect, occurred in a chain, moving through an extended social network. A pioneer family or individual gained access and others followed, often, seemingly with little or no rationale. Being up-to-date related more to relationships within networks than to the utility or otherwise of the new technology. Looking for the primary affordance-in-use is to ignore a palpable prompt in plain view of the researcher—the domestic setting of the computer at the centre of the living space of the family and therefore on display to all visitors. Internet access inertia then requires little or no rationalistic or use-related explanation. The third theme of use also

---

[31]This came to light when the researcher tried, unsuccessfully, to make use of social networking websites in order to contact informants before, during and after the first period of fieldwork. The researcher, at the prompting of Dima, registered for the well-known sites: odnoklassniki.ru, Vkontakte.ru and finally mirtesen.ru. However, Dima, after inputting the minimum of personal information to the profile pages that these sites are predicated on, gave up on them. Very little contact online was achieved with any informants with the exception of Sasha who liked to use the Voice-over-Internet Protocol service Skype. Revealingly though, the researcher remains to date Sasha's only Skype contact. He never turns Skype on because it slows his computer down.

supports this thesis—gaming and leisure use are strongly correlated with existing sociality and the maintenance of social networks. Interest in gaming, or even in the latest movie releases, did not pre-date access in any significant way. With access achieved, leisure uses were embedded in existing temporal and spatial regimes—family evenings, domestic visits, etc. The visit with a new game or to watch a new film becomes an event connected to peer-recognition and status. To download the latest version of the game 'Russian Fishing' is to show that one is 'up to date', not with current affairs as an outsider might impute, but with one's significant peers.[32]

### *The domestic setting: the internet on display as technological wallpaper*

The physical citing of access and the computer's hierarchical positioning as first among other equally visible technical and leisure furnishings within the domestic space underline the importance of 'recognition'. The 'internet', taken here to mean the totality of physical access points including electronic connection devices, is permanently on show in the single living spaces of families, but like other technological wallpaper and status items such as the permanently locked-away Czech crystalware in the dresser, it is a signifying presence more often than it is a utility— whether informational, educational or even in terms of entertainment. The opening ethnographic moment of this essay illustrates this well.

Connectivity in Sasha's family is about showing the frequent kin and other visitors that despite a very meagre income, the household is sufficiently well-run so that disposable income can be spent on a particular type of luxury, even if it remains without a washing machine and the furniture is largely self-built. Technological gadgets—mobile phones, the PC and internet access—are an expense that is not spared; their meaning as an 'affordance' is problematic as they do not significantly widen the social and cultural lifeworld of the users. The adoption and use of the internet can therefore be seen, not as infosumption—and therefore specific to the new media phenomenon—but as a more mundane and well-established technological commodity fetishism and conspicuous consumption in a physically marginal setting, where other such forms of consumption are limited: there are few locally available spaces for public display of wealth and more traditional markers of disposable income are on sale only in the *oblast'* capital.

### *Discussion and conclusion*

One of the ethnographic presuppositions about new media is that the internet must necessarily be socially and culturally embedded in existing structures and spaces and that it would therefore be a mistake to study it as a world apart, dislocated from the rest of people's social lives (Miller & Slater 2000, p. 4). While the transformative civic and political possibilities of the internet and other media are clear, this ethnography has brought agency literally 'back home' in looking at the affordances of virtuality in

---

[32]Among male personal networks, 'show and exchange' of downloaded material as a network affirming practice extended to pornography.

FIGURE 1. DOMESTIC SPACES OF INTERNET ACCESS: THIS ACCESS POINT IS USED BY FIVE DIFFERENT USERS, THREE OF WHOM ALSO USE THIS SPACE AS A BEDROOM.
*Note*: image taken by the author in the field between November 2009 and December 2010.

the everyday Russian context. The setting of technology-in-use—from its position in the domestic space to the structuring of offline communities (both practical and socio-economic)—has a structuring-formative effect on use. In its diffusion to 'ordinary' people and areas it is crucial to understand the meaning and use of 'virtuality' as a 'social accomplishment' (Miller & Slater 2000, p. 6), instead of a 'world apart' from offline sociality. If social research cannot account for the fact that ordinary internet users do not need electronic mail, or even politically variegated online news coverage for that matter, then how can it begin to impute transformative civic potential to a medium? In examining the ordinary Russian internet in ordinary Russian places we find evidence of the assimilation of 'yet another medium into various practices' (Miller & Slater 2000, p. 6) that constitutes the existing, and therefore socially and culturally important everyday. Many of these practices revolve around the domestic sphere— self-provisioning, the maintenance and improvement of living space, hobbies and the sustaining of existing and place-bounded social networks. Such 'ordinary' use and socially related practices link internet use to that of the mobile phone (Lonkila & Gladarev 2008, p. 284); and it is the domestic sphere which is only now in Russia becoming the primary physical connection point to the virtual world, the workplace previously being most favoured. In 2009 the home as the primary space for internet connection passed the threshold of 75%. In contrast, in 2002 less than a third of users connected from home (Galitskii & Sidorova 2009, p. 18). Home use indicates the advent of socio-economically 'ordinary' users (Willis & Tranter 2006, p. 47). As the internet finally comes home in Russia this statistic indicates that the time for a domestic ethnography of the internet has come.

In conclusion, the adoption and use of the internet in the Russian internal margins is explained by separate models of consumption and production, closely related to existing meanings of the domestic space; but these respective modes do not look much like those analysed in the Western context by Bakardjieva. For her, the internet problematised the meaning of the home—it could herald the intrusion of work into the private domestic space with teleworking being an instance of production. The internet could equally enter the home as a 'specific instance of consumption' normally associated with economic relations in the public sphere—shopping, participation in e-commerce, etc. (Bakardjieva 2005b, pp. 24–26). In a Lefebvrian interpretation, new media use would appear to confirm the worst predictions of a technologically determinist position—everyday life is deprived of some of the shelter the domestic sphere is supposed to afford from marketised relations. At the very least, 'previously unchanged characteristics [of everyday life] had to be given up in exchange for the gadgets of technological progress' (Bakardjieva 2005b, p. 51). In the Russian context 'productive' modes of internet use are no less economically determined, but *contra* Neuman, reflect unchanging imperatives of small town life—self-provisioning, the maintenance of the home and updating of skills and knowledge relating to economic priorities. Thus, knitting patterns, diesel engine schemas, allotment advice and DIY furniture plans dominate use. The 'instance of consumption' that the internet brings to the home is inscribed, not in use, but in the acquisition of the new media technology itself. Ironically, the internet, at present, contributes little to market economy-related consumption in the given Russian context, despite the very visible demand for pirated material online, and yet its presence in the home is itself an instance of the marking by informants of their position and potential as consumers. By connecting they 'have arrived', without actually travelling in the sense of enhancing social mobility. Internet inertia means that as one family unit connects, an imperative, without particular recourse to utility, is created in a socially networked unit. If the 'Ivanovs' have it, then so must their social peers, the 'Stepanovs'.

In this case study of a small provincial Russian town with its focus on 'ordinary' users in a 'marginal' setting, Woolgar's first rule of virtuality is fulfilled: 'uptake and use of the new technologies depend crucially on local social context' (Woolgar 2002a, p. 14). Social and geographical imperatives of self-provisioning and difficult economic realities significantly shape use. The internet is integrated into the 'local' by a form of connectivity not predicated on the virtual, but on the physical presence of, and access by others to, the domestic computer-point as an indicator of status. Rules Three and Four—that the 'virtual supplements, rather than substitutes for real activities' and 'the more virtual, the more real'—are also therefore supported by the material presented here (Woolgar 2002a, pp. 16–17).[33] Similarly to an analysis of the impact of the

---

[33]Rule Two states that the 'fears and risks associated with new technology are unevenly socially distributed' (Woolgar 2002a, pp. 15–16). The present research cannot adequately address this hypothesis, but clear indicators of the 'fears' associated with the internet regarding inappropriate underage use were prominent among informants. Rule Five states that the more global a technology, the more local the nuances of its implementation. Again, local relevancies and the absence of interest by informants in communication outside their locale, would support this hypothesis.

FIGURE 2. ACCESS, EVEN THROUGH AN ADSL LINE, CAN BE SLOW IN IZLUCHINO—GIVING MORE TIME FOR NON-VIRTUAL INTERACTION.
*Note*: image taken by the author in the field between November 2009 and December 2010.

telephone in 1940s America or the use of the mobile phone in Jamaica in the 2000s, the internet in provincial Russia can be shown not to 'dramatically affect the localism in small-town life' (Horst & Miller 2006, p. 8). Instead, like Russian mobile phone use, the internet helps maintain close personal networks (persons living in a single city) (Lonkila & Gladarev 2008, pp. 284–85) and, ironically, communication between persons physically in the presence of each other. The internet increases the volume of 'talk', but not virtually. Practices that are by and large socially predicated are enhanced: family and friend leisure pastimes in the domestic sphere, and self-provisioning activities that are often social—gendered hobbies often carried out in the physical company of others: fishing, knitting, tinkering with cars. The internet in Izluchino can be seen to largely reproduce the 'narrowness' of the existing social architecture—placeness, both of the physical citing of access and the user's social milieu, turns out to be a key determiner of the 'virtual'.

*University of Birmingham*

## References

Arestova, O. N., Babanin, L. N. & Voyskunskiy, A. E. (2000) 'Motivatsiya pol'zovatelei Interneta', in Voyskunskii, A. (ed.) (2000) *Gumanitarnye issledovanie v Internet* (Moscow, UDK).
Bakardjieva, M. (2005a) 'Becoming an Internet User in Bulgaria: Notes on a Tangled Journey', *Media Studies/Studia Medioznawcze*, 3, 22.

Bakardjieva, M. (2005b) *Internet Society: The Internet in Everyday Life* (London, New Delhi & Thousand Oaks, CA, Sage).

Bernard, H. R. (2000) *Social Research Methods: Qualitative and Quantitative Approaches* (Thousand Oaks, CA, Sage).

Bradner, E., Kellogg, W. A. & Erickson, T. (1999) 'The Adoption and Use of "BABBLE": A Field Study of Chat in the Workplace', in *Proceedings of the Sixth European Conference on Computer Supported Cooperative Work*, August (Copenhagen, Center for Tele-Information, Technical University of Denmark), pp. 139–58.

Castells, M. (1996) *The Rise of the Network Society* (Oxford, Blackwell).

Castells, M. (2001) *The Internet Galaxy: Reflections on the Internet, Business, and Society* (Oxford, Oxford University Press).

Castells, M. & Kiselyova, E. (1998) 'Russia and the Network Society: An Analytical Exploration', Conference Paper, *Russia at the End of the 20th Century*, Stanford University, 5–7 November.

Cooper, J. (2008) 'The Internet in Russia—Development, Trends and Research Possibilities', in *CEELBAS Post-Soviet Media Research Methodology Workshop*, 28 March, Birmingham, UK.

Croteau, D. & Hoynes, W. (2003) *Media Society: Industries, Images and Audiences* (Thousand Oaks, CA, Pine Forge Press).

Crump, B. & McIlroy, A. (2003) 'The Digital Divide: Why the "Don't-Wants-Tos" Won't Compute: Lessons from a New Zealand ICT Project', *First Monday*, 8, 12, December.

Driesbach, C., Walton, R., Kolko, B. & Seidakmatova, A. (2009) 'Asking Internet Users to Explain Non-Use in Kyrgyzstan', in *Proceedings of International Professional Communication Conference, IPCC '09*, Honolulu, Hawaii, 20–22 July, pp. 1–6.

Eveland, W. P., Jr & Dunwoody, S. (2001) 'User Control and Structural Isomorphism or Disorientation and Cognitive Load?: Learning From the Web Versus Print', *Communication Research*, 28, 1, pp. 48–78.

Fischer, C. S. (1992) *America Calling: A Social History of the Telephone to 1940* (Berkeley & Los Angeles, CA, University of California Press).

Fischer, M., Lyon, S. & Zeitlyn, D. (2008) 'The Internet and the Future of Social Research', in Fielding, N., Lee, R. M. & Blank, G. (eds) (2008) *The Sage Handbook of Online Research Methods* (London, Sage), pp. 519–36.

Galitskii, E. B. (2009a) 'Internet v Rossii: dinamika rosta auditorii', in Oslon, A. A. (ed.) (2009), pp. 13–20.

Galitskii, E. B. (2009b) 'Tipologiya polzovatelei interneta', in Oslon, A. A. (ed.) (2009), pp. 66–90.

Galitskii, E. & Sidorova, A. (2009) *Oprosy 'Internet v Rossii'* (Moscow, Fond obshchestvennoe mnenie).

Gibson, J. J. (1979) *The Ecological Approach to Visual Perception* (Englewood Cliffs, NJ, Prentice-Hall).

Haddon, L. (2006) 'The Contribution of Domestication Research to In-Home Computing and Media Consumption', *The Information Society*, 22, 4, pp. 195–203.

Hammersley, M. & Atkinson, P. (1995) *Ethnography: Principles in Practice*, 2nd edn (London & New York, Routledge).

Hargittai, E. (2007) 'Whose Space? Differences Among Users and Non-Users of Social Network Sites', *Journal of Computer-Mediated Communication*, 13, 1, pp. 276–97.

Hargittai, E. (2010) 'Digital Na(t)ives? Variation in Internet Skills and Uses among Members of the "Net Generation"', *Sociological Inquiry*, 80, 1, pp. 92–113.

Hargittai, E. & Shafer, S. (2006) 'Differences in Actual and Perceived Online Skills: The Role of Gender', *Social Science Quarterly*, 87, 2, pp. 432–48.

Hetherington, K. (1998) *Expressions of Identity: Space, Performance and Politics* (Thousand Oaks, CA, Sage).

Horst, H. A. & Miller, D. (2006) *The Cell Phone: An Anthropology of Communication* (Oxford & New York, Berg).

Howard, P. N., Rainie, L. & Jones, S. (2001) 'Days and Nights on the Internet: The Impact of a Diffusing Technology', *American Behavioral Scientist*, 45, 3, pp. 383–404.

Kline, R. & Pincher, T. (1996) 'Users as Agents of Technological Change: The Social Construction of the Automobile in the Rural United States', *Technology and Culture*, 37, 4, pp. 763–95.

Latour, B. (1987) *Science in Action: How to Follow Scientists and Engineers Through Society* (Milton Keynes, Open University Press).

Lebedev, P. & Galitskii, E. (2009) 'Sotsio-Internet-sreda', in Oslon, A. A. (ed.) (2009), pp. 23–65.

Leonardi, P. M. (2003) 'Problematizing "New Media": Culturally Based Perceptions of Cell Phones, Computers, and the Internet among United States Latinos', *Critical Studies in Media Communication*, 20, 2, June, pp. 160–79.

Lonkila, M. & Gladarev, B. (2008) 'Social Networks and Cellphone Use in Russia: Local Consequences of Global Communication Technology', *New Media & Society*, 10, 2, pp. 273–93.

Miller, D. & Slater, D. (2000) *The Internet: An Ethnographic Approach* (London & New York, Berg).

Neuman, W. R. (1991) *The Future of the Mass Audience* (Cambridge, Cambridge University Press).

Norris, P. (2001) *The Digital Divide: Civic Engagement, Information Poverty, and the Internet Worldwide* (Cambridge, Cambridge University Press).

Ofcom (2008) 'The Nations & Regions Communications Market', available at: http://stakeholders.ofcom.org.uk/market-data-research/market-data/communications-market-reports/cmrnr08/?a=0, accessed 24 December 2011.

Oslon, A. A. (ed.) (2009) *Internet.ru: sotsiologicheskie kontury* (Moscow, Ekoninform).

Pearce, K. E. (2009) 'Explaining Computer, Internet and Social Networking Use in Armenia: The Influence of Skills, Digital Divide Demographics, Language, Innovativeness, Technical Access, Privacy, Social Influence and Collectivism', National Communication Association Conference, Chicago, IL, November.

Schmidt, H. & Teubener, K. (2006) '"Our RuNet"? Cultural Identity and Media Usage', in Schmidt, H., Teubener, K. & Kondradova, N. (eds) (2006) *Control and Shift: Public and Private Usages of the Russian Internet* (Norderstedt, Books on Demand GmbH), pp. 14–20.

Schröder, K. C. (1999) 'The Best of Both Worlds? Media Audience Research between Rival Paradigms', in Alatuusari, P. (ed.) (1999) *Rethinking the Media Audience. The New Agenda* (London, Sage), pp. 38–68.

Söderström, S. (2009) 'Offline Social Ties and Online Use of Computers: A Study of Disabled Youth and their Use of ICT Advances', *New Media & Society*, 11, 5, August, pp. 709–27.

Stern, M. J. (2010) 'Inequality in the Internet Age: A Twenty-First Century Dilemma', *Sociological Inquiry*, 80, 1, February, pp. 28–33.

Wellman, B. (2001) 'Physical Place and Cyberplace: The Rise of Personalized Networking', *International Journal of Urban and Regional Research*, 25, 2, pp. 227–52.

Wellman, B. & Haythornthwaite, C. (eds) (2002) *The Internet in Everyday Life* (Oxford, Blackwell).

Willis, S. & Tranter, B. (2006) 'Beyond the "Digital Divide". Internet Diffusion and Inequality in Australia', *Journal of Sociology*, 42, 1, pp. 43–59.

Woolgar, S. (2002a) 'Introduction', in Woolgar, S. (ed.) (2002b), pp. 1–21.

Woolgar, S. (ed.) (2002b) *Virtual Society? Technology, Cyberbole, Reality* (Oxford, Oxford University Press).

Wyatt, S., Thomas, G. & Terranova, T. (2002) 'They Came, They Surfed, and Went Back to the Beach: Conceptualizing Use and Non-Use of the Internet', in Woolgar, S. (ed.) (2002b), pp. 21–40.

# Co-opting Transmedia Consumers: User Content as Entertainment or 'Free Labour'? The Cases of *S.T.A.L.K.E.R.* and *Metro 2033*

NATALIA SOKOLOVA

## Abstract

The expansion of new media and the advent of Web 2.0 has resulted in dramatic changes in consumers' roles, turning them into 'prosumers': co-participants in cultural production. This essay considers the new situation through analysis of transmedia phenomena produced simultaneously by companies and consumers' activity. The analysis of two Russian transmedial projects: *S.T.A.L.K.E.R.* and *Metro 2033*, reveals a paradox in this phenomenon testifying to the mass expansion and boom in creative practices (rooted, to a great extent, in Soviet cultural traditions) and to their total commodification at the same time.

OCCASIONALLY SOMEONE FROM INTERNET FAN communities of the popular *S.T.A.L.K.E.R.* computer games[1] expresses the opinion that it would be profitable to pay 'modders', or amateur game developers, since their activities encourage a new wave of game modifications and, consequently, bring new profit for companies.[2] The world games industry has witnessed plenty of cases where modders' ideas were successfully employed. For example, another popular and revolutionary windows-based PC game, *Half-Life*,[3] owed much of its enduring popularity and critical acclaim to the fact that game developers provided users with software development kits to create their own content and extensions of the game. And this is true not only of modders; along the same lines we can consider readers' co-authorship in the projects of professional writers who upload their works for multi-authored 'editing' and

I wish to thank the team of the 'New Media in New Europe-Asia' project—Natasha Rulyova, Jeremy Morris and Vlad Strukov—for inviting discussion on new media in the region and Jeremy Morris for his support and comments.

[1]*S.T.A.L.K.E.R.* is a series of three computer games developed by the Ukrainian company GSC Game World (2007–2010).

[2]See, for example, 'Forum po igre *S.T.A.L.K.E.R.*: Zov Pripyati: Mody na ZP. Ponadobyatsya li?', available at: http://www.gscgame.ru/index.php?t = community&s = forums&s_game_type = xr3&thm_page = 3&thm_id = 56&page = 25&sort = DESC&sec_id = 0&offset = -180, accessed 12 May 2009.

[3]*Half-Life* is a popular science-fiction video game developed by Valve Corporation and published by Sierra Studios for PC in 1998.

'writing up' and other fan-based practices which have been used by the media industry. In their free time thousands of people are involved in such activities and this phenomenon has already been both commercially important to producers, and a meaningful form of involvement to consumers. Thus Tapscott and Williams, who introduced the project of 'Wikinomics', or economics for the new media age, believe that the activity of consumers and the 'co-production' of consumers and official producers is one of the main features of the new economics (Tapscott & Williams 2006). This new situation, triggered by the expansion of new media and especially the internet, signifies the implosion of production and consumption, labour and free time, work and entertainment, the author and the audience, the professional and the amateur. Popular culture and entertainment in symbiosis with the internet illustrate this new situation very well.

The activity of 'prosumers', or consumers who have become producers,[4] has recently been the focus of academic discussion. Prosumption is considered as a phenomenon involving both production and consumption and a locus of social change, especially those associated with the internet and Web 2.0 (Bruns 2008; Ritzer & Jurgenson 2009). Many cases in the sphere of popular culture and entertainment media can also be regarded as prosumption. Humphreys, for instance, investigates 'productive players' in online multi-user games (2005), and Sotamaa examines players' productivity with a focus on the multifaceted relations between players and the game industry (2007). Yang considers the new role of 'prosumers' by examining the music-entertainment industry, that integrates fan production, fan promotion and fan consumption all in one (Yang 2009). Martens, in her study of teenagers' participation in collaborative writing projects, concludes that it is quite possible to speak about 'affective labour' (Martens 2011). Peuter and Dyer-Witheford argue that the production of computer games is an ideal example of 'immaterial labour' within the informational economy and stresses that the game industry is inherently dependent on the 'free labour' of game hobbyists (Peuter & Dyer-Witheford 2005). Arvidsson and Sandvik view computer players as a kind of 'immaterial labour' and discuss how this is put to work by the games industry (Arvidsson & Sandvik 2007). The choice of terms in the discussion—'consumers' activity', 'productivity' or 'labour'—itself identifies the researcher's position. A recent conference entitled *Internet as Playground and Factory*[5] demonstrated the importance of the issue and the variety of points of view (Fuchs 2009).

Russian researchers have tended to ignore the phenomenon of prosumption, in particular as applied to popular culture. On the whole, many of them rarely invoke the theme of popular culture and vernacular creativity, and a division between high-brow and low-brow attitudes towards the sphere of this cultural phenomenon tends to relegate it to the realm of 'non-authentic' needs (Birichevskaya & Strelchenko 2006; Savenkova & Frolova 2000). The analysis of various cultural practices relating to prosuming leaves much to be desired, the lack of critical reflection on the issue being especially acute. Though Web 2.0 'business-manifestos', which celebrate prosumers' activity, are becoming increasingly popular among Russian researchers (for example,

---

[4]The term was introduced by Toffler in his book *The Third Wave* (1980).
[5]Held 12–14 November 2009 at The New School, New York City, USA.

the newly published *Wikinomics* by Tapscott and Williams in Russian translation among them (2009)) and while these manifestos have started to influence academic discourse on culture, there is a shortage of research work on the subject.

Analysis of prosumers' activity, as well as the issue of 'co-production' can be successful only if it considers transmedia, the most characteristic phenomenon of the new media age reflecting many of its trends and contradictions. The term 'transmedia' implies the simultaneous promotion of a product via a number of media channels, at least one of which is interactive. In the West the phenomenon is widespread while in Russia it has only recently appeared. This essay examines transmedia in relation to popular culture and more closely the presumption phenomenon. In analysing this phenomenon I use the examples of two successful projects: *S.T.A.L.K.E.R.* and *Metro 2033*. The essay is structured as follows. Firstly, I consider transmedia itself as well as the reasons that enable us to treat it as the defining phenomenon of the new media age. Secondly, I analyse the 'picture' of transmediality in Russia and show why these two projects are archetypically transmedial in comparison with other media projects. Thirdly, the role which consumers play in these projects is defined. I seek to demonstrate that the activity and creativity of consumers, which enable us to treat them as 'prosumers', and their role in the 'co-production' of transmedia, manifests itself not only through actual production of content, ideas and artefacts that contribute to a fictional 'universe' of the projects, but to an even greater extent it manifests itself through production of a special commodity—a brand. I then stress the local dimension of prosumption phenomenon: on the one hand, the role of 'prosumers' is determined by the general logic of contemporary media production and consumption and the spread of new media and the internet, but on the other hand, it is caused by the local specificity and cultural tradition. Finally, I use Marcel Mauss's concept of 'gift-giving' (Mauss 1990) and the Marxist concept of 'labour' in order to consider the two transmedial projects in the context of current debates about prosumption phenomenon (Ritzer & Jurgenson 2009; Terranova 2000; Fuchs 2010). I argue that prosumption in popular culture reveals contradictions: opportunities for mass creativity and self-expression appear at the same time as they are commodified by the companies.

### Transmedia: some theoretical implications

Transmedia is a relatively new cultural and economic phenomenon (Jenkins 2007; Perryman 2008; Scolari 2009). Scholarship does not agree on a single definition but all speak about at least two main characteristics (Jenkins 2007; Perryman 2008; Scolari 2009). Firstly, the term transmedia is a characteristic of the media product itself (although we can only tentatively speak about 'product', since it is a dynamic flexible combination of many media formats rather than a completed artefact). Unlike the widespread advertising practice that delivers content to consumers via numerous channels—so called integrated marketing communications—transmedia assumes the creation of a topical fictional 'world' or 'universe', created by various types of media. Hence, a group of media formats, even if thematically linked, is not yet transmedia. As Jenkins notes, transmedia represents a structure based on the further development of the story world through each new medium, and ideally each medium makes it own unique contribution to the unfolding of the story, the purpose being to create a unified and coordinated entertainment experience (Jenkins 2007). Projects that arose out of the Hollywood film *The Matrix*, the reality TV show *Big Brother*, and the Harry Potter books are considered among the most typical transmedia examples (Evans 2008; Perryman 2008; Vasile 2011).

Another major feature of transmedia is the creation of new cultural practices and experiences for their audiences. Transmedia involves interactive feedback inviting consumers to participate in 'co-production' of a topical fictional 'world' or 'universe'. We can speak of so-called 'interpretive communities' of transmedial projects. The term 'interpretive communities' was coined by Fish (1980) but I employ it to highlight the role of the audience in various media projects.[6] Fish understood readers' interpretations of a text as the process of co-authorship; likewise we can speak about 'collaborative creation' and 'co-production' of the audience in transmedia projects.[7] The 'interpretive community' as related to transmedia is the sum of spectators, readers, listeners or players who are involved in the interpretation of media texts, intensive communication, various cultural practices and the authoring of their own media content and artefacts within a transmedial fictional 'universe'. Such 'interpretive communities' exist in part due to the internet. Therefore transmedia is a cultural phenomenon of the new media, specifically of Web 2.0, with its interactivity, typical for the internet of the 'second generation', and radical changes to a user's role which include user-generated content, user distribution of media content and artefacts which bypass official institutions through blogs and social networks, and online-community formation.

The expansion of transmedia is also caused by the logic of the post-Fordist economy and recent media convergence (Deuze 2006; Hay & Couldry 2011; Jenkins 2004). The emergence of media conglomerates working in multiple industries and carrying out manufacture, distribution and marketing, and active on a global scale and through different media channels, are the social and economic preconditions of transmedia.

### Transmedia in Russia

Russian media and entertainment also has a tendency towards media convergence;[8] however, there are few examples of genuine transmedial projects in the sphere of popular culture. Thus, the huge Russian media-project *Dom-2* (House-2) on the face of it seems to be transmedia. *Dom-2, Gorod Lyubvi* (City of love) is a reality show where participants try to find someone to love and build a house.[9] Created by the TNT channel, it is the most highly rated reality show on television in Russia—extremely popular among young people. It is also the longest-running Russian reality show ever (ongoing since 2004). *Dom-2* is not just a TV show, but a number of other media

---

[6]See, for example, Murley (2008).

[7]See Strukov in this collection.

[8]See, for example, a series of studies on Russian transmedia in *Digital Icons: Studies in Russian, Eurasian and Central European New Media*, 2011, Issue 5, available at: www.digitalicons.org, accessed 2 July 2012.

[9]Programmy TNT, *Dom-2: Gorod Lyubvi*, available at: http://dom2.tnt-online.ru/, accessed 18 May 2010.

formats—a glossy magazine, DVDs, a computer game, content for mobile telephones, several books and a multitude of websites. There are numerous online communities of fans and even 'anti-fans',[10] who, regardless of their particular attitude, work to further disseminate the brand. Despite the plurality of media formats and fan activity, *Dom-2* is not a true transmedial project, however, because of the lack of a topical fictional 'universe' and also an 'interpretive community'. The plot of the reality show is rather lacking in events and details, thus fans do not have much 'food for interpretation' and content creation; websites are orientated towards the discussion of the characters' and hosts' lives. Revealingly, fan art is limited to the creation of computer wallpaper and the most basic animation.[11] There is no fan fiction, fan video, or comics on *Dom-2*.

One might expect that *Dozor* (Watch), the project of fantasy sequels, actively promoted by television channel ORT (Channel One), could lay claim to becoming a transmedial project. The series of books by Sergei Luk'yanenko,[12] films in the genre of supernatural thrillers *Night Watch*[13] and *Day Watch*,[14] were complemented by the release of a computer game.[15] These fantasy worlds have attracted fans: video content based on the films is still being created for YouTube; and there are extreme offline night-time urban role-playing games based on the films.[16] The project had every chance of becoming transmedial, but instead a diffuse, uncoordinated response by consumers resulted in its gradual disappearance as a meaningful transmedial phenomenon.

In contrast the projects *S.T.A.L.K.E.R.* and *Metro 2033* can be characterised as genuinely transmedial. Project *S.T.A.L.K.E.R.* originated as a series of computer games (2007–2010), created by developers from the Ukrainian company GSC Game World.[17] These games are primarily first-person shooters,[18] but unlike many other such games, players are not restricted to a linear plotline and can explore their world as they wish.

*S.T.A.L.K.E.R.* is also the name of a huge fictional transmedial 'universe'. This 'universe' melds the features of the science-fictional 'Zone' from the novella *Piknik na obochine* (Roadside Picnic) by Arkadii Strugatsky and Boris Strugatsky (1972) and the real Chernobyl Zone. The action of the games takes place in Chernobyl in 2012, six

---

[10]Available at: http://domforever.com, accessed 20 August 2011; *Anti-2dom*, available at: http://anti2dom.beon.ru, accessed 14 September 2008.

[11]Fan art is the creation of fans' artefacts on the basis of original artwork.

[12]*Nochnoi dozor*, 1998 (Night Watch); *Dnevnoi dozor*, co-written with Vladimir Vasil'ev, 2000 (Day Watch); and *Sumerechnyi dozor*, 2004 (Twilight Watch).

[13]*Nochnoi dozor*, 2004.

[14]*Dnevnoi dozor*, 2006.

[15]See 'Mir Dozorov', an official site of the game, available at: http://www.dozory.ru/nightwatch.html, accessed 1 September 2010.

[16]See, for example: http://news.samaratoday.ru/news/148000, accessed 12 August 2011.

[17]GSC Game World is a Ukrainian computer games developer, one of the largest in Eastern Europe, which was founded in 1995. The official site of the company is available at: gsc-game.ru, accessed 10 February 2011.

[18]First-person shooters—a type of computer game where a single player, acting on his own, has to destroy all the enemies using weapons. The player follows what is happening in the game through a first-person perspective, so the picture the gamer observes coincides with the 'view' available for the character.

years after a fictional second Chernobyl catastrophe. At that time the Zone is already densely populated, the world of the Zone is full of 'anomalies' and mutants. As envisaged in the Strugatskii book (1972), a stalker transports '*khabar*', or artefacts of alien origin, out of the mystic 'Zone' (Revich 1998). After the Chernobyl disaster of 1986, the name 'stalker' was given to those who took the risk of travelling into the irradiated area to discover the causes of the disaster.

Today *S.T.A.L.K.E.R.* is an easily recognised brand. The series of games is extremely popular in Russia and beyond. For example, it has a very large fan community in Germany and there is a community of foreign gamers on Facebook.[19] The impact of the games is enhanced by many other media formats. *S.T.A.L.K.E.R.* 'literature' (a whole site is devoted to this fan literature—Litstalker.ru) includes a large number of e-books.[20] There are music albums, and content for mobile telephones with content from the games' audio-books;[21] there are plans to produce a science-fiction TV series based on the game styled after 'Men in Black'.[22] There is also a considerable online encyclopaedia containing the database of arms, anomalies and artefacts from 'The Zone'.[23]

The internet abounds in libraries and galleries of fan art based on the games. There is *S.T.A.L.K.E.R.* radio,[24] and a TV channel *STALKER GAME WORLD* on YouTube.[25] Amateur clips (both mini-movies and full-length films) are constantly being uploaded to fan sites and blogs; a substantial series of amateur comics exists. The world of the computer games is adaptable to various media formats and has turned out to be infinitely flexible. The potential of *S.T.A.L.K.E.R.* is far from being exhausted: the developers are working on a new episode of the game: *S.T.A.L.K.E.R.-2*, scheduled for release in 2012 but since scrapped.[26]

The transmedial world of *Metro 2033* developed from, and continues to grow out of, the cult novel of the same name by Dmitry Glukhovskii (2007) which is now a bestseller with translation rights in 23 countries. The fictional 'universe' of *Metro 2033*—like *S.T.A.L.K.E.R.*—is post-apocalyptic, but the action takes place after a nuclear catastrophe in Moscow in the metro system, the biggest atomic shelter in the world. As a result of a nuclear exchange, all large cities have been wiped off the face of the Earth but tens of thousands of surviving Muscovites shelter underground. The rest of the gameworld is very similar to *S.T.A.L.K.E.R.*—there are also 'stalkers' who

[19]Official *S.T.A.L.K.E.R.*, Facebook site, available at: http://www.facebook.com/officialstalker?v=info Official S.T.A.L.K.E.R., accessed 4 December 2010.

[20]Litstalker, available at: http://www.litstalker.ru, accessed 12 May 2009.

[21]Audiostalker, available at: http://www.audiostalker.ru, accessed 2 December 2009.

[22]Kinostalker, available at: http://www.kinostalker.com, accessed 8 September 2010; GSC zanimaetsya s"emkami serial po *S.T.A.L.K.E.R.*, available at: http://vse-stalker.net.ru/news/gsc_zanimaetsja_semkami_seriala_po_s_t_a_l_k_e_r/2010-05-13-78, accessed 13 May 2010.

[23]*S.T.A.L.K.E.R.* encyclopaedia, available at: http://www.stalker-epos.com, accessed 9 April 2010.

[24]See, for example: FSR (*Pervoe stalkerskoe radio*), available at: http://stalker-gsc.ru/news/2009-01-18-311, accessed 18 January 2009.

[25]*STALKER GAME WORLD*, available at: http://www.youtube.com/stalkergameworld, accessed 11 December 2008.

[26]Mini-interv'yu s Olegom Yavorskim, available at: http://www.stalker-zone.info/news/mini_intervju_s_olegom_javorskim/2010-06-22-793, 22 June 2010, accessed 2 July 2012.

come up to the surface of the dead metropolis in search of food; there are anomalies, radiation, mutants and so on.

The project *Metro 2033* can also be classed as transmedial. Taking the fictional universe of novel *Metro 2033* as its foundation, a cycle of spin-off books has been written: the action here takes place outside Moscow, e.g. in the novel by Shimun Vrochek *Metro 2033—Piter* (Vrochek 2010).[27] There is also radio content, a board game and a computer game developed by GSC Game World.[28]

### 'Interpretive communities'

Transmedia can therefore be seen to engage various cultural practices of consumers. The creation of a transmedial project proper is a dual process which involves the specific policies of media companies as well as activity and the creativity of their audience. Both projects discussed possess these features.

Project *S.T.A.L.K.E.R.* is of particular interest because it was mainly formed not by an organised campaign 'from above', via official commercial channels, but rather by the initiatives and activity of ordinary gamers. The role of consumers was vital to the project when it was formed. During the six years of development of *S.T.A.L.K.E.R.: Shadow of Chernobyl* (which was released in 2007) the producer sought fans' reactions to the game's development through forums on the game's internet sites. The first game acquired its fan community and had become an object of cult interest even before it was released (Zlotnitskii 2008), so that there was no need for mass official promotion—compared to the promotion of the films *Night Watch* and *Day Watch*. Then activity proper 'from below' started immediately after the first game was released. Overwhelming success brought about follow-up games, other media formats and made the project enduring. A whole subculture—'stalker culture'—took shape around this game series, including not only online communities, but also events offline, such as conventions and competitions.[29] One example was the international festival *S.T.A.L.K.E.R.-Fest* which took place in Kyiv in 2009.[30] The central city square was turned into a huge exclusion 'Zone', with a replica of the Chernobyl nuclear power station. The festival included a costume contest, a light show, a stunt show and fights in *S.T.A.L.K.E.R.*'s style. A rich internet folklore has sprung up

---

[27]'Mezhavtorskii tsikl "Vselennaya Metro 2033"', available at: http://www.ciscotrain.mirea.rublog. fantlab.ru/work181927, accessed 10 March 2011.

[28]'Ofitsial'noe radio Vselennoi Metro 2033', available at: http://proekt-metro.ru/news/oficialnoe_ radio_vselennoj_metro_2033/2011-07-25-406, accessed 25 July 2011; see also http://www.boardgamer. ru/moi-vpechatleniya-ot-metro-2033, accessed 22 July 2010; *Metro 2033*, official website, available at: http://proekt-metro.ru, accessed 14 September 2010.

[29]See: 'Bol'shaya peintbol'naya igra *S.T.A.L.K.E.R.*', available at: http://www.na-pike.ru/otchet. php?otchet=10, accessed 3 July 2011; 'Straikbol v "Sverdlovskom Stalkere"', available at: http://www. clubstalker.ru/club/playit, accessed 11 August 2011; 'Peintbol'nyi club Vkontakte', available at: http://vkontakte.ru/club4836531, accessed 5 October 2010; 'Peintbol v zone *S.T.A.L.K.E.R.*', available at: http://www.eniki-beniki.in.ua/tour/144-peyntbol.html, accessed 6 June 2010.

[30]Official site of *S.T.A.L.K.E.R.-Fest*, available at: http://stalker-fest.com, accessed 14 June 2010.

around the games, including numerous anecdotes, 'post-nuclear' poetry, 'legends' and even toasts.[31]

The *Metro 2033* project was formed along different lines. Glukhovskii's novel *Metro 2033* (2007) was originally planned as an internet project: the book was uploaded to a website as the author was progressing with it, so during its creation the readers could comment on chapters or even put forward a different narrative move.[32] Glukhovskii's next novel *Metro 2034* (Glukhovskii 2010) was created along the same lines: free-to-view chapter-by-chapter publishing of the book on the official website, although the denouement was only available in the printed version of the book. Readers took part in its production and were acknowledged as the key members of the book's community. Even before the novel was released, the author had been able to provoke immense interest. Later the novel was a publishing success, despite its availability on the internet.

Glukhovskii, who acts not only as the author but also as the project's producer, employed the strategy of 'co-creation' with the reader, which has since become popular among Russian writers.[33] Unlike the majority of them, however, he had planned *Metro 2033* as a multimedia project from the start. For him literature was always a component of a large cross-media experience.[34] Besides the books, he had initially planned to create a game based on *Metro 2033*, available for players via an internet browser, mobile phones and social networks. In his own words, he 'belongs to the first generation of writers who treat computer games as full-value art..., an experience that formats consciousness, determines mentality and creation'.[35] In creating '*Universe Metro 2033*', Glukhovskii purposefully drew on the internet itself as the main means of production for this universe. The web portal 'metro2033.ru' was to turn into 'a virtually real world', providing the opportunity to read new books from the 'Universe', listen to the music, view illustrations, play and even 'live' at one of the stations of the Metro.

Glukhovskii's literary internet projects *Metro 2033* and *Metro 2034* were a success and a large number of readers were given direct feedback. The thematically rich fantasy world of the books was popular with readers. Later, other media formats related to *Metro 2033* appeared and audience interest in them has endured. A computer game was developed by GSC Game World, which produced *S.T.A.L.K.E.R.* Therefore, the *Metro 2033* project was targeted at *S.T.A.L.K.E.R.*'s audience, which brought users of both projects into contact with each other and other different projects.

[31]See, for example, 'Stalkerskie anekdoty', available at: http://www.stalker.by/forum/viewtopic.php?f=68&t=53&sid=68bfcc8555a6aa71d21ffaddccf9bce5, accessed 15 August 2009; 'Stalkerskie pogovorki', available at: http://stalker-book.com/forum/6-1532-1, accessed 18 June 2010.

[32]*Metro, post'yadernyi roman*, available at: http://www.m-e-t-r-o.ru./metro.html, accessed 9 May 2009.

[33]For example, one of the first people who started to upload their novels to the internet was a popular fiction writer Sergey Luk'yanenko. So, while he was writing the chapters of his novel *Draft* (*Chernovik*), readers were invited to introduce their changes to the narration. There was an opinion poll held regarding the style of the book, the gender and age of the protagonist and so on (Chernenko 2009). A similar strategy was employed by Oksana Robski, Andrei Levitskii and others.

[34]See http://mirf.ru/Articles/art4016.htm, accessed 6 July 2012.

[35]See http://mirf.ru/Articles/art4016.htm, accessed 6 July 2012.

Thus, in comparison to *S.T.A.L.K.E.R.* the fan community of the *Metro 2033* project was originally built around the novel's website, a structuring envisaged from the very beginning. Despite the fact that, unlike *S.T.A.L.K.E.R.*, *Metro 2033* was actively promoted (especially on *Runet*) and consumer activity was in large part previously anticipated, constructed 'from above', we still can witness unplanned grassroots prosumption, for example, in the unofficial magazine *Metromir*, with audio attachments.[36] *Runet* can also boast many online communities that actively extend the life of *Metro 2033*,[37] and many events are held offline.

## S.T.A.L.K.E.R. *and* Metro 2033*: the cultural backdrop to transmedial projects*

The logic of official promotion within the two transmedial projects discussed here is widely used the world over but are there any Russia-specific factors in the success of these two projects? First and foremost there is the successful idea behind the projects and their rich topical fictional 'universes'. The *S.T.A.L.K.E.R.* gamers attracted players by an appealing alternative fantasy reality, with many characters and details as well as its dynamic game play and a lively atmosphere. The series is often criticised for its technical details but many users and observers alike think that the developers managed to create 'a uniquely' tangible 'universe'. Users repeatedly refer to the terms 'great game', 'cult game' and 'game-legend' to complete the description.[38] As for *Metro 2033*, according to a number of reviewers, the game based on it is one of the best in Russia—it is 'a perfectly staged adventure, full of suspense' (see, for example, Melekhov 2010).

Certainly, the post-apocalyptic theme has contributed to the popularity of both projects. Their developers highlight the fact that apocalyptic worlds are attractive due to their novelty and explorability, and the possibilities of experimental plotlines and game play (Kazunov 2007). Although post-apocalyptic ideas are universal in culture, both projects have a specifically Russian character.[39] According to Glukhovskii, the post-nuclear setting was chosen because Russians still face the problem of survival.[40] The factor of recognisability and proximity of the post-apocalyptic fictional 'universes' of both projects was also significant in their success. In *Metro 2033* the reader and gamers easily recognise elements of today's Moscow. In *S.T.A.L.K.E.R.* the Zone reproduces features of the real Chernobyl zone. Thus, Claudia Kern, the German author of a *S.T.A.L.K.E.R.* book series, explains why the 'laws of the world' in

[36]See the non-official site of 'Metromir', available at: http://metro2033.livejournal.com/72120.html, accessed 20 October 2011; "Metromir" v sotsial'noi seti "Vkontakte", available at: http://vk.com/metromir_2033, accessed 14 September 2011.

[37]See, for example, Fan-ploshchadka, available at: http://www.metro2033.ru/game/fans/art, accessed 3 October 2009; Fan-portal of 'Universe *Metro 2033*', available at: http://world-metro.ru, accessed 12 August 2011; Portal 'Metro-zone', available at: http://metro-zone.ru, accessed 10 June 2011.

[38]See, for example: *S.T.A.L.K.E.R.*, Zov Pripyati, retsenziya, available at: http://stalker-cs.ru/publ/3-1-0-267, accessed 28 September 2009.

[39]The theme of post-apocalypses is the central issue of a popular project called 'Invasion' which was formed as a transmedia project around the online and print novels of Andrei Levitskii (2011a, 2011b) or the project that started in September 2011—a literature series 'Anabioz' (see http://anabioz.com, accessed 6 July 2012).

[40]See, for example Alekhin (2008).

*S.T.A.L.K.E.R.* seem attractive to a writer: unlike the world of fantasy where the writer creates a completely new world, the world of *S.T.A.L.K.E.R.* is 'real', so a reader can more readily believe that the Zone and its characters can exist in reality (Murygin & Kalintsev 2011). Finally, the criticisms against *S.T.A.L.K.E.R.*'s developers claiming that they turned the tragedy of the Ukrainian people into a brand (Dergunov 2004) are met with the response that their position considers the Chernobyl catastrophe only as historical fact without any political and ideological connotations.[41]

Also central to the popularity of *S.T.A.L.K.E.R.* are its militaristic theme and its utilisation of elements of the Soviet past. Project *S.T.A.L.K.E.R.* today is a cultural mineshaft giving access to a huge seam of resonances and artefacts from the seemingly lost age of the Soviet period. In 'the Zone' of *S.T.A.L.K.E.R.* there are lots of Soviet-era artefacts and cultural referents symbolic of the Brezhnev era, such as Soviet fish preserves labelled *Zavtrak turista* (tourist's breakfast), that, beside functioning as a first-aid and anti-radiation kit, are part of a 'stalker's equipment'.

Thematically *S.T.A.L.K.E.R.* is strongly associated with the military.[42] While *S.T.A.L.K.E.R.* is hardly a sign of nostalgia for Soviet times or the military *per se*, nevertheless such military-style games with their atmosphere of danger, the spirit of brotherhood in war and romanticism demonstrate deep-seated needs for masculine self-realisation and a yearning for national unity. It was no accident that the project clearly attracted a large audience in the former CIS.[43] The military spirit is characteristic of many other games on the post-apocalyptic theme, but in the case of *S.T.A.L.K.E.R.* it particularly unites large numbers of young men who have done military service. Discussions on internet forums, war games and competitions offline, as well as insignia, gift sets and clothing in the *S.T.A.L.K.E.R.* style testify to this.

As mentioned above, the enduring popularity of *S.T.A.L.K.E.R.* and the growth of a whole subculture around it were not connected to a special corporate strategy or to economic actors of importance. Originally the 'stalker culture' around the games was, to a high degree, informal and grassroots-based. Fans' activity is, to a high degree, rooted in an enduring cultural tradition.

In a unique way, project *S.T.A.L.K.E.R.* is connected with the growth and popularity of the science-fiction genre in Soviet times. The period from the end of the 1950s to the 1970s witnessed the appearance of Soviet science-fiction fan clubs, which the authorities saw as culturally undesirable and repeatedly attempted to ban. These clubs organised festivals, conferences and prepared vast bibliographies. *Samizdat* magazines (fanzines) and amateur texts flourished along with translations from other languages, as well as professional Russian-language science fiction which was banned. 'The Golden Age' of science-fiction *samizdat* lasted for a decade from 1988 to 1998,

---

[41]'Sozdatel' kul'tovoi igry *S.T.A.L.K.E.R.* Sergei Grigorovich: "Zov Pripyati" budet gotov k oseni!', *Komsomol'skaya Pravda v Ukraine*, 8 April 2009, available at: http://kp.ua/daily/080409/176054, accessed 12 May 2009.

[42]The connection between popular culture and the military is particularly compelling. See, for example, the case of internet clubs linked to the TV series *Kadetstvo* appealing not only to television-series fans but also to the military and cadets (Sokolova 2009a).

[43]See, for example: 'Stalker.kz. Post-apokalipsis v Kazakhstane', available at: http://stalker.kz/forum, accessed 2 February 2011; Assotsiatsiya stalkerov Belorussii, available at: http://stalkerov.net/interes, accessed 22 June 2011.

until the emergence of the internet (Ivashnikov 2007; Pervushin 2008). The end of the 1970s and beginning of the 1980s saw the onset of live action role-playing games (the first ones were based on Tolkien's novels) that soon acquired a mass character. Stolyarov argues that the development of live action role-playing games was particularly significant in the USSR at a time when computer games were beginning to dominate in the West (Stolyarov 2008).

This science-fiction fan movement provided an enduring fan-orientated cultural and organisational memory. Fan clubs worshiped the creations of Arkadii and Boris Strugatskii, the literary 'fathers' of the stalker's 'profession'; the first stalker, Redrik Shukhart, appeared in their novel *Piknik na obochine* (Revich 1998). There was a fan club of the same name in Sevastopol; a popular magazine *Ural'skii Sledopyt* (Urals pathfinder), which regularly published science-fiction texts, was colloquially called *Uralskii Stalkery*. Science-fiction fans created 'stalker' board games. At the beginning of the 1990s Igor Kovshun, a club member in Odessa, developed a game which featured a map of 'The Zone' and its surrounding areas (Kovshun 1984). In 2001 Boris Strugatsky was involved in the development of a board game *Stalkeriada* (Bugrov 1983; Pervushin 2008). Thus, the large gaming community universally welcomed the computer game development as a logical, long-overdue milestone.

### Prosumers' 'co-production' in transmedia branding

The two cases considered here, while differing in a number of ways, illustrate the important role of 'prosumers'—consumers turned into producers, those who take an instrumental part in the transmedial projects' 'co-production'.

Prosumers make their contribution to projects through creating ideas and practices which are then utilised by official producers or which develop into products or artefacts in their own right. For example, Glukhovskii acknowledges the contribution of readers in creating *Metro 2033* through the internet. According to him, the readers did not affect the artistic side of the novel and its plot, but information provided which helped him to make the book more plausible was invaluable. The readers gave advice on various issues—such as construction of the underground railway and providing detailed descriptions of weapons.[44] The website GSC-Fan.com announced a competition for the best concepts for a new series of games, *S.T.A.L.K.E.R.-2*. The developers claimed that the most creative ideas would be taken into consideration during development.[45] A similar call was made for the development of merchandising clothing,[46] and an online game magazine *Global News of Apocalypse*.[47] A competition

[44]'Dmitrii Glukhovskii: ot kremlevskogo pula do Metro 2033', available at: http://www.aif.ru/culture/article/21533, accessed 2 October 2008.

[45]See, for example: 'Otobrany idei dlya *S.T.A.L.K.E.R.-2*', available at: http://www.stalker-modi.ru/news/otobrany_idei_dlja_s_t_a_l_k_e_r_2/2011-01-21-178, accessed 21 January 2011; 'Itogi sbora idei dlya *S.T.A.L.K.E.R.-2*, kinoproekta i zhurnala', available at: http://gsc-fan.com/?option=com_content&view=article&id=405, accessed 13 September 2011.

[46]'Odezhda dlya stalkera ot GSC', available at: http://www.shadow-of-chernobyl.ru/index.php?topic=11411.0, accessed 23 October 2010.

[47]See http://stalker-gsc.ru/news/nabor_v_redakciju_sajtovskogo_zhurnala/2011-09-16-1527, accessed 16 September 2011.

was also arranged when the company was developing a gift edition of the game *S.T.A.L.K.E.R., Clear sky*, or 'stalker equipment set', which consisted of various collectors' items and many other things beside the game itself.[48] Such competitions have become quite common.

Prosumer activity based on *Metro 2033* has also been widespread. Recently there was a call for users to create their own unique plot of *Metro 2033* for a mobile gaming platform.[49] There was also a video competition,[50] and contests to create some of the tiny details of the universe, such as ammunition used in the weapons.[51]

Apart from prosumers creating content, concepts or artefacts proper, fan art created by them is also of great importance for transmedial projects. Production by fan-art enthusiasts, however obsessive that may seem, contributes to the building of the transmedial 'universe'. It is especially true for cross-genres, when artefacts produced by the fans, and 'the canon' (or the original text that serves as a base for their variations), belong to different genres or media formats. In the case of *S.T.A.L.K.E.R.*, fans produce comics, films and literary pieces based on the computer games, while in the case of *Metro 2033* they have produced video clips based on the book. In fact, fans take centre stage in roles traditionally reserved for media marketing professionals. Thus, GSC Game World arranged several video contests while preparing to shoot a pilot for a TV series of 26 episodes based on *S.T.A.L.K.E.R.* The company elicited concepts and assembled a team of enthusiastic fans. The winners took part in the shooting of the series in 2012. They had produced an amateur film *Sector 62'18*, 2009, which won in the second video contest.[52] Similarly GSC Game World has adopted game modifications created by amateur developers.[53]

The communicative activities of consumers play an increasing role in product promotion and dissemination within transmedial projects, especially with the advent of widespread 'social media' in Russia. Glukhovskii claims that he even has nothing against a pirate version of his book. In his opinion, the text distributed free of charge is an opportunity to advertise the product.[54] However, here the point is not about creating a product proper, though fans, modders and others do create virtual artefacts, widely adopted by the media industry, or which contribute to product dissemination. This issue should be considered in a broader sense: prosumers are crucial in the production of a special commodity—the transmedial project brand.

Nowadays marketing embraces much more than just advertising and promotion of goods and services. A more efficient strategy is to create a communicative–informational

[48]'"Svershilos"!!!!! (Russkaya kollektsionka!!! Kruche zarubezhnoi)', available at: http://allstalk-er.clan.su/news/2008-07-21-12, accessed 21 July 2008.

[49]See http://www.metro2033.ru/news/619855, accessed 10 October 2011.

[50]See http://novalevic.livejournal.com; http://piterhandra.livejournal.com/308751.html, accessed 18 December 2011.

[51]See http://www.gamer.ru/metro-2033-poslednee-ubezhische/ekipiruy-vselennuyu-metro-2033-raboty-finalistov, accessed 22 October 2011.

[52]*Sektor 62'18'—pobeditel' konkursa po S.T.A.L.K.E.R*, available at: http://www.qeek.ru/591-stalker-videofilm-pobeditelya.html, accessed 6 December 2009.

[53]See: 'O S.T.A.L.K.E.R.-2 i ne tol'ko: interv'yu s razravotchikami', available at: http://strelok86.beon.ru/19444-853-o-s-t-a-l-k-e-r-2-i-ne-tol-ko.zhtml, accessed 2 October 2008.

[54]'Chem Dmitrii Glukhovskii shokiruet chitatelei v novom romane *Metro 2034*', *Komsomol'skaya Pravda*, 18 March 2009, available at: http://kp.ru/daily/press/detail/2350, accessed 23 March 2009.

environment around the brand and give consumers opportunities for independent activity and initiatives. Such a strategy fits very well into the modern business thinking, according to which you can achieve effective control over consumers and markets by providing opportunities for maximum consumer creativity (Prahalad & Ramaswamy 2004; Tapscott & Williams 2006; Vargo & Lusch 2004).

Brands are not merely a narrow marketing mechanism but a complex of functional and emotional values, which promise a unique and positive experience for consumers. Arvidsson, one of the best known scholars of branding, argues that a brand today is more a cultural resource than an institution; it is not so much linked with producers or product but rather with the context of consuming (Arvidsson 2005). The model of 'intrusion', when companies decide what people's opinion of a brand should be and implement their ideas in advertising, is being currently replaced by the model of 'involvement' or 'participation', where people are invited to take part in the life of a brand.

This brings us to the issue of 'co-creation' or 'co-production' of a brand by official producers and consumers. Who creates a brand and how is a brand created? Cultural practices as well as the information and communication environment around a brand are taken into account by companies. The same companies can build fan communities, eliciting the interest of consumers in the brand. According to the modern concept of branding, a brand is created by the mutual efforts of companies and consumers. The transmedial projects considered here fall neatly into this category. We can speak about the emergence of the brands *S.T.A.L.K.E.R.* and *Metro 2033* only after fans became involved in activities around them.

The difference between these two projects is that *Metro 2033* was deliberately constructed by Glukhovskii who acted as its author and producer in accordance with contemporary marketing strategies, and so it was well-planned. *S.T.A.L.K.E.R.* to a great degree grew into an unprecedented project due to the initiative and creativity of ordinary gamers; but the projects share lots of features, including dynamic 'interpretative communities', characterised by shared interests, intensive communication and large-scale thematic content creation. In the case of *S.T.A.L.K.E.R.* we can speak about the high degree of self-organisation and the 'integrity' of consumers. Thus, in its circulation, the slogan of *S.T.A.L.K.E.R.* fans: 'Stalkers of the world, unite!', is parallel to Marx's 'Workers of the world, unite!'[55] The fans talk about their community using the discourse of friendship, brotherhood, unity: Community of stalkers,[56] or Brotherhood of stalkers.[57] There are 'statutes' and 'codes of stalkers', describing not only community conventions and rules of behaviour but ethical imperatives as well. An emotional component is of major importance, for example, fans often mention the 'soul' of Stalker.[58] Values of mutuality, a feeling of belonging

[55]'Stalkery vsekh stran, ob"edinyaites'!', available at: http://www.stalker-zone.info/news/stalkery_ vsekh_stran_obedinjajtes/2011-01-13-882, accessed 13 January 2011.

[56]Sodruzhestvo stalkerov, Forum, available at: http://zona-revived.ucoz.ru/forum/6-1-30, accessed 14 July 2009. 'Ustav kluba industrial'nogo turizma "Stalker X—18 Kemerovo"', available at: http://x18-kemerovo.3dn.ru/blog/2009-04-07-57, accessed 21 October 2010.

[57]'Bratstvo ctalkerov na meil ru', available at: http://my.mail.ru/community/becom, accessed 3 February 2011.

[58]This is a name of the project's fan site, see: http://stalker-soul.ru, accessed 12 May 2011.

to the community, meaningful and emotional links between the product and consumers, their interaction with companies as well as with each other—all these features, in the light of the modern concept of a brand, are typical of its development.

Thus, the transmedial brand is produced not only by the traditional official producer but by the consumers as well. However, a paradoxical situation is bound to arise—consumers, or more precisely, prosumers, buy not an isolated product, but a long-lasting relationship with the brand, whose longevity they provide themselves. The higher the degree of their participation in the brand production, the stronger is their feeling of belonging or commitment to these brands.

### Prosumers' 'co-production' as a gift and 'free labour'

The role of consumers (prosumers) in media production and the products created by them clearly require a more detailed analysis. We can divide this into two main strands. The first one sees consumers' 'prosumption' and 'co-production' as the implementation of the 'gift-giving' concept, and the other interprets it along the lines of so-called 'cognitive' (or digital) capitalism and immaterial labour.

The classic concept of the 'gift' phenomenon was proposed by Marcel Mauss. He treated 'the gift economy' as an economic system where goods and services are given rather than sold; it is a system of exchange and redistribution (since ultimately all gifts are returned) that is alien to capitalist society (Mauss 1990). However, recent years have seen various 'manifestos of new media', that class the idea of 'gift-giving' as the basis of post-industrial capitalism. The idea of 'gift-giving' in relation to the new media, specifically, the internet, is quite consistent due to its open structure and free communication. From such a point of view scholars interpret consumers' 'co-production' from the position of companies' and consumers' solidarity and cooperation. These ideas have found clearest expression in Tapscott and Williams (2006) and Jenkins (2006) which treat the issue not only in the context of economics, but in a wider socio-cultural context.

Another approach has been developed by the researchers of the so-called autonomist movement in their concept of 'cognitive' capitalism (Lazzarato 2008; Terranova 2000; Hardt & Negri 2000; Virno 2004). This concept assumes the terminal transformation of modern capitalism as well as emphasising the importance of information technology (Pol're 2008).[59] These authors redefine work and labour with reference to post-Fordist society; the idea of immaterial labour lies at the centre of their neo-Marxist approach. Many phenomena formerly not classed as 'work' belong to immaterial labour, including the creation of such 'products' as fashion, artistic standards, consumption norms, and involvement in advertising and creative industries (Lazzarato 2008). Though autonomists do not consider the idea of immaterial labour

---

[59]The concept of 'cognitive capitalism' (Pol're 2008) is usually elaborated in relation to societies of developed capitalism, however, this is also relevant in examining countries arguably 'catching up' in terms of modernisation. This is characterised by capitalisation of culture and the high-tech sphere as well as by the global nature of the modern media and entertainment industry.

as related specifically to popular culture, it can be easily applied to this sphere, too. In particular, the term 'free labour' (Terranova 2000) can be applied as a way of defining volunteering, unpaid activity by consumers, and also in reference to work that is performed outside the framework of official institutions.

The problem of considering consumers' 'co-production' from the point of view of exploitation is the most controversial issue in recent academic discussion. The adherents of the first approach prefer to speak about the cooperation of official producers and consumers and their mutual benefit. Neo-Marxist researchers such as Terranova argue that any such cooperation is a myth and consider the expansion of consumers' 'co-production' as a new form of exploitation and the beginning of a new upsurge in alienation. In her opinion, 'co-production' does not imply any control on behalf of consumers over the means of production (Terranova 2000).

Detailed analysis of these discussions (Ritzer & Jurgenson 2009; Bonsu & Darmody 2008) is beyond the scope of my essay. However I would argue that the optimism about cooperation between companies and consumers is certainly exaggerated, however much internet activity is based on cooperation and the principle of 'gift-giving'. I agree with Dijck and Nieborg that 'co-creation' or 'cooperation' rhetoric only disguises capital's manipulation of a democratised and collectivist digital space (Dijck & Nieborg 2009). Prosumers' cultural practices, indeed, can be considered in terms of labour: they create ideas and products which are transformed into value, acquired and 'monetised' by capital, but it would be wrong to consider it only in terms of exploitation. This is due to the logic of contemporary cultural production development, where immaterial labour and creative industries play a considerable part. A solely negative evaluation of the prosumption phenomenon and its criticism would be one-sided.

Lazzarato's position may be enlightening on this point. He compares the approaches of Simmel (2004) and Bakhtin (1981) who offer different ways of understanding the relationships between immaterial labour and society (Lazzarato 2008). Lazzarato shows that Simmel, speculating about the phenomenon of fashion, insists on the division between physical and intellectual labour and offers the theory of creativity of intellectual labour. According to Simmel (2004), fashion is created by and for higher classes while lower classes try to imitate it. Lazzarato points out that in his analysis of 'aesthetic production' Bakhtin (1981) instead develops a theory of the social character of creative processes, without accepting the capitalist division of labour as given. Lazzarato finds Bakhtin's position more convincing and notes that in post-industrial society it is necessary to move beyond the understanding of creativity as 'individual' expression, historically the reserve of intellectual elites (Lazzarato 2008). This last point is crucial for a more nuanced reading of consumers' activity. Largely thanks to new media being interactive, we witness an upsurge of 'vernacular' creativity, relatively free of rigid hierarchies typical in 'legitimate' official culture. These mass creative practices are integrated in the general process of culture production as today creativity is considered the underlying principle of existence for many spheres in society—an idea which manifests itself through the famous concept of 'creative class' in the work of Richard Florida (2002).

In the new media age consumers can become co-participants in, and sometimes even real competitors to, traditional media models of production. The shifting roles of consumers, producers, distributors or experts undermine the professional sector monopoly. The need to make new practices legitimate challenges the existing hierarchy of elite producers in culture: culture becomes a battlefield for legitimate production of meanings and norm-setting.[60]

At the same time it would be wrong to overestimate the potential of consumers' (or even prosumers') agency, especially in the sphere of popular culture. Sometimes, the fact of 'co-creation' or 'co-production' is treated as proof that consumers have a chance to control mass media (Goralik 2003; Jenkins 2006). Some authors, moreover, treat this phenomenon in terms of resistance. For instance, Kücklich dwells on the potential for modders' resistance, who, as he rightly notes, have significant symbolic capital. He argues that they cannot build their own counter-culture because they do not fully understand their crucial position in the structure of the game industry. Partly this is because they unthinkingly accept the dominant view of their labour being just 'fun'; he speculates on the prospects of modders' counter-culture development (Kücklich 2005). While this argument has a certain logic, I think that Kücklich simplifies matters when he treats modders and the game industry as opposing parties while in practice they are inseparable. The existence of modders is possible only due to their being an integral part of the game industry; its successful development is now only possible within a model of consumers' participation. Secondly, the protest potential of popular culture is limited since consumers' motivation is connected primarily to consumption entertainment, playing and implementing one's own interests. (Also modders cannot be considered average consumers, because of the level of their technical skills and ways of self-organisation.)

Therefore, the issue of the commodification of consumers' activity and creativity has become more and more urgent. The two transmedial projects discussed in this essay bear witness to this. For example, there is a whole industry of goods and souvenirs on *S.T.A.L.K.E.R.*, such as the clothing collection and bars in the style of the games. The active audience of both projects is ideal for targeted advertising. The creativity and energy of consumers, especially the most active of them, and their community team spirit are used to support brand commitment. Caddell (2009) implies that branding starts with 'courting' already existing communities that are later given the opportunities to act themselves: 'a dollar spent on fans is a dollar invested into keeping, hooking, loyalty and the long-life of the brand'. The idea of 'audience as commodity' developed by Smythe (1977) and elaborated by Fuchs for the new media age as 'internet-prosumers as commodity' (2010), fits well with the notion of immanent and imminent total commodification of the projects.

---

[60]For example, consider the conflict that took place between a Russian TV company AMEDIA and the fan community of the TV series Ne rodis' krasivoi, based on the popular Colombian telenovela *Yo soy Betty, la fea* known in the West as *Ugly Betty*, which took place on the internet-forum dedicated to the series. At all stages of the project promotion the fans were very active and cooperated with the company. When the company decided that the fan activity did not meet necessary norms, it was restricted (first of all, by controlling and moderating the forum). The fans rebelled against the company's policy, arguing that they had a right to solve problems connected with the series promotion. Ultimately they achieved some success (Sokolova 2009b).

## Conclusions

Prosumers' 'co-producing' media in collaboration with official producers is a characteristic tendency of the new media age. The transmedial projects considered in this essay illustrate this tendency and demonstrate the specific features of consumers' creativity in popular culture. Active involvement of consumers in the transmedial projects, which enables us to treat them as 'prosumers' and to speak about their 'co-production' with companies, is found not only in actual production and distribution of the content, and ideas and artefacts that contribute to the creation of the topical fictional 'universe' of the projects, but also in the creation of a communicative environment around the projects' brand. The role of 'prosumers' is determined by the general logic of modern media production and consumption, the media convergence tendency and the expansion of new media and especially the establishment of Web 2.0.

The evolution of *S.T.A.L.K.E.R.* and *Metro 2033* shows that the success of the transmedial projects is connected with a number of factors—social, cultural and marketing. The success of *S.T.A.L.K.E.R.* is substantively rooted in regional cultural traditions, and the analysis of this case brings up a question about transmedia and prosumption phenomenon at the intersection of local contingency and transnational developments.

Transmedia provide more evidence of the complex, controversial nature of popular culture. In this sphere timely reaction to the interests and needs of consumers is imperative for companies' activities, resulting in a complex symbiosis of consumers' wishes and corporate interests. Transmedia involves various creative practices by consumers and the activity of their 'interpretive communities', but commercially it serves to build up a feeling of belonging to a brand, providing its long-term survival. The higher the degree of consumers' 'co-creation' and 'co-production' in transmedial projects, the more acute is the feeling of passionate belonging or loyalty to their brand. Here we face a paradox—on the one hand, consumers receive opportunities for creativity and self-expression, unavailable before, but on the other hand, we witness total commodification of their creativity; moreover, it is implied by the new business model and encouraged by it.

*Samara State University*

## References

Alekhin, A. (2008) 'Apokalipsis dlya shou-biznesa', *Ekspert*, 49, 638, available at: http://www.expert. ru/expert/2008/49/apokalipsis_dlya_shoubiznesa, accessed 15 December 2008.

Arvidsson, A. (2005) 'Brands: A Critical Perspective', *Journal of Consumer Culture*, 5, 2.

Arvidsson, A. & Sandvik, K. (2007) 'Gameplay as Design: Uses of Computer Players' Immaterial', *Northern Lights: Film and Media Studies Yearbook*, 5, 1.

Bakhtin, M. M. (1981) *The Dialogic Imagination: Four Essays* [edited by Michael Holquist, translated by Caryl Emerson & Michael Holquist] (Austin, TX & London, University of Texas Press).

Birichevskaya, O. & Strelchenko, V. (2006) 'Massovaya kul'tura: ontologicheskii smysl i tendentsii kontseptual'noi evolutsii', *CREDO NEW, Teoreticheskii zchurnal*, 3.

Bonsu, S. K. & Darmody, A. (2008) 'Co-creating Second Life: Market Consumer Cooperation in Contemporary Economy', *Journal of Macromarketing*, 28, 4.

Bruns, A. (2008) *Blog, Wikipedia, Second Life, and Beyond: From Production to Produsage* (New York, Peter Lang).

Bugrov, V. (1983) *Sverdlovsk: 'Aelita-83'*, available at: http://www.fandom.ru/convent/26/aelita_1983_12.htm, accessed 24 May 2011.

Caddell, B. (2009) 'Fans are the Future of Digital Marketing', available at: http://whatconsumesme.com/2009/posts-ive-written/fans-are-the-future-of-digital-marketing, accessed 2 March 2010.

Chernenko, E. (2009) 'Vmeste veselo pisat'', available at: http://www.runewsweek.ru/article/26494/?phrase_id=81083, accessed 12 September 2009.

Dergunov, Yu. (2004) 'Komp'yuternye igry i gegemoniya', *Levaya Rossiya. Politicheskii ezhenedel'nik*, 10, 109, available at: http://www.left.ru, accessed 11 November 2009.

Deuze, M. (2006) 'Collaboration, Participation and the Media', *New Media & Society*, 8, 4.

Dijck, J. V. & Nieborg, D. (2009) 'Wikinomics and its Discontents: A Critical Analysis of Web 2.0 Business Manifestos', *New Media Society*, 11, 5.

Evans, E. (2008) 'Character, Audience Agency and Transmedia Storytelling', *Media, Culture & Society*, 30, 2.

Fish, S. (1980) *Is There a Text in This Class? The Authority of Interpretive Communities* (Cambridge, MA & London, Harvard University Press).

Florida, R. (2002) *The Rise of the Creative Class, and How It's Transforming Work, Leisure and Everyday Life* (New York, Basic Books).

Fuchs, C. (2009) 'Conference Report: The Internet as Playground and Factory', *tripleC*, 7, 2.

Fuchs, C. (2010) 'Labor in Informational Capitalism and on the Internet', *The Information Society*, 26, 3.

Glukhovskii, D. (2007) *Metro 2033* (Moscow, Populyarnaya literatura).

Glukhovskii, D. (2010) *Metro 2034* (Moscow, AST).

Goralik, L. (2003) 'Kak razmnozhayutsya Malfoi: Zhanr "fanfik": potrebitel' masskul'tury v dialoge s media-kontentom', *Novyi mir*, 12.

Hardt, M. & Negri, A. (2000) *Empire* (Cambridge, MA, Harvard University Press).

Hay, J. & Couldry, N. (2011) 'Rethinking Convergence/Culture: An Introduction', *Cultural Studies*, 25, 4–5.

Humphreys, S. (2005) 'Productive Players: Online Computer Games Challenge to Conventional Media Forms', *Communication and Critical/Cultural Studies*, 2, 1.

Ivashnikov, K. (2007) 'Kluby lyubitelei fantastiki: opyt dialoga s vlast'yu (1985–1991)', *Novyi istoricheskii vestnik*, 16, 2, available at: http://www.nivestnik.ru/2007_2/29.shtml, accessed 14 October 2011.

Jenkins, H. (2004) 'The Cultural Logic of Media Convergence', *International Journal of Cultural Studies*, 7, 1.

Jenkins, H. (2006) *Convergence Culture: Where Old and New Media Collide* (New York, New York University Press).

Jenkins, H. (2007) 'Transmedia Storytelling', 3 March, available at: http://www.henryjenkins.org/2007/03/transmedia_storytelling_101.html, accessed 18 January 2009.

Kazunov, V. (2007) 'S.T.A.L.K.E.R.—taina semi tochek', available at: http://www.gametech.ru/cgi-bin/show.pl?option=review&id=76, accessed 20 September 2007.

Kovshun, I. (1984) 'Poigraem?', *Ural'ski sledopyt*, 4, available at: http://graymage.narod.ru/text/stalker.htm, accessed 12 October 2011.

Kücklich, J. (2005) 'Precarious Playbour: Modders and the Digital Games Industry', *The Fibreculture Journal*, 5, available at: http://www.journal.fibreculture.org/issue5/kucklich.html, accessed 10 May 2009.

Lazzarato, M. (2008) 'Nematerial'nyi trud', *Khudozhestvennyi zhurnal*, 69, available at: http://xz.gif.ru/numbers/69/nmtrln-trd, accessed 25 May 2009.

Levitskii, A. (2011a) *Nashestvie. Burya mirov* (Moscow, AST, Astrel').

Levitskii, A. (2011b) *Nashestvie. Moskva-2016* (Moscow, AST, Astrel').

Martens, M. (2011) 'Transmedia Teens: Affect, Immaterial Labor, and User-Generated Content', *Convergence: The International Journal of Research into New Media Technologies*, 17, 1.

Mauss, M. (1990) *The Gift: Forms and Functions of Exchange in Archaic Societies* (London, Routledge).

Melekhov, P. (2010) 'Geroi 2010-go', *Igromania.ru*, 31 December, available at: http://www.igromania.ru/articles/137851/Geroi_2010-go.htm, accessed 22 May 2010.

Muniz, M. & O'Guinn, T. (2001) 'Brand Community', *Journal of Consumer Research*, 27, 4.

Murley, B. (2008) 'Interpretive Communities: Beyond "Mass" Communication', available at: http://www.bryanmurley.com/site/wp-content/.../interpretive-communities.pdf, accessed 22 February 2010.

Murygin, V. & Kalintsev, S. (2011) 'Claudia Kern: nemetskii avtor serii S.T.A.L.K.E.R: interv'yu', available at: http://www.gsc-fan.com/?option=com_content&view=article&id=449, accessed 2 October 2011.

Perryman, N. (2008) 'Doctor Who and the Convergence of Media: A Case Study in Transmedia Storytelling', *Convergence*, 14, 1.

Peuter, G. & Dyer-Witheford, N. (2005) 'A Playful Multitude? Mobilising and Counter-Mobilising Immaterial Game Labour', *The Fibreculture Journal*, 5, available at: http://journal.fibreculture.org/issue5/index.html, accessed 11 June 2008.

Pervushin, A. (2008) 'Prokhod zapreshchen: tainstvennye zony', *Mir fantastiki*, 3, 55, available at: http://www.mirf.ru/Articles/art2619.htm, accessed 15 October 2011.

Pol're, B. (2008) 'Kognitivnyi kapitalizm na marshe', *Politicheskii zhurnal*, 2, 179.

Prahalad, C. K. & Ramaswamy, V. (2004) *The Future of Competition* (Cambridge, MA, Harvard Business School Press).

Revich, V. (1998) 'Perekrestok utopiya. Sud'by fantartiki na fone sudeb strany', available at: http://www.lib.ru/RUFANT/REWICH/perekrestok.txt, accessed 17 October 2011.

Ritzer, G. & Jurgenson, N. (2009) 'Production, Consumption, Prosumption: The Nature of Capitalism in the Age of the Digital', *Georgeritzer.com*, available at: http://www.georgeritzer.com/docs/Production%20Consumption%20Prosumption.pdf, accessed 5 July 2010.

Savenkova, E. & Frolova, M. (2000) 'Massovaya kul'tura—kak kul'tura ekrana (amerikanskoe kino)', in Lishaev, S. (ed.) (2000) *Mikhtura verborum-99: Ontologiya, estetika, kul'tura* (Samara, SaGA).

Scolari, C. A. (2009) 'Transmedia Storytelling: Implicit Consumers, Narrative Worlds and Branding in Contemporary Media Production', *International Journal of Communication*, 3, 4.

Simmel, G. (2004) *The Philosophy of Money*, 3rd edn [edited by David Frisby, translated by Tom Bottomore & David Frisby] (London, Routledge).

Smythe, D. (1977) 'Communications: Blindspot of Western Marxism', *Canadian Journal of Political and Social Theory*, 1, 3.

Sokolova, N. (2009a) 'Runet for Television Fans: The Space of/without Politics', *Russian Cyberspace*, 1, available at: http://www.russian-cyberspace.com/issue1/natalia-sokolova.php, accessed 5 July 2009.

Sokolova, N. (2009b) *Populyarnaya kul'tura Web 2.0:k kartografii sovremennogo medialandshafta* (Samara, Samarskii universitet).

Sotamaa, O. (2007) 'Let Me Take You to the Movies: Productive Players, Commodification and Transformative Play', *Convergence: The International Journal of Research into New Media Technologies*, 13, 4.

Stolyarov, A. (2008) *Osvobozhdennyi Edem* (Moscow, AST).

Stroeva, K. (2005) 'Ne prosto fantastika', *Novoe literaturnoe obozrenie*, 73.

Strugatsky, A. & Strugatsky, B. (1972) 'Piknik na obochine: Fantasticheskaya povest', *Avora*, 7, pp. 27–43; 8, pp. 38–51; 9, pp. 38–51; 10, pp. 42–51.

Tapscott, D. & Williams, A. (2006) *Wikinomics: How Mass Collaboration Changes Everything* (London, Atlantic Books).

Tapscott, D. & Williams, A. (2009) *Vikinomika. Kak massovoe sotrudnichestvo izmenyaet vse* [translated by Pavel Mironov & Galina Vasilenko] (Moscow, BestBusinessBooks).

Terranova, T. (2000) 'Free Labor: Producing Culture for the Digital Economy?', *Social Text*, 63, 18.

Toffler, A. (1980) *The Third Wave* (New York, Bantam Books).

Vargo, S. & Lusch, R. (2004) 'Evolving to a New Dominant Logic for Marketing', *Journal of Marketing*, 68, 1.

Vasile, A. (2011) 'Decoding a Transmedia Classic, Enter the Matrix', *Transmedialab.org*, available at: http://www.transmedialab.org/en/case-study/etude-de-cas-decryptage-d%E2%80%99un-classique-du-transmedia-enter-the-matrix-2, accessed 17 May 2011.

Virno, P. (2004) *A Grammar of the Multitude: For an Analysis of Contemporary Forms of Life* (New York, Semiotext[e]).

Vrochek, S. (2010) *Metro 2033* (St Petersburg, Piter.AST& Astpel).

Yang, L. (2009) 'All for Love: The Corn Fandom, Prosumers, and the Chinese Way of Creating a Superstar', *International Journal of Cultural Studies*, 12, 5.

Zlotnitskii, D. (2008) 'Seriya "S.T.A.L.K.E.R"', *Mir Fantastiki*, 64, available at: http://www.mirf.ru/Articles/art3201.htm, accessed 20 January 2009.

# Spatial Imagining and Ideology of Digital Commemoration (Russian Online Gaming)

## VLAD STRUKOV

*Abstract*

I trace the impact of digital economy on the cultural production on *Runet* by examining the Russian online gaming industry and focusing on a massively multiplayer online role playing game, *Allods Online*. I explore the game as a new form of cultural commemoration that goes beyond established discourses of nostalgia, war and trauma. I utilise Henry Lefebvre's concept of the production of space in my examination of the ludic environment of *Allods Online* as an ideological space. I call for a renegotiation of the Russian mediascape in its post-broadcast phase for it to include new ludic spaces of production of cultural symbols.

> Visual space in its specificity contains an immense crowd, veritable hordes of objects, things, bodies. These differ by virtue of their place and that place's local peculiarities, as also by virtue of their relationship with 'subjects'. Everywhere there are privileged objects which arouse a particular expectation or interest, while others are treated with indifference. Some objects are known, some unknown, and some misapprehended. Some serve as relays: transitory or transitional in nature, they refer to other objects. (Lefebvre 1992, p. 29)

IN THIS ESSAY I ADDRESS A SERIES OF INTERRELATED theoretical concerns. Firstly, I trace the impact of the digital economy on cultural production in the Russian-speaking section of the internet and I put forward a framework for thinking about the transformational impact of digital technologies on cultural experiences in networked ludic spaces. Secondly, I explore the impact of new cultural industries on the digital infrastructures in post-communist societies, the case of the Russian Federation, and how they perform, collaborate and share information in the global space. I call for a renegotiation of the Russian mediascape in its post-broadcast phase for it to include new ludic spaces of production of cultural symbols and value. Thirdly, I explore new forms of cultural commemoration that go beyond established discourses of nostalgia, war and trauma.[1] Finally, I engage in the discussion of ludic spaces by focusing on what Henri Lefebvre (1992) identifies as elements of the production of space—'objects, things, bodies'—and explore new paradigms of thinking about ludic environments as ideological spaces. These concerns are brought together by the logic of space which is understood in the essay in a number of ways: as a local place which is mobilised

---

[1] For a representative range of approaches, see Boym (2001), Etkind (2009), Oushakine (2000, 2010) and Trubina (2010).

politically in the age of globalisation; as a locus of cultural imagining and a sphere of negotiation of cultural heritage, that is a place of commemoration; as a surface that functions as a display and apparatus for constructing identities; and as a product of cultural and economic interrelations that produce an ideology of space. In my conceptualisation, space is viewed as a promise of multiplicity, heterogeneity that is always in process and never a closed system as it relies on collaborative practices of cultural engagement.

*Space of augmenting: context and tools*

Since 1991, Russian society has undergone a series of dramatic changes, the outcome of which has been frequently manifested spatially: for example, the dissolution of the USSR resulted in the emergence of the term 'near-abroad' (for the former Soviet republics) that reconfigured the perception of geopolitical, social and private (that is familial) space for many Russians who found themselves 'stranded' abroad; the conflict in Chechnya has posed a threat to the Russian territorial integrity; the economic boom of the mid-1990s and then mid-2000s reshaped the Russian cityscape with new areas of commerce and leisure replacing industrial sites and producing some of the most extravagant examples of contemporary architecture and sculpture, especially in Moscow. The imaginary landscape of the nation has undergone equally aggressive restructuring with waves of modernisation and retrograde commemoration creating a troubled vista of the nation's psyche and producing an agonising revision of its own past.

In the digital realm, *Runet*, in spite of its geopolitical connotations as the 'Russian internet', has come to signify the transnational, deterritorialised environment of the Russophone World Wide Web; that is the online activity of Russian-speaking communities in Russia proper, the near-abroad, Israel, the USA and other countries. The transition of the top-level country domain name from '.su' to '.ru' in 1994 and most recently to '.рф' (which in Russian stands for 'the Russian Federation'), is a manifestation of reorienting the country socio-politically as well as technologically and culturally. It is undeniable that *Runet* has emerged as a space of intense political engagement, especially for the Russian diaspora, as well as of cultural production. For example, in the 1990s, *Runet* was one of the spaces which gave birth to net.art, an artistic movement that appeared as a reaction to the collapse of the USSR and that revolutionised the perceptions of political and public space, especially the space of the museum and the practice of public display. The works of net.artists Olia Lialina[2] and Alexei Shulgin[3] contributed to the creative process of reimagining space as a volatile cultural construct in relation to new interactive, deinstitutionalised areas of the internet, and thus defined a new arena of commemoration.[4]

While in the early days of *Runet*, artists imagined the internet as a utopian space of ultimate freedom and democracy, since 2000 *Runet* has been an area where different political, economic and cultural interests are represented, staged and contested. *Allods*

[2]See http://www.teleportacia.org, accessed 25 June 2012.
[3]See http://www.easylife.org, accessed 25 June 2012.
[4]See Paul (2003) for an account of net.art.

*Online* (2009–present) embodies the peculiar mix of socio-political and cultural changes I refer to above, as well as post-Soviet expanses of nostalgia for the past, which are manifested primarily through the spatial organisation of the environment in which the project operates.

Figure 1 provides an example of the re-appropriation of Stalinist iconography in the gaming space of *Allods Online*. It shows the mediation of the planned Palace of the Soviets designed by Boris Iofen, Vladimir Geil'frekh and Vladimir Shchuko in 1934, in the landscape of Khadagan.

*Allods Online* (AO hereafter) is a massively multiplayer online role playing game (MMORPG) that employs a complex visual apparatus enabling the player to interact with other players as well as the gaming environment, constructed specifically to emulate the Russian and Soviet pasts.[5] The players navigate across an imaginary world of Allods, an ancient civilisation, forming alliances with other players and taking part in open combat and various quests. The three-dimensional virtual world of AO carries the recognisable markers of Russian classical cultural heritage as well as of the grand Soviet style. Russian history is constructed out of structural elements presented in the game sometimes nostalgically and sometimes ironically. National (fictional) history is itemised as a series of objects, details, patterns, references and

FIGURE 1. A SCREENSHOT OF THE GAME SHOWING A VIRTUAL REPLICA OF THE PROPOSED PALACE OF SOVIETS IN MOSCOW.
*Source*: Vlad Strukov.

[5]See http://www.allodsonline.com, accessed 25 June 2012.

practices that appear in the space[6] of the game and simultaneously enable its ludic assembly.

There has been a tendency in the popular and academic discourse on MMORPGs to view them as spaces lying outside national territories and to consider the actual virtual worlds they create as states with their own economy, laws and culture. This essay brings the discussion back to the realm of the national cultural tradition by exploring the construction of space in AO in relation to Russian popular cultural forms and geopolitics of cultural representation.[7] I examine the cultural apparatus that celebrates extraterrestrialism and the constitutive role of space, especially its role in forming, maintaining and reforming cultural identities.[8] My focus is on the double use of computer gaming as a form of commemoration of Russia's past and as a means of imagining and myth-making. Therefore, the essay paves the way for the discussion of the role of space and spatial practices in Russian computer games as well as advances research in digital technologies in post-communist states that so far has predominantly focused on debates around censorship and other forms of government control.[9]

The essay approaches AO through a combination of socio-historical interpretative models and critical aesthetics; the latter is a means to interpret the production of meaning whereby the game is perceived as both a work of art[10] and a cultural artefact. The study is based on the analysis of a corpus of 400 images that includes concept art, game advertising and screengrabs of game footage.[11] While being cognisant of

---

[6]By focusing on space, I do not mean to suggest that time is not important, or that time and space are in opposition to each other; rather in this essay, I am concerned with space as a product of interrelations—architectural, economic, social and primarily aesthetic as presented in the design of the game space—whereby I argue for the possibility of the existence of multiplicity thanks to contemporaneous plurality. In other words, I explore heterogeneity of space as a metaphor for offline intersections of global capital and local production. For an understanding of time in computer games see the following: Eskelinen (2001), Crogan (2003), Juul (2004), Lindley (2005), Hitchens (2006) and Tychsen and Hitches (2008).

[7]Raymond Williams (1974) identified electronic media as important agents in the production of social and cultural identities. He particularly argued that media are *sites* of collective self-imagining that occur as part of everyday life.

[8]In my discussion of *Allods Online* I do not employ the concept of 'chronotope' because I am not concerned with the issue of genre, that is a specific mode of speaking that—as conceptualised by Bakhtin in his 1937 essay *Forms of Time and of the Chronotope in the Novel*—contains a specific worldview and defines generic boundaries. This is not to say that such a consideration would not be productive; on the contrary, it might elucidate—among other things—a connection between the literary forms used in *Allods Online* and the typological structure of the game. In fact, traces of Bakhtinian thought are present—but not explicitly acknowledged—in the general classification of games and game design proposed by Elverdam and Aarseth (2007). Furthermore, Bakhtin's insistence on the role of time in defining literary genres has been taken up by game scholars via Huizinga's writing on game culture, particularly his view that a game takes place in its own boundaries of time and space (Huizinga 1955).

[9]As a good example, see a series of studies in the special issue of *Digital Icons: Studies in Russian, Eurasian and Central European New Media* entitled 'Between Big Brother and the Digital Utopia: e-Governance in Post-Totalitarian Space', available online at: http: http://www.digitalicons.org/issue03/, accessed 15 July 2011.

[10]For a discussion of video games as art see Tavinor (2009, especially ch. 9).

[11]I built the corpus over 10 months in 2010; the corpus comprises of three groups of images of approximately the same size: (a) art work produced by Nival and available for free download on the internet; (b) images that document my own experience of playing *Allods Online*; and (c) images

possible multiple ways of exploring AO—such as the creation of online identities, the role of social bonds and online groupings, corporate ideology, war and the re-staging of colonial and post-colonial narratives and settings, circulation of gaming capital in the offline world, and many other aspects—I specifically concern myself with the task of analysing AO as a medium of visual imagery and cultural memory as well as an exercise in the production of space, that is to say I examine AO as a spatial practice. I achieve this aim by situating AO in Russian computer gaming culture and the national visual tradition and by employing the concept of collaborative aesthetics as a discursive tool. My investigation also implicitly raises a number of meta-issues related to game-world analysis, including the role of user-modifiable content, the notion of limits in a limitless space, the relationship between game mechanics and game worlds, and computer games as an intersection of narrative structures and socio-cultural practices, and finally, the issue of materiality and immateriality in both constructed and reconstructed persistent worlds. While occasionally employing terminology from film studies, I, however, do not explore the general notion of a game world in comparison with other fictional worlds such as the filmic world;[12] instead, I focus on the geopolitics of space of AO in relation to game design, play and critical aesthetics.

The essay consists of two parts; the first part situates AO culturally as well as methodologically making links to the wider discourse of Russian cultural industries and its ideological apparatus. The second part focuses on the spatial construction of AO and explores the imaginary world of the game both in its diachronic progression and synchronic manifestation. These strands of analysis are linked together by the concept of space understood as 'transitory or transitional' (Lefebvre 1992, p. 209) and utilised to account for the complex network of transformation in the field of Russian cultural production. In my game analysis, I utilise Lefebvre's distinction between 'spatial practice', 'represented space' and 'representational space' (Lefebvre 1992); the last—read as a combination of fictional, play and social spaces towards an existential whole—provides the foundation for my analysis and is further advanced as a space of cultural imagining.

*Space of transition: discourses of gaming and cultural value*

As an extension of dominated spaces, leisure spaces are arranged at once functionally and hierarchically. They serve the reproduction of production relations. (Lefebvre 1992, p. 384)

Since the dissolution of the USSR, Russia has experienced a dramatic transfiguration of its economies of work and leisure, with new infrastructures of enjoyment, sociality

---

produced by other gamers that present their views on the gaming experience. In all instances the main criterion for selecting the images was their representation of the space; interactive elements of the game were not considered for the purpose of this study. The visual elements were analysed and classified according to the used representational strategies, particularly in terms of spatial design. Due to space constraints only two images are displayed in this essay. Further examples of the images derived from or based upon AO can be found at: http://allods.mail.ru, accessed 14 June 2012.

[12]For a study of the relationship between Russian film and computer games, see the special issue of *Digital Icons: Studies in Russian, Eurasian and Central European New Media*, issue 8, available online at www.digitalicons.org, accessed 25 June 2012.

and consumption emerging in the country and redefining its urban and rural spaces. Such a transfiguration has resulted in the production of new functionalities and hierarchies that Lefebvre identifies in relation to leisure industries. The position of online gaming on the Russian cultural horizon is that of transition, whereby it is simultaneously in a position of cultural dislocation and transience. While online gaming has become one of the most widespread and cheapest entertainment forms in the country's urban centres as well as on its numerous external and internal peripheries, the attitude to computer gaming is mixed, which indicates a discrepancy between the actual social practice and the value system. At one end of the spectrum we find condescending and occasionally degrading comments made in the popular press; their authors assign computer gaming to 'asocial teenagers' and 'miserable husbands' and put forward a notion of computer gaming as a crime or as a threat to the traditional nuclear family.[13] These views reflect Russian technophobia in contrast to the celebration of new technologies that we find at the other end of the spectrum: in this sense President Medvedev epitomises Russia's new digital modernity whereby his ideological discourse builds on the premise of Russia's competitive edge in the digital realm. From the cultural perspective, Russia is one of the few countries in the world that officially recognises computer gaming as a sport. The Russian Computer-Sport Federation (*Federatsiya komp'yuternogo sporta Rossii*) was registered in 2000 with the purpose of promoting computer gaming. Since the victory at the Seoul global gaming competition in 2002 (Labastov 2002), Russian interest in computer gaming has grown exponentially, which is manifested in the proliferation of specialised publications, news coverage and representations of gaming and gamers in popular television and film productions.[14]

In spite of the extraordinary enthusiasm of the gamers, there has been insufficient research into Russian computer gaming culture. In the West, game studies, or ludology, has become a burgeoning field of research, finding itself at a cross-roads of academic disciplines. In turn, video-game theory is gaining acceptance in academia as part of new media theory and communication studies.[15] In Russia, game theory is still viewed as an odd subject for professional pursuit, which was characteristic of Anglo-American academia in the 1980s and early 1990s.[16] A similar, if not more paradoxical situation can be found in the field of (Slavic) Cultural Studies. It is a commonplace view that computer games are an important medium for conveying cultural values and memory, yet there is inadequate discourse concerning the socio-political and cultural implications of gaming in post-communist societies in Cultural Studies, or its Russian

---

[13]See, for example, opinions expressed in the *Zhenskie strasti* online journal and blog (*Zhenskie strasti*, 2008, available at: http://www.passion.ru/faq/s.php/1924.htm, accessed 1 September 2010).

[14]See, for example, Nurbek Egen's television series *Set'* (The net) (2007) and Pavel Sanaev's filmic diptych *Na igre* (Hooked) (2009, 2010).

[15]For a concise overview of the history of the study of video games, see Wolf and Perron (2003a, pp. 2–13).

[16]I intend to emphasise that such a difference in perception is not a mark of backwardness of Russian academia but rather of a different configuration of intellectual enquiry and academic discipline: I anticipate that in Russia game studies will evolve, not as part of new media or communication studies, but rather as part of film studies, a field that is currently taking centre stage in Russian universities.

half-sister, '*kul'turologiya*'.[17] While Russian gamers often feature in ludological publications,[18] there has been no dedicated research exploring any aspect of computer gaming in Russia. Furthermore, it is generally rare that researchers focus on national configurations of computer games, or consider localisation and engagement with the national cultural tradition.[19]

*Space of production: the Russian games industry as an emerging space*

In strategic spaces resources are always localized. (Lefebvre 1992, p. 356)

In her study of the *Uru* gaming community, Pearce (2009) conceives of identity as place, determined by the logic of transculturation, reclamation and self-determination. In this section of the essay, I argue that place might be considered as identity, that is an area of open ongoing production which has an inherent connection to the expanding (Russian) modernity conceived in spatial terms and in relation to a specific subjectivity. I trace this emerging identity by considering the extent of the Russian gaming market, the global flow of capital associated with game production and the labour conditions of such production. In my case study, online gaming on the Russian-speaking internet is a matter of 'strategic space' (Lefebvre 1992) perceived as an economy of cultural symbols in the global flow of information.

The Russian games industry (*igrovaya industriya*), which includes design, production and distribution of video, mobile, online and other games, is one of the fastest growing cultural industries in the country (Ruutu *et al.* 2009). Online games in Russia take up 18% of the game market (IKS-Consulting 2007), and in 2009 the value of the Russian market for computer games was estimated at about £3 billion with the share of online games of £173 million (Finam.ru 2010).[20] On average, a brand new Russian-made video game costs R250–350 (£5.27–7.39) while old games can be purchased for about R100 (£2.12) (Mikhailov 2007); at the same time pirated copies of games can be obtained for as little as R25 (£0.53) a piece.[21] In large cities access to the internet in '*igrovye salony*'

---

[17]In this essay I hope to ensure the argument does not follow the logic of 'othering'. For a discussion of the importance of national reading of gaming cultures, see Shaw (2010).

[18]See, for example, Adrienne Shaw's study of video games and homosexuality (2009), and also Silvia Lindtner and Paul Dourish's study of Chinese adaptations of games created in Russia (2011).

[19]As a good example of an exception proving the rule, see Dal Yong Jin and Florence Chee's study of the Korean online game industry (2008).

[20]However, all estimates might be inaccurate for a number of reasons. Firstly, the very diversity of products and complexity of the industry itself confound any exact estimate. Secondly, many Russian game producers develop products for Western and Eastern game distributors who then sell their products to consumers in Russia, thus distinguishing the Russian market from its foreign counterparts is problematic. The important role of Russian developers in the global game market has been acknowledged at numerous occasions (see for example, BBC documentaries on computer games (Temple 2004)).Thirdly, a lot of products are sold in Russia illegally as a result of widespread piracy, obscuring the real cash flow of the market.

[21]For example, in August 2010, Russian online store ozon.ru, an analogue of amazon.com, sold brand new computer games for as much as R2,499 (£52.75). Such a high initial price for a product was not an indication of the inflation or price growth but rather of the fact that the seller aimed to make a profit before the product would be resold illegally on the pirate market.

(gaming salons) and internet cafes is between R40 and R80 (£0.84–1.69) per hour, whereas the monthly fee for unlimited access to broadband internet with such providers as Beeline and MTS is about R450 (£9.51). Thus, with more companies providing broadband internet at competitive prices, more players engage in online games from their homes rather than from '*igrovye salony*'. The transition of gaming from '*igrovye salony*' to people's homes signifies a transition from public to private spaces as well as the establishment of a sustainable industry focused on individual consumption. At the same time the low price of computer games and the damage caused by piracy encourage Russian game makers to focus on online gaming because it utilises different forms and types of payments and reduces resale of games on pirate markets.

AO is one of many exported and locally produced MMORPGs available in the country.[22] *Allods Online*[23] (2009–present) was developed by the Nival Group whose diverse portfolio includes such games as MMORPGs *Dragon World* (*Drakonika*), *Cabal Online*, multiplayer racer *Level-R*, online karaoke *Super Star* (*Stan' zvezdoi*), and strategy games *Prime World* and *King's Bounty*.[24] Nival Group was founded in 1996 by Sergei Orlovskii (born in 1972) and has emerged as Russia's most active and powerful player on the market. The group specialises in the development and marketing of online games and social network gaming applications for users in Russia and other countries. In order to maintain its global appeal, Nival has two development studios in Moscow and Minsk, directly employing over 200 developers. The group also has an office in California, representing the company's interests abroad. In addition, Nival functions as a maker of franchised produce for many Western and Eastern companies, and it also outsources substantial amounts of work to smaller companies in Russia and the former Soviet republics, thus positioning itself at a specific junction in the global network of computer game producers.

In March 2010, the group was acquired by Mail.ru, Russia's leading internet and communications portal (Komissarova 2010). One of the reasons for the acquisition was to strengthen the position of the game maker and its produce on the network of Russian-speaking gamers around the globe. The acquisition resulted in the creation of a large online gaming group (owning over 50% of the market share).[25] The merger of Mail.ru and Nival Group opened new channels for collaboration between Russian and global producers, exporting Russian-made products and beta testing and localising exports for domestic use,[26] thus securing multichannel, multidimensional

---

[22]Specifically Russian popular MMORPGs include *Legenda—nasledie drakonov* (The legend—the legacy of dragons; http://w2.dwar.ru); *Haddan* (http://haddan.ru); *Fragoriya* (http://fragoria.mail.ru); *Piratiya* (World of pirates, http://www.piratia.ru); *Nochnoi dozor* (Night watch, http://www.dozory.ru); *Gorod tantsa* (The city of dance, http://www.parapa.ru), and many others (sites accessed 25 June 2012).

[23]*Allods Online* was produced by a team of game developers with Aleksandr Mishulin serving as the project's creative director. The game saw four years of development and cost about $12 million to produce.

[24]See http://www.nival.com/, accessed 25 June 2012.

[25]It has more than 40 types of games on offer in addition to successful international franchises such as *Silkroad Online*, *Lord of the Rings Online*. Mail.ru also owns a 50% stake in another large Russian game producing company Nikita.Online (http://www.nikitaonline.ru, accessed 25 June 2012); this makes Mail.ru a significant player in the Russian and CIS market for games and online applications.

[26]For example, *Pandora Saga* from Rosso Index K.K. and *Jade Dynasty* from Beijing Perfect World. The full list of authored and franchised online games can be accessed on the group's portal available at

collaborative work in the global networked game industry with a complex architecture of creative processes and capital distribution. In fact, game developers have their professional associations, with game designers from Russia and around the world holding an annual conference in Moscow.[27] There is a thriving game advertising industry with a great number of online and offline publications specialising in writing about the industry, its products and consumers.[28] These publications vary in their focus, audience, circulation and quality.[29]

Thus, Russia enjoys a well-developed cultural space of computer gaming, which benefits from a growing production sector as well as a thriving industry of mediation, circulation and consumption with both the producers and players perceiving themselves as part of a larger, global network of gamers. The cartography of the gaming industry in Russia indicates that it operates as a multimodal interface with a global dimension particularly as regards the flow of products, and to a lesser degree of labour. There is evidence to make a case for Russian re-narrativisation of local capital, expansion into adjoining markets and attempts to overcome geographical embeddedness. Finally, the analysis of the Russian gaming industry calls for a spatial difference to be convened as a temporal sequence: I aim to examine AO as temporal progress in relation to the development of technology.

### Space of asynchrony: evolution of Allods

*Global space* established itself in the abstract as a void waiting to be filled, as a medium waiting to be colonised. (Lefebvre 1992, p. 125)

In this section of the essay I explore AO as a media space that, as Lefebvre suggests, is waiting to be determined while the cartography of the digital economy is being constantly redrafted. The design history of *Allods Online*—understood here as social spatiality of the medium—goes back to 1996 when the newly formed Nival Group had an ambition to create a MMORPG but lacked the necessary experience. Moreover, at that time access to the internet in Russia had its technological and financial limitations, i.e. the majority of users relied on very expensive dial-up connections,[30] thus preventing the popular appeal of online games. In many respects, the evolution of

---

http://corp.mail.ru/games/ru/portfolio/publishing, accessed 12 July 2010. The information on games for social networks is available at http://corp.mail.ru/games/ru/portfolio/social, accessed 12 July 2010.

[27]See http://www.kriconf.ru, accessed 25 June 2012.

[28]For example, *Games Star* (http://gamestar.ru, accessed 25 June 2012), *Igromaniya* (Game mania, http://www.igromania.ru, accessed 25 June 2012), *Virtual'nyi zhurnal geimera* (Gamers' virtual journal, http://vgplayer.net, accessed 25 June 2012), *Zhurnal geimera* (Gamer's journal, http://virtgamer.com, accessed 25 June 2012).

[29]For example, *Strana igr* (The land of games) is published in Moscow twice a month, it has a circulation of 80,000 copies distributed in Russia and abroad, and its main focus is on new games and industry events; *Navigator igr* (Game navigator) is a monthly publication with a circulation of 50,000 copies sold only in Russia; the journal provides in-depth analysis of new games and industry developments. Understandably, Russian online publications make a profit not from the sale of copies but from advertising and product placement.

[30]For an overview of internet development in Russia, see Strukov (2009, pp. 208–15).

AO reflects the development of the very technology of the internet and its social and artistic components: dial-up connection is eventually replaced with broadband internet, single-player PC-based games turn into massively multiplayer online games, and the 'digital intelligentsia' (Strukov 2006) is displaced by the ordinary internet user, respectively.

Similarly to net.art, discussed in the introduction to this essay, the first single-player *Allods* was developed by a consortium of artists and technology experts.[31] It was released in 1998, and was distributed by Russia's software giant 1C. A new version of the game called *Allody: pechat' tainy* (Allods: the secret seal) was released in 1998 both in Russia and in the West. The game proved to be extremely successful with players all over the world, and so Nival launched an add-on game called *Allody-2: povelitel' dush* (Allods-2: the lord of souls) in 1999, and a multi-language sequel *Proklyatye zemli* (Cursed lands) in 2000. In 2000–2006, Nival produced a number of games, including the very popular *Silent Storm* and *Night Watch*. This period was also characterised by the group's expansion in Russian regions and the acquisition of small, developing companies in Russia and abroad.

From the commercial point of view, *Allods* exemplifies the success story of a software company that survived the turmoil of the financial crisis of 1998 and Russia's exposure to global markets in the 2000s, and presents a convincing case of post-industrial integrity and expansion. Artistically, while AO derives some of its imagery from the first single-player *Allods*, it is a completely different gaming environment. First, the spatial construction of AO incorporates a set of spaces and practices that are specific to MMORPGs, such as traditional Player *versus* Computer (PvC) and Player *versus* Player (PvP) environments, as well as quests, mini-games and other types of activities that enhance the player's standing in the group and advance her capabilities. AO presents a contiguous space which is a fictional construct and relies on a fantasy of a world filled with objects, natural and architectural patterns, and populated with various beings. The exploration of this world is one of the main aims of the games. In turn, the online practices correlate and engage not only with the fictional world of the game, but also with the world of the players, i.e. the virtual realm that all of their online activities constitute, and the virtual environment of online networks in which the game is positioned, acted out and enjoyed. Therefore, the gamer continuously considers the physical, ludic, emotional and ideological engagement with these territories of gaming and congealed social practices with the help of which a vision of the Russian past is constructed.

### *Space of historical imagining: fictional mythology of* Allods Online

Spatial practice is neither determined by an existing system, be it urban or ecological, nor adapted to a system, be it economic or political . . . . On the contrary, thanks to the potential energies of a variety of groups capable of diverting homogenized space to their own purposes, a theatricalized or dramatized space is liable to arise. (Lefebvre 1992, p. 391)

[31]Script-writer Anna Klimenkova, artists Elena Rachkova, programmers Iurii Blazhevich and Georgii Osipov, among others.

MMORPGs are concerned with creating specific environments—virtual worlds—whose boundaries are negotiable within the game design and reflect the social aspect of online interaction.[32] In fact, as Celia Pearce argues, virtual worlds are the "'collective creation of belief," since virtual worlds are, by definition, social constructions' (Pearce 2009, p. 17). The fabric of such worlds is fundamentally fictional, even if they attempt to recreate the world outside the game. At the same time, these worlds display all the qualities of (social) reality, whereby shared visions of fantasy and/or of memory are localised through the experience of play. Furthermore, the spaces of such worlds are progressively dramatised by players and therefore resist homogenisation.

AO is set in a fictional world of Sarnaut, initially a planet populated by peace-loving and prosperous societies of humans and other beings. Two ancient territories were particularly prominent, the country of Iul and the country of Dzhun. The first was the birthplace of humankind on Sarnaut, and it is perceived by the game designers and players as a repository of human experiences and values. The second country, organised as an empire, eventually disappeared. Thus, from the very outset, AO privileges the binary structure of the game world, with clear divisions between human and non-human, prehistoric and modern, rational design and mystical forces, as well as between more existential concepts of life and death, continuation and termination, purity and contamination.[33] The impetus for the fictional world of AO is provided by a cataclysmic event. This violent upheaval causes the world of Sarnaut to break up into fragments—allods—resulting in the extinction of a number of beings. The apocalyptical event is perceived to have origins in the very nature of Astral, the matter that fills in the space and determines natural and social developments. The myth of the cataclysm creatively renders the dissolution of the USSR and the emergence of independent states—allods—in the 1990s.

After the cataclysm, two conflicting parties emerge causing the allods to gravitate into one of the two spheres of influence, the League and the Empire. The first, also known as Kaniya, is a free association of allods that does not have a ruler. The second, Khadagan, is ruled by an authoritative leader Nezeb. While Kaniya is an agrarian society, Khadagan is an industrial powerbase with its population residing in large urban centres. The game mythology asserts that citizens of Khadagan are raised in the belief that an authoritative state is the only legitimate form of social organisation; they are wholeheartedly devoted to their empire and no matter what their trade all of them are warriors. Khadagan and Kaniya are constantly at war with each other, thus players are expected to take sides in the confrontation. Therefore, the mythology of *Allods Online*, and especially its dystopian visions, provides a creative account of Russia as a state, land and universe, with Kaniya symbolising the pre-modern era in the country's development with its purportedly strong democratic traditions and

---

[32]See Castronova (2001) for an overview of a large-scale study of virtual worlds from an economic point of view.

[33]In its dualistic structure and mystical concerns, *Allods Online* is similar to the fictional world of Sergei Luk'yanenko's book series *Nochnoi Dozor* (Night Watch) and its film adaptation produced by Timur Bekmambetov (2004, 2006). Both the Night Watch series and *Allods Online* depict a world that is divided by mystical alliances with the boundaries between them blurred and spatial and cultural transgressions imminent (Strukov 2010).

Khadagan epitomising Soviet totalitarian modernity. The associations are further enhanced on the visual level explored in detail below (see Figure 2).

According to AO mythology, the land of Khadagan was initially populated by nomads who are described in the game as 'dark in complexion and with black hair'; the populace of Khaniya, on the contrary, are of fair complexion with blonde hair. The bodily characteristics of the game's two main races and the names of the two opposing countries allude to Russian early history, with Khadagan referring to the Asian connection (Khadagan—Khaganate), and Kaniya indicating a link to Western Europe (-iya being a common ending used in the names of Western countries in Russian, for example '*Angliya*' (England)). Therefore, AO constructs an imaginary world that examines the Russian past as a set of cultural dichotomies, with Kaniya and Khadagan representing not only the West–East ideological divide but also a set of symbols and narratives of a prehistoric and historic past, tradition and modernisation, (pagan) ritual and (Soviet) ideology, high and low fantasy, history and myth, with players' characters taking a special place in this mythological order, the game's lore and cultural positions as they bring their own interpretations of histories and cultural signifiers.

### Space of confrontation: Kaniya versus Khadagan

The second logic embodied in this spatialization is a logic (and strategy) of metaphor—or, rather, of constant metaphorization. Living bodies, the bodies of 'users'—are caught up not only in the toils of parcellized space, but also in the web of what philosophers call 'analogons': images, signs and symbols. (Lefebvre 1992, p. 98)

FIGURE 2. A GAMER'S AVATAR OCCUPYING THE IDYLLIC SPACE OF KHANIYA.
*Source*: http://allods.mail.ru/, accessed 14 June 2012.

Juul (2004) has argued for a concept of modelling game time in the form of layers; Hitchens (2006) has redefined the concept by analysing the correlation between the player's multi-layered experience of time in the gaming world. Similarly, in MMORPGs space is constructed as a set of layers, each of which is transformed by the player—through comprehension, interaction and action—into places, whereby the subjective gaming experience establishes a sense of location. Therefore, space is constructed as a set of tensions—confrontations—that are designated depending on the player's (adopted) identity, action and memory. In the remainder of the essay, I analyse the spatial structure of AO in accordance with these principles, focusing on the last aspect that I consider as 'analogons', to borrow Lefebvre's terms, that is as impulses of cultural memory codified in the symbolic imagery of the game. Furthermore, I explore the relationship between space and cultural memory as a form of presence in the virtual world. Heeter distinguishes between three forms of presence in the virtual world: personal presence, 'the extent to which and reasons why you feel like you are in a virtual world'; social presence, 'the extent to which other beings (living or synthetic) also exist in the world and appear to react to you'; and environmental presence, 'the extent to which the environment itself appears to know that you are there and react to you' (1992, p. 2). I utilise her categories of presence as modes of interpretation of the spatial layers of AO in their relation to Russian culture and history.

The player's engagement begins with a series of choices that result in the construction of a specific identity, which in turn determines the spatial perception of the environment. AO puts forward historical narratives as a matter of choice and not predeterminations:[34] the player inscribes self on the space of the games, and each choice results in locating self in a specific layer of the game: on the side of Kaniya or Khadagan, as belonging to a composite of peoples and races,[35] and as demonstrating gaming allegiances by navigating a specific environment. Unlike in other MMORPGs, in *Allods Online* the distinction between places is not racial but rather ideological. While freely mixing beings derived from fantasy and science fiction, AO maintains a high degree of separation between its two principal gaming territories and the values they stand for. The ideological distinction between layers of the space is carried out as a performative act in that the player conducts sundry activities codified as labour and positions the player as a performing element within the spatially constructed discourse.[36]

The actions of the characters help structure the space of the game, whereby sub-tiers of the game—for example, use of the mechanics of popular games of poker and blackjack and slot machines—provides not only detours from the main objective that expand the experience of the gamer but also the narrative perspective instead of seeing

---

[34]By contrast, Bekmambetov's *Night Watch* presents an illusion of choice whereas in reality a character's position on the side of good or evil is predetermined (by making it a blood-line related event, the director completes the task of essentialising his subjects); the director overcomes the dichotomy of being by blurring the boundaries between the very elements of these dichotomies.

[35]With two warring factions, *Allods Online* has six races and 28 classes, creating a nearly endless source of adventure and excitement.

[36]Characters conduct seven types of activities that are divided into two main categories: the gatherers and industrial producers.

the player in a fixed position. The game participants' active role results in the perception of the game space as the place of production: in AO this is literally manifested in the practice of making objects out of available parts. Therefore, gamers can customise their avatars and consequently create a more personalised gaming environment. For that purpose, the world of AO is filled with materials and objects that can be used for making gadgets, jewellery, weapons and so forth, enabling a malleable environment in which the gamers interact with the fictional world and other gamers.

This creative 'materiality' of AO makes a valuable addition to the standard rankings, levels, military structures and nomenclatures available in other MMORPGs. Moreover, the opposition of the rural as a symbol of the good and the technological equating the evil, a division that is normally derived from fantasy landscapes, is challenged in that characters are able to move between different allods and occupy multiple spaces, making the conceptualisation of 'home' problematic. By culturally 'uprooting' the gamer, AO destabilises geospatial and ideological adjacencies within the world of the game and problematises a sense of belonging. Hence the game retells the story of modernity through spatialisation and through challenging connections and discontinuities. Overall, the virtual world of AO has a simultaneous function of self and other, with the locus of the player determining its precise role. Players are not required to be familiar with Russian history in order to succeed; however, their participation inevitably constructs a specific vision of Russian history. While the divided world of AO represents the broken history of Russia, the gaming experience is akin to a multi-layered journey through the country's history.[37] Similarly, as in many rhizomatically organised narrative spaces, AO creates a tension between the (assumed) linear as well as the nonlinear narrative of the past that the gaming experience constructs by mixing chronological periods, real and fictional characters and persona, as well as transcending 'real' and fantastic spaces. Russian history and history in general are perceived, not as a continuous flow of historical events, but rather as a sequence of periods, each of which displays its own inner logic and purpose.[38] The confusing world of AO maintains its integrity thanks to the binary construction of space—a distinction between allods and the astral (grounded and ephemeral spaces with historical and suspended time, respectively) and between Kaniya and Khadagan, or imperial and Soviet periods of Russian history. Because playing AO involves acting on behalf of one of them, the gamer's movements through various territories function as a means of transcending boundaries and distinctions and engaging in a ludic reassessment of the country's political past and cultural heritage by interacting with objects and the architectural space of the game world. As a result, the player's actions

[37]A similar technique of superimposing layers of historical and cultural allusions and signifiers is used in Aleksandr Sokurov's 2002 *Russkii kovcheg* (Russian Ark). Such layering of history becomes possible thanks to the filmmaker's use of a specially designed digital camera and it creates an effect of the film operating as a first-person shooter, i.e. a genre of a computer game.

[38]In fact, in both *Russian Ark* and *Allods Online*, the representational systems include external, fixed coordinates as well as internal, shifting referents associated primarily with the memory of the subject, and finally time-based systems that enable the viewer/player, respectively, to follow lines of movement and the narrative sequence.

enable the emergence of information in the physical sense as objects carrying meaning, i.e. as digital artefacts with their political, economic and cultural significance.

The cultural significance of space is realised through the recreation of a style perceived as an ideological zone of play. The space of Khadagan is rendered in the architectural style of Stalinist Moscow; at the centre is a building reminiscent of the Palace of the Soviets designed by Boris Iofen, Vladimir Geil'frekh and Vladimir Shchuko in 1934. This building functions as a spatial point of reference for the players and an anchor of cultural memory. The spatial design of Khadagan emulates not only the (unrealised) architectural grandeur of Stalin's architecture but generally its specific neo-classical style that has become known as 'stalinskii ampir' (Stalin's imperial style).[39] For example, the game contains details of the building design with its characteristic use of sculpture as part of the decorative system. It presents an ostentatious display of Soviet ideological paraphernalia, namely the red banners, star and state symbols. In its careful use of granite wall cladding, colour scheme and contrast between huge windows on the higher level and castle-like windowless walls of the lower level—in the style of the KGB building in Lubyanka Square—the building is no less imposing, hence the depicted disproportion between the building and the gamer's avatar. The dates engraved on the facade are not arbitrary and have a complex function as cultural referents. On one level, they refer to the period when Palladian style became prominent in Europe and to its uses and abuses in Stalin's Moscow.[40] On another level, they refer to the early period of the construction of the Russian empire—roughly from the rule of Ivan the Terrible to the foundation of St Petersburg—in other words, a period of centralisation of power and Moscow's prominence as the imperial capital that was repeatedly mythologised under Stalin.[41]

The game also dutifully replicates the industrial power of the Soviet empire. Life in Khadagan depends on extraordinary power generators; their overpowering size and their primary location inside the city's buildings indicate the primacy of the industrial complex in the USSR and a longing for the lost industrial power in contemporary Russia.[42] The energy produced by the generators is necessary for travel and defence purposes; in other words, it symbolises the response to the threat of globalisation perceived by some members of the Russian public.

The space of Khadagan is also filled with Soviet social practices, and so it continues the cultural legacy of the Soviet regime. For example, players need to attend a combat training centre designed as a typical Soviet 'voenkomat' (military commissariat), a body of the government responsible for drafting recruits and providing training. The training centre is set in 'the Red Steppe', a territory controlled by Khadagan; the virtual 'voenkomat' is ringed with a concrete fence; its inner court has a geometric layout; tall poplar trees and strategically placed street lights mimic the classic columns of the main building. The totalising symmetry of the centre creates an overall effect of strict discipline, control and subjugation. Ferocious orks command the centre; they

---

[39]For a detailed discussion of the style and its spatial ideology, see Cooke (1997).

[40]For example, Palladio's Loggia del Capitaniato in Vicenza of 1572 inspired Ivan Zholkovskii's 1934 apartment building at 13 Mokhovaya ulitsa.

[41]For example, Sergei Eisenstein's *Ivan the Terrible* (Part I, 1944).

[42]See Zyuganov (2012) for calls to rebuild the Soviet power base.

wear stylised military uniforms with exaggerated epaulettes signalling their nasty nature. Their facial features are a combination of stereotypical vampires (the enormous fangs and red eyes) and Russian police (shaven head and heavy lower jaw), thus parodically conveying the recent discourse about corruption among Russian police, especially road police ('GAI'), which 'suck blood' by demanding bribes.[43]

Verbal interactions and sound art used in the game maintain the association with the Soviet cultural discourse. For example, Khadagan's motto is reminiscent of Soviet slogans: 'Motherland, Leader, Victory; Life is a struggle, Death is a hindrance; Happiness is unity, the Empire is the motherland, the League is an atavism, Astral is the reality'. As in others MMORPGs, AO provides players with instructions on how to achieve their goals. These instructions combine elements of Soviet military jargon, known to average Soviet and post-Soviet citizens from popular literature and war films,[44] as well as elements of official jargon used in pioneer games from the Soviet period.[45] Moreover, the game designers included popular Soviet songs such as 'Aviamarsh'[46] as part of the game soundtrack which incidentally was recorded by the Central Symphony Orchestra of the Russian Ministry of Defence.[47] This explicitly signposts the ideological intentions of the game designers and also accounts for the process of commodification of the Soviet past presented in the game in the form of digital artefacts and practices.

The dazzling display of authoritarian symbols, combat tools and means of control conveys the world of Khadagan as predominantly an urban militaristic masculine world. It contrasts with rural, complaisant, feminine Kaniya, a distinction rendered even on the level of the nouns used as names for the territories (grammatically speaking, Kaniya is a feminine noun and Khadagan is masculine). If Khadagan controls 'the Red Steppe', Kaniya rules over the ore rich 'Siveriya' with the deep 'Vertysh' flowing through its domain; thus, Kaniya evokes the familiar landscape of Siberia with the river Irtysh as an imperial appendix to the country's core. Kaniya, in its winter setting, displays a barren snow-covered terrain with a masterfully rendered glow of polar lights. It is the world of the Varangian Rus with characters clad in carefully reconstructed Viking armour roaming its frost-bound lands. By contrast Kaniya in the summer is a pastoral paradise with its mighty trees, shady grounds, bright skies and intense sunlight. This landscape makes a connection to Russia's southern historical influences manifested in the masterfully developed Orthodox style of churches and houses.

---

[43]In 2010, Russian motorists staged protests against the notoriously corrupt Russian traffic police by attacking their use of privileges on the road, i.e. motorists would fix blue buckets onto the roofs of their vehicles that would mock the signal system used by the road police and other sections of Russian government that gives them special privileges on the road. These protests were mobilised with the help of blogs and social media on the internet (see, for example, the site of 'the society of blue buckets' (*assotsiatsiya golubykh vederok*), available at: http://sineevedro.ru, accessed 25 June 2012).

[44]For example, '*razvedka, zasedanie shtaba*' (intelligence, the meeting of the command staff).

[45]For example, '*zastava, otriad, vmeste veselee*' (outpost, team, it's fun to do things together).

[46]'*My rozhdeny, chtob skazku sdelat' byl'u …*' (We were born to turn the fairy tale into reality …), composed by Iurii Khait, lyrics by Pavel German, 1922.

[47]Part of the soundtrack is available online at: http://files.tarakanov.net/mp3/AllodsOnline-MainTitle.mp3, accessed 25 June 2012.

The game pedantically reconstructs not only the imaginary landscape of Medieval Rus but also its everyday practices, elements of interior design and utilitarian objects, like 'the red corner' (*krasnyi ugol*) in a Russian home, the place where icons are displayed and prayers are performed, the samovar, the horse-driven carriage and many others. The game celebrates the now lost '*byt*' (everyday life) of rural Russia as well as reproduces the familiar (pseudo-) Russian cityscape: it is the world of the Russian fairy tale—the '*tridevyatoe tsarstvo*' (the far away kingdom)—with its house on chicken legs, the moving stove, the cross-roads, the *leshyi* (forest spirit) and so forth. The players construct their quest through the fairy tales, replicating magic scenarios and situations. For example, as they enter the ancient city, the game shows a view of a cobbled street with a perspective over a church through decorative gates, a real view of the Russian museum in Red Square through the Voskresenskie (resurrection) gate in Moscow, which provides an ironic comment not only on the current trend in romanticising pre-modern Russia, but also on the position of the official historian whose role is symbolised by the building of the museum.

In contrast to the official debates over Russian history,[48] AO delegates the production of historical knowledge to the players by exploiting their memory of the past—the world of Khadagan—and by utilising cultural imagination of more distant past—Kaniya—which is part of the myth-making tendency found in contemporary Russian film as well as literature. The spatial organisation of the enchanted world of AO is derived from the literary heritage. For example, the game features a giant oak tree growing at a river bank or on a sea shore, which is a visual reference to the prologue from Alexander Pushkin's poem *Ruslan and Lyudmila* (1820). As the gamer continues her quest she comes across the witch from Nikolai Gogol's short story *Vii* (1833). If the gamer succeeds in her battle with the witch, she will come face-to-face with a monster who demands she lifts up his eyelids. The quest is informed by the setting and plot of Gogol's story with the evil spirit of Vii terrorising the inhabitants of a small Ukrainian village.

AO borrows from literary sources, fairy tales and (the mythology of) Soviet life, and adheres to the vision of Medieval Rus and Soviet Russia both in terms of the landscape and architectural design as well as in terms of the organisation of the visual space of the game. AO exploits not only the gamer's knowledge of the visceral space but also her familiarity with the cultural space, particularly with Russia's mythological and mythologised heritage. This creative strategy enables the game to assimilate and to assemble various localities and characters into a cohesive narrative and ludic space, rendering the Soviet historic past *en par* with the country's literary and mythological heritage. Conversely, the Soviet past is perceived as a source of various legends and mysterious events, and the game resolves the real confrontations over Russian history, and particularly, the legacy of the Soviet regime, by providing an interactive place where these issues can be negotiated. Therefore, the space of AO is organised as a commemorative environment whereby historical narratives are presented as game layers (Juul 2004): the player re-enacts her location and thus engages with the national

---

[48]In 2009, President Medvedev established a government commission to counteract 'falsification' of Russian history; the move was perceived in the liberal media as an attempt to control historical knowledge (see for example Ivanova 2009).

history constructed as a set of tensions. In the game, the symbolic imagery of the game—given as 'analogons' (Lefebvre 1992)—signposts Russian cultural memory and realises it as different forms of presence.

### Space of collaboration: conclusions, or an illusion of closure

In the concluding section of the essay, following Dean Keith Simonton's use of the term in his sociological research on Oscar-nominated films (2002), I employ the notion of 'collaborative aesthetics' as a process of creating an environment by means of purposeful collaboration between different members of the public. Participation and collaboration are inherent to the networked digital medium (Paul 2003; Pearce 2009), which supports and relies on a constant exchange and flow of information, and are important elements in multi-user environments such as online gaming. In relation to MMORPGs, the collaborative model helps explicate the creative process itself. It requires a complex collaboration between artists, programmers, designers, researchers and other participants, whose role may vary from that of a consultant to a full collaborator. As the analysis of the AO business model demonstrates, this work process is fundamentally different from an industrial scenario where producers hire people to build components for their work according to elaborate instructions; rather, it is a collaborative process that requires complex aesthetic decisions that are made at various stages of the process. In this regard, the collaborative model of online gaming is comparable with the film industry, since creating computer games nowadays requires input from graphic designers, composers, scriptwriters, animation artists, game designers and, in some cases, live actors. The audiovisual aspect of creating games as well as playing creates a multimedial business surrounding each game. In addition, the collaborative aesthetics implies the use of gamers' input, which constitutes another level of participation or collaboration. Online games are ultimately software systems in which the creation of meaning relies on the content provided by the public.[49]

Furthermore, I apply the notion of collaborative aesthetic to the idea of shared fantasy. This traditionally refers to the practice of role play and gender play, which is a much-researched internet phenomenon, taking place in such different online environments as chat groups, MUDs,[50] and online computer games (Turkle 1997; Yee 2006). However, in my analysis, I extend the notion to involve the ideas of commonality expressed through the phenomenon of shared fantasies of nationhood, represented in online games through game mythology, construction of specific environments and providing avatars. I argue that this shared fantasy of nationhood, along with the use of 'contiguous, explorable, inhabitable and persistent worlds, embodied persistent identities and consequential participation' (Pearce 2009, pp. 18–20) serves as a prerequisite for the existence of the virtual world of AO. It takes the form of a spatial characteristic that embraces textual, graphical and audio representations as well as displays imaginary completeness and narrative cohesion as was demonstrated in the previous two sections of this essay. Spatiality functions as

---

[49]See Sokolova's essay in this collection on transmediality and fan labour.
[50]Multi-user dungeons.

an element that produces a sense of coherence and continuity, both socially and aesthetically, appealing to a variety of gamers.

By definition, communities involved in playing MMORPGs are predominantly trans-local, and often trans-national; in the case of AO and other Russian-made games, the global appeal of the product is enhanced with the help of a simultaneous launch of the game in other languages, in this case Italian, French, German, Spanish and English. At the same time the distinctive feature of AO is that it is aesthetically rooted in local, national culture. The game displays a strong centripetal force towards Slavic, Russian and Orthodox culture whereby the Soviet aspect of AO reveals a more centrifugal dynamic structure based on the common, shared memory of the recent past that requires collaborative assemblage to maintain its meaning. The binate logic of AO implies that, on the one hand, gamers engage in restoration of the lost culture with the purpose of exploring, understanding and reconfiguring the loss in contexts that simultaneously appear nostalgic and ironic. On the other hand, they get involved in creating a collective nostalgia for a past that never was, or what media theorist Norman Klein calls 'collective imaginary' (Klein 1997, p. 29). The architectonics of such imaginary offers emotional and immediate experiences that create a sense of affiliation. This sense is manifested through the spatial construction of the game, the elements of which acquire shared meanings of artefacts.

According to Benedict Anderson, nations as imagined communities occurred as a result of print capitalism (2006); the digital worlds of MMORPGs provide a space for renegotiating occurrences of commonality and question the validity of the term 'nation' in the post-national, trans-national condition facilitated by ludic digital environments with greater emphasis on the participatory aspect of identity construction. The procedural nature of MMORPGs renders space as an eco-system that continues to evolve as a result of social networking as well as the developers' interventions with various add-ons expanding the universe of AO and their imaginary playfield. Therefore, AO simultaneously provides a sense of stability through its spatial orchestration and a sense of uncertainty rooted in the lucid qualities of the environment. Open-ended, unstructured, creative play is similar to the work of imagination that occurs in worlds regulated by communication protocols, interfaces and transaction mechanisms. AO magnifies the complexity of Russia's post-Soviet transition and provides gamers with a social, collaborative space where the processes of remembering and forgetting can be played out. Thus, Russian game designers and players, like those based in Silicon Valley, are engaged in the production of technological innovation as well as their own ideology. In relation to the American industry, Barbrook and Cameron famously called it 'Californian Ideology', which emerged at the intersection of techno-utopianism, free market economy and capitalist democracy (2001). The new ideology of AO—as a symbol of the Russian games' industry with its digital tools, immersive interfaces and whole industry of cultural mediation—advances technological narratives of progress and provides a space for negotiating the laws of free market economy. AO is simultaneously engaged in a critical revision of the Russian past and as a result operates in a multimodal cultural regime in multiple ideological spaces.

*University of Leeds*

## *References*

Anderson, B. (2006) *Imagined Communities: Reflections on the Origin and Spread of Nationalism* (London, Verso).

Baigarova, P. (2010) '"Electronic Russia": Reality or (Empty) Promises? (Interview with Ivan Ninenko)', *Digital Icons: Studies in Russian, Eurasian and Central European New Media*, 3, pp. 103–6.

Barbrook, R. & Cameron, A. (2001) 'The Californian Ideology', in Ludlow, P. (ed.) (2001) *Crypto Anarchy, Cyberstates, and Pirate Utopias* (Cambridge, MA, MIT Press), pp. 363–87.

Bogost, I. (2007) *Persuasive Games: The Expressive Power of Videogames* (Cambridge, MA & London, The MIT Press).

Boym, S. (2001) *The Future of Nostalgia* (New York, Basic Books).

Castronova, E. (2001) *Virtual Worlds: A First-Hand Account of Market and Society on the Cyberian Frontier*, CESIfo Working Papers, available at: http://papers.ssrn.com/sol3/papers.cfm?abstract_id¼294828%20, 12 December 2010.

Chumachenko, A. (2007) 'Tochnykh dannykh … Rossiiskii rynok razvitiya komp'uternykh igr', Online conference, *Finam.ru*, 9 October, available at: http://www.finam.ru/analysis/conf00001001D3/default.asp, accessed 4 August 2010.

Cooke, C. (1997) 'Beauty as a Route to "the Radiant Future": Responses of Soviet Architecture', *Journal of Design History*, 10, 2, pp. 137–60.

Crogan, P. (2003) 'Gametime: History, Narrative and Temporality in Combat Flight Simulator 2', in Wolf, M. & Perron, B. (eds) (2003b), pp. 275–301.

Elverdam, C. & Aarseth, E. (2007) 'Game Classification and Game Design: Construction Through Critical Analysis', *Games and Culture*, 2, 1, pp. 3–22.

Eskelinen, M. (2001) 'Towards Computer Game Studies', *Digital Creativity*, 12, pp. 175–83.

Etkind, A. (2009) 'Post-Soviet Hauntology: Cultural Memory of the Soviet Terror', *Constellations: An International Journal of Critical and Democratic Theory*, 16, 1, pp. 182–200.

Finam.ru (2010) 'Industriya komp'uternykh igr: potentsial rosta i problemy piratstva', 28 May, available at: http://www.finam.ru/analysis/conf0000100331/default.asp, accessed 1 September 2010.

Heeter, C. (1992) 'Being There: The Subjective Experience of Presence', *Presence, Teleoperators and Virtual Environments*, 1, 2, pp. 262–71.

Hitchens, M. (2006) 'Time and Computer Games or "No, That's Not What Happened"', in Wong, K. K. W., Fung, L. C. C., Cole, P. & Pisan, Y. (eds) (2006) *Proceedings of the CGIE Conference* (Perth, Western Australia, Murdoch University), pp. 44–51.

Huizinga, J. (1955) *Homo Ludens: A Study of the Play Element in Culture* (Boston, MA, Beacon Press).

IKS-Consulting (2007) 'Rossiiskii rynok tsyfrovykh igr', 7 November, available at: http://www.iks-consulting.ru/search/262939.html, accessed 4 August 2010.

Ivanova, U. (2009) 'Pechal'naya istoriya: Medvedev sozdal komissiyu po bor'be s iskazheniem istorii', *kasparov.ru*, 19 May, available at: http://www.kasparov.ru/material.php?id=4A1263D299255, accessed 14 February 2012.

Jin, D. Y. & Chee, F. (2008) 'Age of New Media Empires: A Critical Interpretation of the Korean Online Game Industry', *Games and Culture*, 3, 38, pp. 38–58.

Juul, J. (2004) 'Introduction to Game Time', in Wardrip-Fruin, N. & Harrigan, P. (eds) (2004) *First Person: New Media as Story, Performance, and Game* (Cambridge, MA, MIT Press), pp. 131–42.

Klein, N. (1997) *The History of Forgetting: Los Angeles and the Erasure of Memory* (London & New York, Verso).

Komissarova, V. (2010) 'Mail.ru Completes its Acquisition of Astrum Nival', 25 March, available at: http://www.nival.com/eng/news/2010/document3966.shtml, accessed 12 July 2010.

Kuz'min, A. (2007) 'Ochen' prosto', *Rossiiskii rynok razvitiya komp'uternykh igr*, Online conference, 9 October, available at: http://www.finam.ru/analysis/conf00001001D3/default.asp, accessed 4 August 2010.

Labastov, S. (2002) 'Rytsari klavyatury', *Itogi*, 50, 17 December, available at: http://www.itogi.ru/archive/2002/50/103804.html, accessed 2 August 2010.

Lefebvre, H. (1992) *The Production of Space* [translated by Donald Nicholson-Smith] (Oxford, Blackwell Publishers).

Lindley, C. (2005) 'The Semiotics of Time Structure in Ludic Space as a Foundation for Analysis and Design', *Game Studies*, 5, 1, available at: http://www.gamestudies.org/0501/lindley, accessed 6 January 2012.

Lindtner, S. & Dourish, P. (2011) 'The Promise of Play: A New Approach to Productive Play', *Games and Culture*, 6, 5, pp. 453–78.

Livingston, S. (2009) 'Private Lives, Public Connections, on the Mediation of Important Things', *Public Lecture*, University of Leeds, 3 December.

Lyskovskii, A. (2007) 'Vot dannye ...', *Rossiiskii rynok razvitiya komp'uternykh igr*, Online conference, 9 October, available at: http://www.finam.ru/analysis/conf00001001D3/default.asp, accessed 4 August 2010.

Mikhailov, A. (2007) 'K sozhaleniyu ...', *Rossiiskii rynok razvitiya komp'uternykh igr*, Online conference, 9 October, available at: http://www.finam.ru/analysis/conf00001001D3/default.asp, accessed 4 August 2010.

Miloslavskii, M. (2010) 'V Kazani zakryli igrovoi salon', *Komsomol'skaya Pravda*, 13 August, available at: http://kazan.kp.ru/online/news/720227, accessed 1 September 2010.

Oushakine, S. (2000) 'In the State of Post-Soviet Aphasia: Symbolic Development in Contemporary Russia', *Europe-Asia Studies*, 52, 6, pp. 991–1016.

Oushakine, S. (2010) 'Totality Decomposed: Objectalizing Late Socialism in Post-Soviet Biochronicles', *Russian Review*, 69, 4, pp. 638–69.

Paul, C. (2003) *Digital Art* (London, Thames and Hudson).

Pearce, C. (2009) *Communities of Play: Emergent Cultures in Multiplayer Games and Virtual Worlds* (Cambridge, MA & London, The MIT Press).

Ruutu, K. Panfilo, A. & Karhunen, P. (2009) *Cultural Industries in Russia* (Copenhagen, Nordic Council of Ministers).

Shaw, A. (2009) 'Putting the Gay in Games: Cultural Production and GLBT Content in Video Games', *Games and Culture*, 4, 3, pp. 228–52.

Shaw, A. (2010) 'What is Video Game Culture? Cultural Studies and Game Studies', *Games and Culture*, 5, 4, pp. 403–24.

Simonton, D. (2002) 'Collaborative Aesthetics in the Feature Film: Cinematic Components Predicting the Differential Impact of 2,323 Oscar-Nominated Movies', *Empirical Studies of the Arts*, 20, 2, pp. 115–25.

Strukov, V. (2006) 'Actors and Agents: Russian Internet and Civil Society', lecture presented at St Anthony's College, University of Oxford, 9 June.

Strukov, V. (2009) 'Russia's Internet Media Policies: Open Space and Ideological Closure', in Beumers, B., Hutchings, S. & Rulyova, N. (eds) (2009) *The Post-Soviet Media: Conflicting Signals* (New York & London, Routledge), pp. 208–23.

Strukov, V. (2010) 'The Forces of Kinship: Timur Bekmambetov's Night Watch Cinematic Trilogy', in Goscilo, H. & Hashamova, Y. (eds) (2010) *Cinepaternity: Fathers and Sons in Soviet and Post-Soviet Film* (Bloomington, IN, Indiana University Press), pp. 191–217.

Tavinor, G. (2009) *The Art of Videogames* (Oxford, Wiley-Blackwell).

Temple, M. (2004) 'Tetris: From Russia with Love', *BBC*, documentary, available at: http://www.bbc.co.uk/bbcfour/documentaries/features/tetris.shtm l, accessed 10 January 2011.

Terranova, T. (2006) 'Of Sense and Sensibility: Immaterial Labour in Open Systems', in Krysa, J. (ed.) (2006) *Curating Immateriality: The Work of the Curator in the Age of Network Systems* (New York, Autonomedia), pp. 27–39.

Trubina, E. (2010) 'Past Wars in the Russian Blogosphere: On the Emergence of Cosmopolitan Memory', *Digital Icons: Studies in Russian, Eurasian and Central European New Media*, 4, pp. 63–85.

Turkle, S. (1997). *Life on the Screen: Identity in the Age of the Internet* (New York: Simon & Schuster).

Tychsen, A. & Hitches, M. (2008) 'Game Time: Modelling and Analyzing Time in Multiplayer and Massively Multiplayer Games', *Games and Culture*, 4, 2, pp. 170–201.

Volkov, D. (2009) 'Rynok kom'uternykh igr: vremya novykh modelei', *iBusiness*, 31 March, available at: http://www.ibusiness.ru/markets/414652/, accessed 4 August 2010.

Williams, R. (1974) *Television: Technology and Cultural Form* (London, Fontana).

Wolf, M. (2003) 'Abstraction in the Video Game', in Wolf, M. & Perron, B (eds) (2003b), pp. 47–67.

Wolf, P. & Perron, B. (2003a) 'Introduction', in Wolf, M. & Perron, B. (eds) (2003b), pp. 1–24.

Wolf, M. & Perron, B. (eds) (2003b) *The Video Game Theory Reader* (New York & London, Routledge).

Yee, N. (2006) 'The Demographics, Motivations and Derived Experiences of Users of Massively-Multiuser Online Graphical Environments', *PRESENCE: Teleoperators and Virtual Environments*, 15, pp. 309–29.

Zyuganov, G. (2012) 'Politicheskie debaty: Zyuaganov protiv Prokhorova' [video], 6 February, available at: http://www.youtube.com/watch?v=LK5lNPss1Hs, accessed 20 February 2012.

# Rebranding Russia's Capital City on Selected Social Media

## GRAHAM ROBERTS

'In the city there's a thousand things
I want to say to you'
The Jam, *In the City*

### Introduction

Of all the changes to have taken place in Russia since the collapse of the USSR in 1992, surely one of the most striking – certainly the most visible – must be the metamorphosis of Moscow.[1] First there were the statues, both those that came down (such as the monument to Felix Dzerzhinsky, founder of the Soviet secret police, on Lyubanka Square), and those that went up, for example Tsereteli's controversial memorial to Peter the Great. There was also a raft of name changes, to streets, squares and even certain metro stations ('Dzerzhinskaya' is now 'Lyubyanka', while 'Lenino' has become 'Tsaritsyno', for example). Next came the new buildings, including the high-rise apartment blocks with grandiose names such as 'Egoist' and 'Triumph Palace', the gigantic shopping malls called 'Crocus City' or 'Vegas', and the downtown business district known as 'Moscow-City'. Then the night-clubs arrived, along with the fleshpots, the exclusive restaurants, and the new museums, such as the Lumière Brothers' Centre for Photography, opened in March 2010 in the former Krasnyi Oktyabr chocolate factory. Below the ground, the Moscow metro has significantly expanded, gaining not only a number of new stations but even an entire new line. (The extent of Moscow's post-Soviet transformation has even prompted Clowes (2011) to argue that the city is currently traversing a profound 'identity crisis'.)

A number of the city's parks have also undergone significant changes, none more so than the most iconic of all Moscow's green open spaces, Gorky Park (see for example Weaver 2011). It now offers open-air activities such as aerobics and yoga, as well as beach volleyball in the summer, and snowboarding in the winter. An open-air cinema and contemporary art gallery were recently opened in the

---

[1]On the different aspects of Moscow's post-Soviet transformation, see for example Boym 2001, Beumers 2005, Kondakov 2008, Roberts 2008, Lemon 2009, Goldstein 2011 and Goscilo 2011.

park (Stathaki 2012), while it has even begun to host rock concerts (such as the 'Subbotnik' festival on 6 July 2013, headlined by UK band Arctic Monkeys). It is perhaps not surprising, then, that it was here that the latest project designed to change Moscow's image, 'Я ❤ Москву' ('I love Moscow') was launched by Sergey Kapkov, erstwhile general manager of the park and current Head of Moscow City Council's Culture Department, in the summer of 2013. On 11 July, the first of a series of monuments was installed in the middle of the park. This consisted of an angular white 'tick'-shaped structure (a stocky/stubby version of the Nike 'swoosh', or a nod towards the city's constructivist heritage?), approximately 3 metres high and 4 metres in diameter, set on a low, 4-metre-wide plinth. 'Ya' appears as a stencilled letter in the shorter, left-hand leg, with 'Moskvu' similarly stencilled in the other, longer leg, positioned at right angles to it. Nestling in the angle between the two legs is a 1m-high bright red heart, so that the entire structure can be read, from left to right, as saying 'Ya lyublyu Moskvu'.

Kapkov's initiative is intended to be a major municipal rebranding exercise. The project's website, 'moscowiloveyou.ru', makes this crystal clear:

> Unfortunately, recent years have witnessed a persistent tendency to run down ['rugat"] Moscow. People have become reluctant or ashamed of saying how much they love this city. We have decided to express our love [for Moscow] and help all those who wish to do the same. This is why we designed the logo 'Ya lyublyu Moskvu'. Declaring one's love for one's home city has become a well-established global trend. 'Ya lyublyu Moskvu' is destined to become a world-famous global brand like 'I ❤ NY', 'IAmSterdam', 'cOPENhagen' and others.[2]

What is most significant about this exercise is not the attempt to reposition Moscow as just another Western metropolis. Neither is it the fact that a city should be trying to change its public image. After all, countless cities have already done this, even before the 'I love NY' campaign of the 1970s mentioned here. East European cities to have engaged in recent rebranding include Kiev, Minsk and Budapest (on the last of these, see Szondi 2011, Smith and Puczko 2010, and Kavaratzis 2011).[3] Many other Russian cities have engaged in official 'rebranding' exercises since the end of Communism, such as Perm (Gavrilova 2012). In 2008, this self-styled 'easternmost city of Europe' began to reposition itself as a centre for contemporary arts to rival Moscow. The year 2010 saw the launch of the city's new logo, a red Cyrillic 'П' (for 'Пермь'), designed by Artemy Lebedev (Gavrilova 2012).[4] The Permian Gate, a 12-metre-high triumphal arch in the shape of the city's new logo, designed by Nikolay Polissky and made entirely from the branches of pine trees,

---

[2]'O proekte "Ya lyublyu Moskvu"': http://moscowiloveyou.ru/page/about.html (accessed 14 January 2014). It is noteworthy that no other Russian city is mentioned here, despite the fact that a number of them have undergone rebranding exercises of their own. The reference to Moscow as a 'brand' is itself ironic: as Yurchak (2005) notes, in the Soviet era, the word was used by Russians to refer exclusively to *foreign* brands.

[3]For an incisive discussion of the logos used by various cities of the CIS in their recent rebranding exercises, see 'Azbuka gorodov. Vypusk No. 1' (31 January 2013), available at: http://citybranding.ru/alpha1/ (accessed 14 January 2014).

was formally opened in the city in June 2011 (Gavrilova 2012). The same month saw the first 'White Nights of Perm' ('Belye nochi v Permi') annual arts festival. All this has been accompanied by a communication campaign on specific social media networks. The city of Perm has a Vkontakte page entitled 'Ya lyublyu Perm !!' which currently has over 9,000 followers. There are also a number of institutional Twitter accounts linked to the city of Perm, such as 'Zhivaya Perm' (over 1,500 followers), 'Belye nochi v Permi' (1,710 followers), Perm Opera (over 2,000 followers) and the Perm Museum of Modern Art (nearly 2,300 followers). Perm's rebranding example has been followed by other Russian cities such as St Petersburg, Nizhny Novgorod and Ulyanovsk (Shafranskaya 2012; see also Bertrand 2012). As for Moscow, it is not the first time that the idea of rebranding the city has been floated. In 2010, the online Russian newspaper *The Village* organised a competition to redesign the 'brand of Moscow', and six different designers entered (Trabun 2010a, 2010b). In 2012, the mayor's office announced a competition to find a new 'brand' for Moscow (Po 2012). At the time of writing (February 2014), the winner has yet to be announced.

What is most noteworthy, perhaps, is the use that the 'I love Moscow' campaign has made of social media. Perm's experience notwithstanding, this is the first time that a public, municipal body has invested so heavily in Web 2.0 applications such as Facebook, Vkontakte, Instagram, Twitter and YouTube.[5] Social media are purportedly at the heart of the project, notably those characterised by what Kaplan and Haenlein (2010) call medium levels of social presence and media richness—social networking sites such as Vkontakte and Facebook and content communities such as Instagram and YouTube. It is easy to understand the attraction. As Kaplan and Haenlein put it (2010, p. 67):

> Social Media allow firms to engage in timely and direct end-consumer contact at low cost and higher levels of efficiency than can be achieved with more traditional communication tools. This makes Social Media not only relevant for large multinational firms, but also for small and medium sized companies, and even non-profit and governmental agencies.

---

[4]See also 'Permskii tsentr razvitiya dizaina ostalsya bez deneg' (19 April 2010), available at: http://www.sostav.ru/news/2010/04/19/cod3/ (accessed 30 October 2013). Lebedev's design studio was founded in 1995, and is now one of the most sought after creative agencies in Russia. With its motto 'Design will save the world', it has offices in Moscow, Kiev and New York. Recent projects undertaken by the studio include the website of Russian search engine Yandex, the logos for Gorky Park and the city of Yaroslavl, and the new Moscow metro map (in 2013). Details concerning these and other projects undertaken by the studio can be found at the company's website, available at: http://www.artlebedev.ru (accessed 3 February 2014).

[5]We follow the definition of 'Web 2.0' as:

> a collection of open-source, interactive and user-controlled online applications expanding the experiences, knowledge and market power of the users as participants in business and social processes . . . [supporting] the creation of informal users' networks facilitating the flow of ideas and knowledge by allowing the efficient generation, dissemination, sharing and editing/refining of informational content. (Constantinides and Fountain 2008, quoted in Brennan and Croft 2012, p. 113)

The importance to the project of social media is made clear on the homepage of moscowiloveyou.ru:

> By the end of July there will be 'Ya lyublyu Moskvu' structures in 14 of the city's biggest parks, in order for as many people as possible to participate in the 'Ya lyublyu Moskvu' project. In order to take part, simply have a photo of yourself taken next to the structure and post it on your social media networks with the hashtag #yalyublyumoskvu. The brightest, most positive and most joyful photographs will be entered for the 'photo of the day' on the home page of the project's web site.[6]

In the rest of this chapter we propose to look at the use Moscow is making of this new media. Before we do so, however, we need to take a closer look at the academic literature that has grown up around the concept of city branding itself, and also explain our methodology.

### *Literature review and data collection*

The 'I Love NY' campaign, with which Kapkov explicitly compares his own initiative, was launched nearly 40 years ago, in 1976. City branding as a practice has been around far longer, however – very possibly as long as cities themselves (Stigel and Frimann 2006, Dinnie 2011). On the other hand, the *theory* of, and literature on, city branding emerged from the pioneering work on nation branding by Anholt (1997). Anholt's initial use of the term 'nation brand' was, he claimed, in recognition of the growing importance of reputation for cities in meeting the particular challenges – social, economic, cultural, political, etc. – facing them. It was also part of a growing interest among scholars in the latter part of the twentieth century in the relationship between 'place' (global, national, regional, urban, domestic etc.), social relations and historical change (see for example Anderson 1983/2006, Giddens 1990 and Gieryn 2000).

At first a sub-theme of place branding (Anholt 2002), city branding has in the last ten years emerged as an academic subject in its own right, in a context of a changing global economic environment, accelerating urban decay and increasing competition for ever scarcer resources. There is now even a burgeoning industry of 'how to' books that address the practical, policy issues behind city branding (see for example Kotler *et al.* 1999, Anholt 2007 and Moilanen and Rainisto 2008). As well as policy makers, the subject has attracted scholars from across a broad range of disciplines, including urban studies, sociology, anthropology, political science, economics, marketing and semiotics.[7] Themes that have attracted particular

---

[6]The choice of many of these photographs suggests that Kapkov's initiative should be seen in the broader context of President Putin's on-going attempt to rebrand Russia herself. For example, one Instagram image chosen as 'photo of the day' in late September 2013 featured two Russian athletes brandishing the official Olympic flame, celebrating the forthcoming 2014 Winter Olympics scheduled to take place not in Moscow, but over a thousand kilometres away, in the Black Sea resort of Sochi. On the (re)branding of Russia as *nation*, see for example Klyukanov *et al.* (2008), Miazhevich (2010), Ostapenko (2010) and Simons (2011).

[7]More extensive surveys of the city branding literature can be found in Lucarelli and Berg (2011) and Martinez (2012).

attention, and which are all to a greater or lesser extent relevant to the 'Ya lyublyu Moskvu' campaign include: the specificities of branding cities, as opposed to the branding of regions or nations (Caldwell and Freire 2004); the differences between city *marketing* and city *branding* (Kavaratzis 2004); 'creativity' and city branding (Healey 2004); the relationship between corporate branding and city branding (Trueman, Klemm and Giroud 2004, Hankinson 2007, Kavaratzis 2009); how best to generate a single brand vision around the interests of the city's multiple private, public and quasi-public stakeholders (Virgo and de Chernatony 2006); the use of sports events to brand a city (Zhang and Zhao 2009, Belloso 2011, Fola 2011); and, most recently, the role bloggers can play in helping to boost, or contest, a city's official brand image (Florek 2011, Hvass and Munar 2012, Larsen 2014). As Anholt (2009) has observed, one of the recurrent themes in the literature is the opposition between two radically opposed schools of thought; on the one hand there are those who regard cities as they might products or corporations, to be promoted using catchy slogans and eye-catching logos, while on the other hand there are those who take a holistic view of city branding, seeing it as part of a much broader attempt to address real social and economic problems (see also Ashworth and Kavaratzis 2009 and Ashworth 2010).

In recent years, two areas have proved particularly popular in the city branding literature. The first of these concerns cities in countries currently undergoing profound economic transition – those defined by the IMF as 'emerging' economies, such as Hungary (on Budapest see Smith and Puczko 2010, Kavaratzis 2011 and Szondi 2011), Turkey (on Ankara see Hayden and Sevin 2012), Kazakhstan (Gaggiotti, Low Kim Cheng and Yunak 2008) or China (Zhang and Zhao 2009, Larsen 2014). The second has been the role played by social media in city branding. This latter trend was marked in particular by the special issue of *The Journal of Travel and Tourism Marketing* in 2013 devoted specifically to social media in tourism and hospitality (see in particular the article by Leung *et al.* here (Leung *et al.* 2013), and also Hvass and Munar 2012). As this last example suggests, however, the focus of research into city branding has tended to be on outsiders (be they tourists or sources of Foreign Direct Investment such as corporations), and on making the city attractive to them, rather than on the interests and perspectives of residents themselves. The needs of this latter group – who are the focus of Kapkov's initiative in Moscow – remain considerably under-researched in the city branding literature (see Aitken and Campelo 2011). Another issue still relatively overlooked is the use that cities themselves make of social media; most scholars who look at social media do so from the point of view of consumers/bloggers (see for example Ketter and Avraham 2012, or Tussyadiah and Zach 2013). The aim of the present chapter, then, is to focus on the 'I love Moscow' project, and to use it as a case study to analyse its implications both for city branding, and for social media branding.

Before we begin our analysis, however, we need to introduce a concept which is becoming increasingly important, both for city branding and for social media branding (and indeed for branding in general). This is the notion of 'brand co-creation'. A number of scholars have begun to look at what is generally referred to as 'co-creation' of the city brand by the consumer (see for example Aitken and Campelo 2011, Ketter and Avraham 2012, and Tussyadiah and Zach 2013). This notion goes back to the work of theorists such as Prahalad and Ramaswamy (2004) who were among the first to argue that interaction between brands and consumers could help create value for the brands themselves (see also Foster 2007).[8] What is important to note, however, is the insistence Prahalad and Ramaswamy make on the role in this process, not so much of individual consumers, but of consumers who come together to form *communities*. The phrase 'brand community' was first coined by Muñiz and O'Guinn (2001). Taking their cue from a raft of nineteenth- and twentieth-century social scientists and philosophers – and especially theorists of identity such as Durkheim (1893) and of community such as Tonnies (1887), Janowitz (1952) and Anderson (1983/2006) – they argued that in today's fragmented, post-modern consumer society, it is now brands, rather than physical places or religious groups, that create a sense of belonging for people. It is worth quoting in full their definition of brand community, since it makes explicit not just the despatialised, mediated and commercial aspects of these communities, but also their moral dimension (Muñiz and O'Guinn 2001, p. 412):

> A brand community is a specialized, non-geographically bound community, based on a structured set of social relationships among admirers of a brand. It is specialized, because at its center is a branded good or service. Like other communities, it is marked by a shared consciousness, rituals and traditions, and a sense of moral responsibility. Each of these qualities is, however, situated within a commercial and mass-mediated ethos, and has its own particular expression. Brand communities are participants in the brand's larger social construction and play a vital role in the brand's ultimate legacy.

For Muñiz and O'Guinn, by no means the least interesting aspect of these communities is that they can help reinforce the emotional bond between consumer and brand, and thereby ultimately strengthen the brands themselves (on this point see also McAlexander, Schouten and Koenig 2002, Cova 2006 and Stokburger-Sauer 2010).

As might be expected, the original concept has somewhat evolved since its first formulation – especially since the advent of Web 2.0, which has generated a marked increase in scholarly interest in online brand communities. Amine and Sitz (2007) were among the first to seek to develop Muñiz and O'Guinn's original definition to the realm of online branding in an article analysing three online consumer-led discussion groups, devoted to Nikon, Canon and Apple's now defunct Newton personal assistant respectively. A number of scholars have followed suit (see for example Schau, Muñiz and Arnould 2009, Bonnin and Odou 2010, Healy and McDonagh 2013 and Wirtz *et al.* 2013). Indeed, Web 2.0 lends itself particularly well to the construction of certain types of brand community, where the link between consumers may be just as, if not more, important than the goods or services on offer

---

[8]While brand co-creation is related to the notion of 'prosumption', the two should not be confused. 'Prosumption' has been an important concept in marketing and branding theory for nearly 30 years. Kotler (1986) traces the notion back to the assertion by Toffler (1980) that in the Industrial Age, people produce more and more of their own goods and services. As Ritzer and Jurgenson (2010) argue, however, user-generated online content is an increasingly important form of 'prosumption'. For further discussion of the ways in which social media are contributing to a blurring of the distinction between 'consumption' and 'production', see Cova and Cova (2012), Sokolova (2012) and Fuchs (2014).

from the brand itself (indeed, Bonnin and Odou suggest that in the case of the online dating agency Meetic, the service *is* the link: 2010, p. 28). The research agenda has slowly shifted to take into account not just consumer-led discussion groups, but also brands' official community websites and social media pages. Habibi, Laroche and Richard (2014) are among an increasing number of scholars who look at how communities form around brands' social media pages, arguing that it is precisely the social and networked nature of such pages that makes them an ideal environment in which such communities can develop (see also Wallace, Buil and de Chernatony 2012 and Zaglia 2013). At the same time, some scholars have begun to question the type and nature of the role online brand followers can effectively play in brand creation (see for example Quinton 2013). As we have seen, Kapkov's project is introduced on its website in terms which explicitly encourage Muscovites both to collaborate actively in the creation of the Moscow city brand, and to imagine themselves as a community of like-minded individuals: 'We have decided to *express* our love [for Moscow] and help *all those who wish to do the same*' [our emphasis].[9] The most important question, then, is not whether Kapkov succeeds in rebranding Moscow; rather, it is: to what extent does he succeed in creating a community around the Moscow 'brand'?

To answer this question, we propose a three-stage methodology. To begin with, we conducted a statistical analysis of the images which appear on certain of the project's social media. This is because Kapkov's project, as we have seen, lays such emphasis on the pictorial representation of Moscow. It is also because, as Kavaratzis has cogently argued (Kavaratzis 2004, quoted in Miles 2010, p. 41), the primary way in which we apprehend the post-industrial city is via the visual image. Indeed, Kapkov's project confirms (if such confirmation were still needed) that we are now in the age of 'visual consumption' (Schroeder 2002). During the visual analysis stage, we focussed on the project's Vkontakte page for a number of reasons. First, Vkontakte currently has more than three times the number of unique users in Russia than Facebook (40 million as opposed to approximately 12 million: Fernandes 2012). Second, the project itself has almost 50% more followers on Vkontakte than on Facebook (at the time of writing – February 2014 – this is just over 18,000, as opposed to approximately 12,500). Third, all the images posted on the Vkontakte page remain accessible indefinitely, whereas the majority of those on the Facebook page are removed over time, thereby rendering impossible any attempt to give a satisfactory statistical account of activity on its page (it should be noted that all the images currently viewable on the project's Facebook page can also be found on its Vkontakte page). We then analysed the textual interactions between followers of 'I love Moscow', filtering them for relevance to our chosen theme (brand community). Our methodology is essentially 'observational netnography' (Kozinets 2006), including historical analysis, visual analysis, textual analysis and

---

[9]On the use of this kind of emotional appeal to solidarity in other areas of branding in post-Soviet Russia, see Morris (2007) and Roberts (2014).

statistical analysis. While Kozinets suggests that netnographic research should ideally be multi-method (2006, p. 132), we took only a limited part in online discussions, in order to avoid the possibility of soliciting biased responses, following McQuarrie, Miller and Phillips (2013). Fourth, and finally, because the comments on the project's official social media may be censored or even removed by brand community managers, we looked at the images posted on Instagram, using the hashtag #yalyublyumoskvu (for the methodology used, see below).

### The 'I love Moscow' project: a 'brand community'?

At the time of writing (February 2014), the images which appear on the project's Vkontakte page (in a total of 1,229 posts) can be broken down into the following categories, in decreasing order of frequency:

- Photos taken of people standing in front of or next to the 'I love Moscow' statue (including an album of 255 such photos at the top of the page). The majority of these (approximately two-thirds) are women, aged between approximately 18 and 25, and usually photographed on their own;
- Panoramic views of the city (these are very often included at the beginning of the day, with the caption 'Good Morning, Moscow!', or at the end, with 'Good Night, Moscow', and many of them juxtapose both modern and more ancient buildings);
- Historical monuments and buildings dating from before the Soviet era, many of which are closely associated with Moscow, such as St Basil's, the Kremlin, GUM, or the Novodevichy complex;
- Cultural/sporting events of various sorts: these are as diverse as they are numerous, and include the arrival of the Olympic flame in the city ahead of the 2014 Winter Olympics, celebrations for Russian Flag Day (22 August), the 'Active City' festival of 31 August, with mayor Sobyanin[10] promoting cycling, a collective pillow fight in the city centre described as a 'flashmob', a Robotics exhibition in Sokolniki Park, and Jamaican athlete Usain Bolt DJ-ing in Gorky Park during the World Athletics Championships of August 2013;
- Parks/outdoor places/nature. The main parks featured include Gorky Park, Sokolniki and VDNKh, now called 'VVTs' – 'Vserossiiskii Vystavochnyi Tsentr'. There are posts on the best places for a picnic, the best children's playgrounds in Moscow, or the fact that people can now marry in Kolomenskoe Park;
- Photos of the city from the Soviet era. These are particularly diverse and include: the (recently demolished) Rossiya hotel in 1966, the first motorised bus (a British Leyland) in the 1920s, peasant women celebrating victory over the Nazis on Red Square in May 1945, volunteers during the construction of Moscow State University, Mayakovsky's Renault car (taken in 1929) and the Opening Ceremony of the 1980 Olympic Games;[11]

[10]As this is the only image of Sobyanin on the project's Vkontakte page, it would appear that Kapkov has been careful to distance himself and his project from the city's mayor.

[11]Twenty of these photos are posted not by the project's administrator, but rather by a community member, a certain 'Slus Tabl'.

- Modern, post-Soviet buildings, monuments, structures and statues. Many are of the Moscow-City business complex – shown so often that on 17 December one community member is moved to complain 'haven't you any other nice things to show us?' – while others are of less well-known, more modest structures, including statues of Pushkin and his wife, Rostropovich, Adam and Eve, Soviet-era circus clown Yury Nikulin, Baron Munchhausen and even Sherlock Holmes and Dr Watson;
- Young children and families in a variety of settings;
- Various city centre scenes, such as pedestrianised streets, bridges or ponds. Some of these serve to introduce themes such as 'How to sleep cheap in Moscow', or 'Moscow for Lovers';
- Historical monuments and buildings from the Soviet era, such as the Ukraine Hotel (and various other examples of Stalinist 'wedding cake' architecture), the Lenin sports stadium at Luzhniki, the Ostankino TV tower, or Vera Mukhina's Worker and Peasant statue in front of VDNKh;
- Cultural centres, such as art galleries, museums, theatres and cinemas;
- The Moscow metro, with images ranging from historical photographs to up-to-date shots of carriages featuring the special 'Sochi 2014' livery, on 10 January 2014;
- City photographs from the pre-Soviet period, such as the Maly Theatre taken in the 1890s, or the great flood of 1908;
- Outdoor sports such as skating, snowboarding, mountain-biking and even beach volleyball;
- Other miscellaneous images, such as: bars, clubs, restaurants and cafes (including photos of plates of food, from sushi to steak, and from fruit pie to pizza); photos of the 'I Love Moscow' statue with nobody standing in front of it; animals (mainly 'cute', often on a 'good morning' or 'good evening' post); a wide variety of shops and markets; transport routes, maps of the metro, etc.; depictions of medieval Russia, such as Pavel Ryzhenko's painting of the Battle of Kulikovo of 1380 (accompanied by a brief note which explains that the battle reinforced Moscow's role in the unification of the Russian state); photos of souvenirs on sale as part of Kapkov's project; plans of unrealised Soviet architectural projects; images of famous people linked with Moscow (such as Repin's portrait of Pavel Tretyakov, who gave his name to Moscow's Tretyakov Gallery); railway stations; hotels; and libraries.

How might one characterise the 'brand' that emerges from these pages? Perhaps most obviously, Moscow is presented as a dynamic, family-friendly city, a modern metropolis with a proud historical tradition and rich cultural diversity.[12] One could argue that Kapkov's extensive use of popular imagery to produce a coherent, linear brand narrative helps transform Russia's capital into the kind of 'iconic' brand discussed by Holt (2004).[13] (One might add that, like most iconic brands, Kapkov's Moscow exploits

---

[12]As with all 'brands', so those following 'I love Moscow' may have multiple, and even conflicting, loyalties. During a series of private exchanges in which we took part, one individual repeatedly confided that although she lived in the capital, she was actually born in St Petersburg, and preferred her native city to Moscow.

[13]'Iconic' brands discussed by Holt include Levi's, Pepsi, Marlboro and Jack Daniel's.

popular memory to construct a myth in which all dissenting voices – what Thompson and Tian (2008) refer to as 'countermemories' – are silenced.) At the same time, the many different ways in which the site links the city's present to its past, suggesting that history (albeit of a very sanitised kind) is an integral part of today's Moscow, is typical of many so-called 'heritage' brands (see Urde, Greyser and Balmer 2007).[14] Indeed, to paraphrase Heilbrunn (2006), Kapkov's Moscow functions like so many other brands which enshrine contradictory principles, such as the past and the present, the very distant and the here and now. But to return to the issue raised in our introduction, are we dealing here with a genuine brand *community*?[15] In order to answer this question, we need to look at the responses to Kapkov's campaign by Russian bloggers themselves – first, the comments to posts on the project's Vkontakte page, and second, the images on Instagram using the projects dedicated hashtag.[16]

We take first the comments on Vkontakte. As we saw earlier, Muñiz and O'Guinn (2001) argued that communities are characterised by, first, shared consciousness, second, rituals and traditions, and third, a sense of moral responsibility. The first of these, a shared consciousness, appears at a number of moments: members engage in discussion of how to ease transport problems in the city, for example (18 December, 29 comments); on another occasion, they angrily defend Moscow against a blogger who dismisses the city as 'a dirty little village full of crooks and paedophiles', accusing him of being a provincial who knows nothing about their city (1 November, 11 comments). An awareness of shared rituals and traditions comes across in, for example, a post on 25 December, when in reaction to a photograph of the statue of Yury Dolgoruky (the founder of Moscow) dressed in a red and white Father Christmas costume, one community member angrily asks: 'Why do we have to have these Santa Claus costumes everywhere? Where is our RUSSIAN [in capitals in the text] Grandfather Frost in dark blue and light grey?' (although this outburst meets with no response). At another moment, an invitation to vote for one's favourite film about Moscow elicits 13 responses (all of the films mentioned are from the Soviet era). Finally, a sense of moral responsibility is present in a discussion on 30 October, following the announcement that volunteers are to begin patrolling the city's streets and courtyards in search of individuals smoking in places where it is forbidden to do so (an exchange which also contains much evidence of shared consciousness). Participants express views on, for example, the rights and wrongs of volunteering, the morality (and practicality) of forbidding people from smoking in public places, and the question of whether miscreants should be fined.

There are, then, examples of each of the characteristics of a typical community as identified by Muñiz and O'Guinn (2001). It should be said, however, that even those examples that we found are remarkably few in number, when one considers

---

[14]As well as private corporations, many brands discussed in the growing literature on 'heritage' brands are public institutions, such as the British monarchy (see Hudson and Balmer 2013).

[15]While the campaign refers to itself on Vkontakte as a 'community', this is standard practice on this particular social medium not just for brands but for all sorts of groups.

[16]The corresponding url is: iconosquare.com/tag/ялюблюмоскву/ (accessed 28 July 2014).

both the number of official community members (in itself small for a city with an official population of over 12 million) and the length of time the page had been operational at the time we carried out our analysis. The discussions also tend to be extremely brief (the exchange concerning smoking contains a total of 12 comments, only three of which are responses to points made by other community members). Moreover, the majority of the discussions on the project's Vkontakte and Facebook pages involve people merely venting their anger about such issues as public transport price rises, or asking for information about the location of a particular tourist attraction or event. Consequently, while a brand community may be said to emerge, it is extremely fragile.[17] Indeed, Amine and Sitz (2007) argue that there are three stages in the construction of an online brand community: first, the creation of a dedicated community space (a stage which includes regular interactions between community members); second, the construction of a collective identity (with the emergence of rules governing interaction, and collective strategies to solve conflict); and third, stabilisation of hierarchical structures, a stage during which members arrange actual, physical meetings between each other. The first of these three stages is clearly present here, while there is also some evidence of the second; during a discussion on 6 November on the rights and wrongs of city centre parking charges, one member tells another he is a degenerate and accuses him of participating in discussion groups on pornographic sites, at which point another community member tells them, 'stop arguing!'. As for the third stage, no attempt is made by any other of the community members to actually meet up – something which comes across particularly clearly in the 21 responses to the question posed by brand managers on 25 December, namely 'How are you planning to celebrate New Year?'[18]

The fragility of the brand community becomes even clearer if we examine the ways in which the campaign's official hashtag, yalyublyumoskvu, is used on Instagram. Although we looked at as many images as possible, because of the large number of pictures involved (over 25,000 at the time of writing), we decided to focus on certain specific days, notably 10 September 2013, 10 October 2013, 10 November 2013, 10 December 2013 and 10 January 2014. Statistical analysis of the images for these dates revealed the following principle differences between the official Vkontakte page and the Instagram account:

– a far higher incidence of portraits, of the self or of friend(s), many of which are taken in front of famous monuments, and resemble typical 'tourist' photographs;
– a lower proportion of images showing people in front of an 'I love Moscow' statue, animals, children and families, historical monuments, statues, or modern buildings;

---

[17]This fragility is particularly evident on the project's Twitter account, which at the time of writing (February 2014) has a mere 323 followers – hardly impressive when one considers that there are currently over 2.3 million Twitter users in Russia (Kelly *et al.* 2012).

[18]It is of course perfectly possible that individual community members make precisely such arrangements by contacting each other on their personal 'walls', rather than on the main page. However, this in itself does not enable the emergence of a shared consciousness – quite the opposite, we would contend, since any exchanges and/or arrangements would be private, rather than public.

– the virtual absence of images relating to the kind of cultural event which appears so frequently on the project's official Vkontakte and Facebook pages.
– the presence of elements totally absent from the official pages, including: 'off-the-wall' images with little or not apparent link to Moscow, such as screenshots of homework schedules taken from the 'notes' page of an iPhone, or an image of a collection of wine bottle corks, arranged so as to fill every square millimetre of the screen; photographs that are difficult to decipher, including several taken from inside a car on a rainy day, the raindrops on the car window making it impossible to make out what we are supposed to be looking at; and photographs that appear to be 'spam' (images posted by professional models, florists, cake shops or manicurists; these – sometimes appearing in a series of nine or ten at a time – are among the most popular posts, regularly attracting over 150 comments, far higher than the average, which is three).

What is especially noteworthy is the self-conscious performativity of many of the Instagram (self-)portraits. Some of the images are strikingly reminiscent of work by contemporary photographers such as Larry Sultan, Hiromix, Nan Goldin or Juergen Teller, who have legitimised the diary style of photography (Cotton 2009). This points to the fact that posts in Instagram serve first and foremost to display the self, and only secondly (if at all in some cases) the brand, with users inserting multiple hashtags in order to be seen by as many different people as possible (see Winter 2013). What this suggests perhaps more than anything else is that encouraging consumers to post photographs on Instagram is by no means the best way of building a sense of 'brand community'. Indeed, when discussion does emerge, it is generally extremely short (far shorter than on Vkontakte); for example, a photograph of the awning covering the front of iconic city centre toyshop, Detskii Mir, currently undergoing complete reconstruction, produces the following exchange, remarkable for its succinctness:

– natashking: It's a real shame.
– chuchundra_mc: But it doesn't matter what it looks like from the outside, what's important is what it was like inside . . . the merry-go-round, the passageways, the lifts, fascinating staircases.
– natashking: Such memories! It takes the breath away.

The kind of brand community that emerges from the project's Instagram page, then, is a community of individuals seeking first and foremost to brand *themselves*, rather than express their support for the Moscow 'brand'. In this respect, 'I love Moscow' is not alone; self-promotion by 'fans' is a problem experienced by many brands which try to encourage online communities, as Cova (2006) has noted in the case of Nutella, for example (see also Belk 2013).

## Conclusion

As Clarke (2003, p. 190) has aptly commented, quoting Bauman (2001), globalisation has spawned a number of attempts by city managers and urban planners throughout the world to assert some kind of local identity by foregrounding local landmarks, historical events or cultural specificities. Kapkov's initiative is typical of this growing trend. The question remains, however (especially given our point about the use of the hashtag on Instagram): can he generate a genuine community of followers around his 'I love Moscow' brand? In principle, there is no reason why not. After all, for Anderson, whose work on nationalism has proved the starting point for so much work on brand communities, of both the off- and online variety (see for example Muñiz and O'Guinn 2001, and Cayla and Eckhardt 2008), communities ultimately exist not in any physical sense, but rather in the minds of their various members. As he puts it: 'all communities larger than primordial villages of face-to-face contact (and perhaps even these) are imagined. Communities are to be distinguished, not by their falsity/genuineness, but by the style in which they are imagined' (Anderson 2006, p. 6). This may be good news for Kapkov and Moscow's brand managers, especially since physical distanciation makes places such as cities increasingly 'phantasmagoric', with actual contact between inhabitants increasingly rare (as Giddens noted nearly a quarter of a century ago: 1990, pp. 18–19).

Anderson's most important theoretical contribution, however, has been not his insistence on the quintessentially imagined nature of communities, but rather the idea that historically, the most important factor behind both the secularisation of modern society and the consciousness (and indeed legitimation) of that particular imagined community called 'the nation' has been the rise of what he calls 'print capitalism'. By this he means the combination of the rise of an economic system based on private ownership of the means of production, distribution and exchange, on the one hand, and the invention of that particular form of information technology known as the newspaper, on the other. He makes this point in a passage which we propose to reproduce here in full since it is particularly relevant to our discussion. Referring specifically to the collective reading of the morning or evening edition (what he describes as a 'mass ceremony') in the late nineteenth and early twentieth centuries, Anderson continues (2006, pp. 35–36):

> The significance of this mass ceremony [. . .] is paradoxical. It is performed in silent privacy, in the lair of the skull. Yet each communicant is well aware that the ceremony he performs is being replicated simultaneously by thousands (or millions) of others of whose existence he is confident, yet of whose identity he has not the slightest notion. Furthermore, this ceremony is incessantly repeated at daily or half-daily intervals throughout the calendar. What more vivid figure for the secular, historically clocked, imagined community can be envisioned? At the same time, the newspaper reader, observing exact replicas of his own paper being consumed by his subway, barbershop, or residential neighbours, is continually reassured that the imagined world is visibly rooted in everyday life.

Much of what Anderson says here aptly describes how consumers use social media today. There is the reference to 'silent privacy', for example, mention of the 'ceremony [. . .]' replicated by thousands (or millions) of others [. . .] at daily or half-daily intervals', or the idea of '[an] imagined world [. . .] visibly rooted in everyday life'. If we substitute, then, 'social media follower' for 'newspaper reader' in the above quotation, we can begin to see the potential of Web 2.0 to produce precisely the kind of community that would add value to Kapkov's own project. Indeed, Anderson himself makes the crucial point later on in his book that advances in communications technology, such as radio and television, serve the same community-building function as the newspaper (he describes them as 'allies unavailable [to print] a century ago': 2006, p. 135). Social media are, we would argue, even more of an 'ally' in this respect than radio or television, since they facilitate direct (and indeed immediate) communication between users – whether by 'liking' somebody else's post, sharing it, commenting upon it, or even responding directly to it, either textually or visually, via an embedded photo or video.[19] Indeed, as we have seen, scholars from Amine and Sitz (2007) to Habibi, Laroche and Richard (2014) argue that Web 2.0 is particularly conducive to the formation of online brand communities.

Yet on the whole, those brand communities which do emerge online tend to be rather fragile, whether the brand in question is Kapkov's Moscow or Ferrero's Nutella (Cova 2006). This is in stark contrast to what happens when individuals (and families) consume popular culture and produce cultural memory in an environment which, rather than being predominantly online, is entirely physical. In a remarkable article on the National Western Stock Show and Rodeo in South Dakota, Peñaloza (2001) argues that this annual event is a quintessential site of cultural meaning-making (which makes it in ways reminiscent of 'I love Moscow', although in a very different context, naturally). Referring both to Anderson (1983) and de Certeau (1984), Peñaloza discusses the various modes of participation in the show – strolling through exhibition stands, sampling food, watching the rodeo, participating in role-play, interacting both with other consumers and with cattle producers (not least by buying produce to take home), etc. She argues persuasively that via activities such as these, visitors consume history, produce their own cultural meanings and reaffirm their common heritage. In doing so, Peñaloza argues, they both assert the uniqueness of the American mid-West, and construct their own social – and communal – identity. As Peñaloza's study suggests, the negotiation of cultural meaning and memory between consumers (and indeed between consumers and producers) at the Stock Show and Rodeo is both multifaceted and multi-dimensional. It involves several different types of consumer behaviour (entertainment, education, shopping), various levels of situational positioning (personal, historical, etc.), several kinds of cultural interactions (adults/children, humans/animals, ranchers/visitors, etc.), and different modes of market interaction (involving marketers, producers and industry representatives). As such, it is richer, deeper and more diverse than would ever be possible for the 'I love Moscow' initiative as it is currently designed online. And it is precisely this richness, depth and diversity that helps make the community 'imagined' by consumers at the show so durable.

This does not necessarily mean that Kapkov's own project is destined to fail, however. In his fascinating study on the city, Alan Blum makes the following remark (2003, p. 231; quoted in Miles 2010, p. 52):

> [T]he city [. . .] can be theatricalised as an expanded public stage in which all participants are reciprocally performers and audience, two-in-one, seeing each other and being seen in an environment enlarged to illuminate its varied choices and restricted to keep dark the possibility that those differences make no real difference.

---

[19]Of course, it might be objected that unlike one's fellow newspaper readers, one does indeed have a clear notion of the identity of other online community members. Yet this is merely an illusion, since while the person I exchange views with online may claim to be a 15-year-old schoolgirl from Moscow, he may in fact be a 52-year-old paedophile from Murmansk.

Clearly, the 'I love Moscow' project, and others like it, suggests that city branding is moving into a new phase, one in which the use of social media turns those living in the city from passive viewer to active participant; from spectator to spectacle. Thanks, at least in part, to Kapkov, Moscow has indeed become the kind of 'theatricalised [. . .] expanded public stage' which Blum so eloquently describes. Exploiting consumers' increasing desire to put themselves in the picture – literally to make spectacles of themselves – Kapkov's project has, as we noted earlier, generated over 25,000 followers on Instagram alone. This 'spectacularisation' of consumption is typical of so-called postmodern trends in branding encapsulated by projects such as Nike Town – the US sportswear manu-facturer's original concept store/museum founded in Portland, Oregon, in 1990. As we have seen, the 'I love Moscow' construction itself bears an uncanny resemblance to the Nike swoosh (even if it is a little chunkier). The connections go far deeper, however. As Peñaloza has argued elsewhere, Nike Town promotes a form of 'spectacular' consumption designed to offer the subject – and especially the male subject – the opportunity to ease what she calls his 'existential anxiety about [his] place in the hegemonic order' (1998, p. 437). At the risk of sounding reductivist, such anxiety is, we would argue, precisely what Moscow – and Russia in general – has been experiencing ever since losing its Soviet empire (Clowes 2011, Brundny 2013). This loss of power has been compounded recently by people 'running down' the city – something which, as we saw earlier, the project's website mentions quite explicitly. What better way, then, to defuse this anxiety than by presenting one's ideal image, offering oneself up as a spectacle to be consumed, not only physically, but also – and primarily – in the virtual reality of cyberspace. Furthermore, this goes not just for the city itself, but also for its millions of subjects, eager to exploit whatever 'cultural capital' (Bourdieu 1984) the city may have left to offer.

As Arvidsson aptly puts it, although in a different context (2006, pp. 191–192), 'when Nike puts its swoosh on Berlin playgrounds, the idea is that the brand can feed off the sociality, meanings and experiences that people produce on those playgrounds'. In a similar way, Kapkov appears to be hoping that the Moscow brand can benefit from goodwill generated by the installations he has placed in so many of the city's parks. However, if Kapkov really does want Moscow to become a 'global brand', he will need to strengthen considerably the fragile brand commu-nity that has begun to emerge around his project and work together with members of that community. This in turn will require a number of things, including: first, promoting genuine dialogue with bloggers by embedding (moderated) 'Discussion Boards' on the project's Vkontakte page, in line with the practice of many corporate brands in Russia (such as Levi's, who currently have 27 such boards, including a 'Truth Box' ['Budka Glasnosti']); second, engaging with bloggers on LiveJournal, possibly by creating an official LiveJournal account; third, differentiating 'I love Moscow' from the plethora of other current Moscow branding initiatives (see Trabun 2010a, 2010b, and Po 2012); and fourth, identifying further ways of transforming the city's cultural capital into economic capital (Bourdieu 1984), both for the city's residents and for foreign visitors. 'I love Moscow' is an interesting start – but whether it can help Kapkov achieve his long-term objectives remains to be seen.

## Bibliography

Aitken, R. and Campelo, A. (2011) 'The four Rs of place branding', *Journal of Marketing Management*, 27, 9–10, August, pp. 913–33.

Amine, A. and Sitz, L. (2007) 'Émergence et structuration des communautés de marque en ligne', *Décisions Marketing*, 46, April–June, pp. 63–75.

Anderson, B. ([1983]/2006) *Imagined Communities: Reflections on the Origin and Spread of Nationalism*, 2nd edn (London and New York, Verso).

Anholt, S. (1997) 'Africa needs brand aid', *Monocle*, 6, 1, September, pp. 56–57.

—— (2002) 'Foreword to the special issue on place branding', *Journal of Brand Management*, 9, 4–5, April, pp. 229–39.

—— (2007) *Competitive Identity: The New Brand Management for Countries, Regions and Cities* (Basingstoke, Palgrave Macmillan).

—— (2009) 'Editorial', *Place Branding and Public Diplomacy*, 5, 1, February, pp. 1–4.

Arvidsson, A. (2006) 'Brand value', *Brand Management*, 13, 3, February, pp. 188–92.

Ashworth, G. J. (2010) 'Should we brand places?', *Journal of Town and City Management*, 1, 3, pp. 248–53.

Ashworth, G. J. and Kavaratzis, M. (2009) 'Beyond the logo: Brand management for cities', *Journal of Brand Management*, 16, 8, July–August, pp. 520–31.

Bauman, Z. (2001) 'On globalization: Or globalization for some, localization for some others', in Beilharz, P. (ed.), *The Bauman Reader* (Oxford, Blackwell), pp. 298–311.

Belk, R. W. (2013) 'Extended self in a digital world', *Journal of Consumer Research*, 40, 3, October, pp. 477–500.

Belloso, J. C. (2011) 'The city branding of Barcelona: A success story', in Dinnie, K. (ed.) *City Branding: Theory and Cases* (Basingstoke, Palgrave Macmillan), pp. 118–23.

Bertrand, M. (2012) 'Le "city branding" débarque en Russie', 21 August, available at: http://www.lecourrier derussie.com/2012/08/21/le-city-branding-nouvelle-mode-en-russie/, accessed 30 October 2013.

Beumers, B. (2005) *Pop Culture Russia! Media, Arts, and Lifestyle* (Santa Barbara, Ca, Denver, Co and Oxford, ABC-Clio).

Blum, A. (2003) *The Imaginative Structure of the City* (Montreal, Kingston and London, McGill-Queen's University Press).

Bonnin, G. and Odou, P. (2010) 'Les communautés imaginées, un territoire d'action marketing? Le cas de l'entreprise de rencontre en ligne Meetic', *Décisions Marketing*, 58, April–June, pp. 27–36.

Bourdieu, P. (1984) *Distinction: A Social Critique of the Judgement of Taste* (Cambridge, MA, Harvard University Press).

Boym, S. (2001) *The Future of Nostalgia* (New York, Basic Books).

Brennan, R. and Croft, R. (2012) 'The use of social media in B2B marketing and branding: An exploratory study', *Journal of Customer Behaviour*, 11, 2, Summer, pp. 101–15.

Brundny, Y. M. (2013) 'Myths and national identity choices in post-communist Russia', in Bouchard, G. (ed.) *National Myths: Constructed Pasts, Contested Presents* (London and New York, Routledge), pp. 133–56.

Caldwell, N. and Freire, J. R. (2004) 'The differences between branding a country, a region and a city: Applying the Brand Box Model', *Journal of Brand Management*, 12, 1, September, pp. 50–61.

Cayla, J. and Eckhardt, G. M. (2008) 'Asian brands and the shaping of a transnational community', *Journal of Consumer Research*, 35, 2, August, pp. 216–30.

Clarke, D. B. (2003) *The Consumer Society and the Postmodern City* (London and New York, Routledge).

Clowes, E. W. (2011) *Russia on the Edge: Imagines Geographies and Post-Soviet Identity* (Ithaca and London, Cornell University Press).

Constantinides, E. and Fountain, S. J. (2008) 'Web 2.0: Conceptual foundations and marketing issues', *Journal of Direct, Data and Digital Marketing Practice*, 9, pp. 231–44.

Cotton, C. (2009) *The Photograph as Contemporary Art*, 2nd edn (London, Thames and Hudson).

Cova, B. (2006) 'Développer une communauté de marque autour d'un produit de marque: L'exemple de my nutella The Community', *Décisions Marketing*, 42, April–June, pp. 53–62.

Cova, B. and Cova, V. (2012) 'On the road to prosumption: Marketing discourse and the development of consumer competencies', *Consumption, Markets and Culture*, 15, 2, June, pp. 149–68.

de Certeau, M. (1984) *The Practice of Everyday Life* (Berkeley, CA and London, University of California Press).

Dinnie, K. (ed.) (2011) *City Branding: Theory and Cases* (Basingstoke, Palgrave Macmillan).

Durkheim, E. ([1893]/1933) *The Division of Labor in Society* (New York, Free Press).

Fernandes, I. (2012) 'Vkontakte is Facebook's formidable rival in Russia', 12 June, available at: http://adage.com/article/global-news/vkontakte-facebook-s-formidable-rival-russia/235331/, accessed 17 September 2013.

Florek, M. (2011) 'Online city branding' in Dinnie, K. (ed.) *City Branding: Theory and Cases* (Basingstoke, Palgrave Macmillan), pp. 82–90.

Fola, M. (2011) 'Athens city branding and the 2004 Olympic Games', in Dinnie, K. (ed.) *City Branding: Theory and Cases* (Basingstoke, Palgrave Macmillan), pp. 112–17.

Foster, R. J. (2007) 'The work of the new economy: Consumers, brands, and value creation', *Cultural Anthropology*, 22, 4, pp. 707–31.

Fuchs, C. (2014) 'Social media and the public sphere', public lecture given at University of Westminster, February 19.

Gaggiotti, H., Low Kim Cheng, P. and Yunak, O. (2008) 'City brand management (CBM): The case of Kazakhstan', *Place Branding and Public Diplomacy*, 4, 2, May, pp. 115–23.

Gavrilova, D. (2012) 'The Perm cultural revolution', available at: http://www.iwm.at/read-listen-watch/transit-online/the-perm-cultural-revolution/, accessed 20 September 2013.

Giddens, A. (1990) *The Consequences of Modernity* (Stanford, CA, Stanford University Press).

Gieryn, T. F. (2000) 'A space for place in sociology', *Annual Review of Sociology*, 26, pp. 463–96.

Goldstein, D. (2011) 'Hot prospekts: Dining in the new Moscow', in Goscilo, H. and Strukov, V. (eds) *Celebrity and Glamour in Contemporary Russia: Shocking Chic* (London and New York, Routledge), pp. 255–78.

Goscilo, H. (2011) 'Zurab Tsereteli's exegi monumentum, Luzhkov's largesse, and the collateral rewards of animosity', in Goscilo, H. and Strukov, V. (eds) *Celebrity and Glamour in Contemporary Russia: Shocking Chic* (London and New York, Routledge), pp. 221–54.

Habibi, M. R., Laroche M. and Richard, M-O. (2014) 'Brand communities based in social media: How unique are they? Evidence from two exemplary brand communities', *International Journal of Information Management*, 34, 2, April, pp. 123–32.

Hankinson, G. (2007) 'The management of destination brands: Five guiding principles based on recent developments in corporate branding theory', *Journal of Brand Management*, 14, 3, February, pp. 240–54.

Hayden, C. and Sevin, E. (2012) 'The politics of meaning and the city brand: The controversy over the branding of Ankara', *Place Branding and Public Diplomacy*, 8, 2, May, pp. 133–46.

Healey, P. (2004) 'Creativity and urban governance', *Policy Studies*, 25, 2, June, pp. 87–102.

Healy, J. C. and McDonagh, P. (2013) 'Consumer roles in brand culture and value co-creation in virtual communities', *Journal of Business Research*, 66, 9, September, pp. 1528–40.

Heilbrunn, B. (2006) 'Cultural branding between utopia and a-topia', in Schroeder, J. E. and Salzer-Mörling, M. (eds) *Brand Culture* (London and New York, Routledge), pp. 103–17.

Holt, D. B. (2004) *How Brands Become Icons: The Principles of Cultural Branding* (Harvard, Harvard Business School Press).

Hudson, B. T. and Balmer, J. M. T. (2013) 'Corporate heritage brands: Mead's theory of the past', *Corporate Communications: An International Journal*, 18, 3, pp. 347–61.

Hvass, K. A. and Munar, A. M. (2012) 'The takeoff of social media in tourism', *Journal of Vacation Marketing*, 18, 2, April, pp. 93–103.

Janowitz, M. (1952) *The Community Press in an Urban Setting* (Glencoe, IL, Free Press).

*Journal of Travel and Tourism Marketing* (2013) [Special issue on social media] 30, 1–2, January.

Kaplan, A. M. and Haenlein, M. (2010) 'Users of the world, unite! The challenges and opportunities of Social Media', *Business Horizons*, 53, 1, January, pp. 59–68.

Kavaratzis, M. (2004) 'From city marketing to city branding: Towards a theoretical framework for developing city brands', *Place Branding*, 1, 1, pp. 58–73.

—— (2009) 'Cities and their brands: Lessons from corporate branding', *Place Branding and Public Diplomacy*, 5, 1, February, pp. 26–37.

—— (2011) 'The dishonest relationship between city marketing and culture: Reflections on the theory and case of Budapest', *Journal of Town and City Management*, 1, 4, pp. 334–45.

Kelly, J. *et al.* (2012) *Mapping Russian Twitter*, Berkman Center for Internet and Society at Harvard University, March 20, available at: http://cyber.law.harvard.edu/publications/2012/mapping_russian_twitter, accessed 20 September 2013.

Ketter, E. and Avraham, E. (2012) 'The social revolution of place marketing: The growing power of users in social media campaigns', *Place Branding and Public Diplomacy*, 8, 4, November, pp. 285–94.

Klyukanov *et al.* (2008) 'Nation branding and Russia: Prospects and pitfalls', *Russian Journal of Communication*, 1, 2, Spring, pp. 192–222.

Kondakov, E. (2008) *Russkaya seksual'naya revolyutsiya* (Moscow, Octopus).

Kotler, P. (1986) 'The prosumer movement: A new challenge for marketers', *Advances in Consumer Research*, 13, 1, pp. 510–13.

Kotler, P., Asplund, C., Rein, I. and Haider, D. (eds) (1999) *Marketing Places Europe: How to Attract Investments, Industries, Residents and Visitors to Cities, Communities, Regions and Nations in Europe* (London, Pearson Education).

Kozinets, R. V. (2006) 'Netnography 2.0', in Belk, R. W. (ed.) *Handbook of Qualitative Research Methods in Marketing* (Cheltenham and Northampton, MA, Edward Elgar), pp. 129–42.

Larsen, H. G. (2014) 'The emerging Shanghai city brand: A netnographic study of image perception among foreigners', *Journal of Destination Marketing* (forthcoming).

Lemon, A. (2009) 'The emotional lives of Moscow things', *Russian History*, 36, pp. 201–18.

Leung, D., Law, R., van Hoof, H. and Buhalis, D. (2013) 'Social media in tourism and hospitality: A literature review', *Journal of Travel and Tourism Marketing*, 30, 1–2, January, pp. 1–2.

Lucarelli, A. and Berg, P. O. (2011) 'City branding: A state-of-the-art review of the research domain', *Journal of Place Management and Development*, 4, 1, pp. 9–27.

McAlexander, J. H., Schouten, J. W. and Koenig, H. F. (2002) 'Building brand community', *Journal of Marketing*, 66, 1, January, pp. 38–54.

McQuarrie, E. F., Miller, J. and Phillips, B. J. (2013) 'The megaphone effect: Taste and audience in fashion blogging', *Journal of Consumer Research*, 40, 1, June, pp. 136–58.

Martinez, N. M. (2012) 'City marketing and place branding: A critical review of practice and academic research', *Journal of Town and City Management*, 2, 4, pp. 369–94.

Miazhevich, G. (2010) 'Sexual excess in Russia's Eurovision performances as a national branding tool', *Russian Journal of Communication*, 3, 3–4, Fall–Winter, pp. 248–64.

Miles, Steven (2010) *Spaces for Consumption* (London, Sage).

Moilanen, T. and Rainisto, S. K. (2008) *How to Brand Nations, Cities and Destinations: A Planning Book for Place Branding* (Basingstoke, Palgrave Macmillan).

Morris, J. (2007) 'Drinking to the nation: Russian television advertising and cultural differentiation', *Europe-Asia Studies*, 59, 8, December, pp. 1387–1403.

Muñiz, A. M. Jr. and O'Guinn, T. C. (2001) 'Brand community', *Journal of Consumer Research*, 27, 4, March, pp. 412–31.

Ostapenko, N. (2010) 'Nation branding of Russia through the Sochi Olympic Games of 2014', *Journal of Management Policy and Practice*, 11, 4, pp. 60–63.

Peñaloza, L. (1998) 'Just doing it: A visual ethnographic study of spectacular consumption behaviour at Nike Town', *Consumption, Markets and Culture*, 2, 4, pp. 337–400.

—— (2001) 'Consuming the American West: Animating cultural meaning and memory at a stock and rodeo show', *Journal of Consumer Research*, 28, 3, December, pp. 369–98.

Po, Alisa (2012) 'U Moskvy budet svoi turisticheskii brend', 20 March, available at: http://www.the-village.ru/village/city/city/112481-brand, accessed 20 September 2013.

Prahalad, C. K. and Ramaswamy, V. (2004) 'Co-creation experiences: The next practice in value creation', *Journal of Interactive Marketing*, 18, 3, Summer, pp. 5–14.

Quinton, S. (2013) 'The community brand paradigm: A response to brand management's dilemma in the digital era', *Journal of Marketing Management*, 29, 7–8, May, pp. 912–32.

Ritzer, G. and Jurgenson, N. (2010) 'Production, consumption, presumption: The nature of capitalism in the age of the digital "prosumer" ', *Journal of Consumer Culture*, 10, 1, March, pp. 13–36.

Roberts, G. (2008) 'Signes extérieurs de richesse: Argent et représentations dans la Russie post-soviétique', in Vatanpour, S. (ed.) *L'Argent et la Monnaie: Représentations et Concepts* (Lille, Presses Universitaires de Lille-3), pp. 85–94.

—— (2014) 'Message on a bottle: Packaging the Great Russian Past', *Consumption, Markets and Culture*, 17, 3, June–August, pp. 295–313.

Schau, H., Muñiz, A. Jr. and Arnould, E. (2009) 'How brand community practices create value', *Journal of Marketing*, 73, 5, September, pp. 30–51.

Schroeder, J. E. (2002) *Visual Consumption* (London and New York, Routledge).

Shafranskaya, I. (2012) 'Guest article: Russian place branding – we start with the alphabet', available at: http://blog.inpolis.com/2012/05/18/russian-place-branding-we-start-with-the-alphabet/, accessed 20 September 2013.

Simons, G. (2011) 'Attempting to re-brand the branded: Russia's international image in the 21st century', *Russian Journal of Communication*, 4, 3–4, Fall–Winter, pp. 322–50.

Smith, M. and Puczko, L. (2010) 'Out with the old, in with the new? Twenty years of post-socialist marketing in Budapest', *Journal of Town and City Management*, 1, 3, pp. 288–99.

Sokolova, N. (2012) 'Co-opting transmedia consumers: User content as entertainment or "free labour"? The cases of S.T.A.L.K.E.R. and *Metro 2033*', *Europe-Asia Studies*, 64, 8, October, pp. 1565–83.

Stathaki, E. (2012) 'Garage centre pavilion designed by Shigeru Ban in Gorky Park, Moscow', 22 October, available at: http://www.wallpaper.com/architecture/garage-centre-pavilion-designed-by-shigeru-ban-in-gorky-park-moscow/6140, accessed 3 February 2014.

Stigel, J. and Frimann, S. (2006) 'City branding – All smoke, no fire', *Nordicom Review*, 27, 2, pp. 245–68.

Stokburger-Sauer, N. (2010) 'Brand community: Drivers and outcomes', *Psychology and Marketing*, 27, 4, April, pp. 347–68.

Szondi, G. (2011) 'Branding Budapest', in Dinnie, K. (ed.) *City Branding: Theory and Cases* (Basingstoke, Palgrave Macmillan), pp. 124–30.

Thompson, C. and Tian, K. (2008) 'Reconstructing the South: How commercial myths compete for identity value through the ideological shaping of popular memories and countermemories', *Journal of Consumer Research*, 34, 5, February, pp. 595–613.

Toffler, A. (1980) *The Third Wave: The Classic Study of Tomorrow* (New York, Bantam Books).

Tonnies, F. ([1887]/1957) *Gemeinschaft und Gesellschaft* (East Lansing, Michigan State University Press).

Trabun, Daniil (2010a) 'Perestroika: brend goroda – chast' 1', 14 September, available at: http://www.the-village.ru/village/city/people/105023-perestroyka-brend-goroda-2010-09-13-16-40-27, accessed 20 September 2013.

Trabun, Daniil (2010b) 'Perestroika: brend goroda – chast' 2', 14 September, available at: http://www.the-village.ru/village/city/people/105031-perestroyka-brend-goroda-2, accessed 20 September 2013.

Trueman, M., Klemm, M. and Giroud, A. (2004) 'Can a city communicate? Bradford as a corporate brand', *Corporate Communications: An International Journal*, 9, 4, pp. 317–30.

Tussyadiah, I. and Zach, F. (2013) 'Social media strategy and capacity for consumer co-creation among destination marketing organizations', in Cantoni, L. and Xiang, Z. (eds) *Information and Communication Technologies in Tourism 2013* (Berlin and Heidelberg, Springer), pp. 242–53.

Urde, M., Greyser, S. A. and Balmer, J. M. T. (2007) 'Special issue papers: Corporate brands with heritage', *Journal of Brand Management*, 15, 1, September, pp. 4–19.

Virgo, B. and de Chernatony, L. (2006) 'Delphic brand visioning to align stakeholder buy-in to the City of Birmingham brand', *Journal of Brand Management*, 13, 6, July, pp. 379–92.

Wallace, E., Buil, I. and de Chernatony, L. (2012) 'Facebook "friendship" and brand advocacy', *Journal of Brand Management*, 20, 2, December, pp. 128–46.

Weaver, C. (2011) 'Abramovich brings Midas touch to Gorky Park', 28 July, available at: http://blogs.ft.com/beyond-brics/2011/07/28/abramovich-brings-midas-touch-to-gorky-park/#axzz2cb71k5oa%29, accessed 3 February 2014.

Winter, J. (2013) 'Selfie-Loathing. Instagram is even more depressing than Facebook. Here's why', 23 July, available at: http://www.slate.com/articles/technology/technology/2013/07/instagram_and_self_esteem_why_the_photo_sharing_network_is_even_more_depressing.html, accessed 14 January 2014.

Wirtz, J., *et al.* (2013) 'Managing brands and customer engagement in online brand communities', *Journal of Service Management*, 24, 3, pp. 223–44.

Yurchak, A. (2005) *Everything Was Forever, Until It Was No More: The Last Soviet Generation* (Princeton, NJ, Princeton University Press).

Zaglia, M. E. (2013) 'Brand communities embedded in social networks', *Journal of Business Research*, 66, 2, February, pp. 216–23.

Zhang, L. and Zhao, S. X. (2009) 'City branding and the Olympic effect: A case study of Beijing', *Cities*, 26, 5, October, pp. 245–54.

# Index

1C (Russian software giant) 274
9/11 events 2

'active audiences' 5
Agadamov, Rustem 96
*Allods Online* (AO) computer game 266–9,
    272–3, 273–6, 277–8, 280, 282–3
*Allody: pechat' taimy* (Allods: the secret seal)
    274
*Allody-2: povelitel' du* (Allods-2, the lord of
    souls) 274
American Standard Code for Information
    Interchange (ASCII) 168
'analogons' term 277
Anderson, Benedict 287
Arab Spring 2, 118, 139
Arctic Monkeys 287
Arendt, Hannah 74
Aronczyk, Melissa 147
'ASCII Greek' 168
'asocial teenagers' 270

Bakhtin, Mikhail 211, 260
Balibar, Etienne 182
Baltic states (Estonia, Latvia and Lithuania)
    148
Baron Cohen, Sacha 148, 158
'basic computer literacy' (Russian political
    leaders) 99
'Be Proud, Write in Cyrillic' slogan
    (Macedonia) 177
Belarus 152
belarus.by 152
Belykh, Nikita 74, 94, 100, 110
Berkman Centre for Internet and Society
    (Harvard University) 7
*Big Brother* (TV programme) 249
'blog': definition 94; term 94
blog Medvedev—public consent: blog
    without a blogger 79–81; blogosphere
    77–9; conclusions 87–9; introduction 72–3;

network state 75–7; politics in blogging 73–5;
    pressures 83–5; symbolic empowerment
    81–3; visual effectiveness 85–7
Blog-Medvedev (BM) 73, 76–7, 79–80, 81–2,
    84, 85–7, 88–9
blogging central Asia to citizen media, from—
    *neweurasia* blog project: convergence,
    amalgamation and bridge blogging 66–9;
    description 54–5; development 62–6;
    history 55–62; introduction 52–4
blogging central Asia to citizen media, from—
    *neweurasia* blog project—development:
    contextual users 62–3; logistics and
    organisation 63–5; personnel 65–6
'blogging challenge' 67
blogging governors in Russia: activity
    104–5; audience response 105;
    description 96–7; federal districts 101;
    growth 100–1
blogging the Other—national identities in
    blogosphere: China/Chinese in Russian
    mass media 207–9; Chinese in Russian
    blogs 216–18; conclusions 223–4; genre
    in blogs 218–23; introduction 205–7;
    Russians in Chinese blogs 212–16; theory
    and methods 209–12
blogging for the president—online diaries of
    Russian governors: allegiance—why are
    there so many blogs? 98–102; efficient
    statement blogs 107–9; how? 102–5;
    internetchik *blogs* 110–11; introduction
    92–4; overview 94–8; political impact
    105–6; PR blogs 106–7
blog.mail.ru 210
Blum, Alan 299–300
*Blut und Boden* nationalism 146
Bolter, Jay David 146
book: objectives 7; summary 5–11
*Borat: Cultural Learnings of America
    for Make Benefit Glorious Nation of
    Kazakhstan* 158

305